NAMBU

A Foreteller of Modern Physics

New Edition

World Scientific Series in 20th Century Physics

Published

Vol. 24 Subnuclear Physics — The First 50 Years: Highlights from Erice to ELN
by A. Zichichi
edited by O. Barnabei, P. Pupillo and F. Roversi Monaco

Vol. 25 The Creation of Quantum Chromodynamics and the Effective Energy
by V. N. Gribov, G. 't Hooft, G. Veneziano and V. F. Weisskopf
edited by L. N. Lipatov

Vol. 26 A Quantum Legacy — Seminal Papers of Julian Schwinger
edited by K. A. Milton

Vol. 27 Selected Papers of Richard Feynman (with Commentary)
edited by L. M. Brown

Vol. 28 The Legacy of Léon Van Hove
edited by A. Giovannini

Vol. 29 Selected Works of Emil Wolf (with Commentary)
edited by E. Wolf

Vol. 30 Selected Papers of J. Robert Schrieffer — In Celebration of His 70th Birthday
edited by N. E. Bonesteel and L. P. Gor'kov

Vol. 31 From the Preshower to the New Technologies for Supercolliders — In Honour of Antonino Zichichi
edited by B. H. Wiik, A. Wagner and H. Wenninger

Vol. 32 In Conclusion — A Collection of Summary Talks in High Energy Physics
edited by J. D. Bjorken

Vol. 33 Formation and Evolution of Black Holes in the Galaxy — Selected Papers with Commentary
edited by H. A. Bethe, G. E. Brown and C.-H. Lee

Vol. 35 A Career in Theoretical Physics, 2nd Edition
by P. W. Anderson

Vol. 36 Selected Papers (1945–1980) with Commentary
by Chen Ning Yang

Vol. 37 Adventures in Theoretical Physics — Selected Papers with Commentaries
by Stephen L. Adler

Vol. 38 Matter Particled — Patterns, Structure and Dynamics — Selected Research Papers of Yuval Ne'eman
edited by R. Ruffini and Y. Verbin

Vol. 39 Searching for the Superworld — A Volume in Honour of Antonino Zichichi on the Occasion of the Sixth Centenary Celebrations of the University of Turin, Italy
edited by S. Ferrara and R. M. Mössbauer

Vol. 40 Murray Gell-Mann — Selected Papers
edited by H. Fritzsch

Vol. 41 Many-Body Theory of Molecules, Clusters, and Condensed Phases
edited by N. H. March and G. G. N. Angilella

Vol. 42 Selected Papers of K. C. Chou
edited by Y.-L. Wu

Vol. 43 Nambu: A Foreteller of Modern Physics
edited by T. Eguchi and M. Y. Han

For information on Vols. 1–23, please visit http://www.worldscientific.com/series/wsscp

World Scientific Series in 20th Century Physics **Vol. 43**

NAMBU

A Foreteller of Modern Physics

New Edition

Editors

T. Eguchi

Rikkyo University, Japan

M. Y. Han

KAIST, Korea

World Scientific

NEW JERSEY • LONDON • SINGAPORE • BEIJING • SHANGHAI • HONG KONG • TAIPEI • CHENNAI

Published by

World Scientific Publishing Co. Pte. Ltd.
5 Toh Tuck Link, Singapore 596224
USA office: 27 Warren Street, Suite 401-402, Hackensack, NJ 07601
UK office: 57 Shelton Street, Covent Garden, London WC2H 9HE

Library of Congress Cataloging-in-Publication Data
Nambu : a foreteller of modern physics / editors T. Eguchi, Rikkyo University, Japan, M.Y. Han, Korea Advanced Institute of Science and Technology, Korea.
pages cm -- (World scientific series in 20th century physics ; v. 43)
Includes bibliographical references.
ISBN 978-981-4578-14-1 (hardcover : alk. paper) -- ISBN 978-981-4583-05-3(pbk) (alk. paper)
1. Particles (Nuclear physics) 2. Nambu, Yoichiro, 1921– I. Eguchi, T. (Tohru), 1948– editor of compilation. II. Han, M. Y., editor of compilation. III. Series: World Scientific series in 20th century physics ; v. 43.
QC793.2.N35 2014
539.7'2--dc23

2013041116

British Library Cataloguing-in-Publication Data
A catalogue record for this book is available from the British Library.

The editors and publisher would like to thank the following publishers of the various journals and books for their assistance and permissions to include the selected reprints found in this volume:

American Physical Society
Elsevier Science Publishers B.V.
Gordon & Breach Publishing
Physical Society of Japan
Scientific American
The Nobel Foundation
University of Chicago News Office

Printed in Singapore by B & Jo Enterprise Pte Ltd

ACKNOWLEDGMENTS

We would like to express our gratitude to Professor K. K. Phua, Chairman of the World Scientific Publishing Co., for initiating the project of publishing this volume.

We are also thankful to Professor Yoichiro Nambu who personally translated into English the article of reminiscences first published in Japanese for the Bulletin of Japan Physical Society.

One of us (M.Y.H.) would also like to thank Professor Sung-Mo Kang, President of the Korea Advanced Institute of Science and Technology (KAIST), for providing opportunity to work on this project while in Korea.

Lastly, we both are thankful for the efforts of Ms. Chionh, an editor of the World Scientific Publishing Co.

PREFACE

In 1995 a book called *"Broken Symmetry, Selected Papers of Y. Nambu"* was published that contained 40 seminal papers of Nambu selected from 148 of his publications. The book was edited by T. Eguchi and K. Nishijima.

Since then a few important events took place.

In 2008, Nambu was awarded the Nobel Physics Prize for his work on the spontaneous symmetry breaking. Many physicists considered this to be an award long overdue.

In 2009, Kazuhiko Nishijima passed away at the age of 82. Nishijima was one of Nambu's closest lifelong friends who was one of the editors of the book *"Broken Symmetry"*. He is best known for the discovery of the strangeness quantum number that formed the basis of the Gell-Mann–Nishijima formula.

Coming into the 21st century, developments in three major areas of elementary particle physics, the electroweak theory with the Higgs mechanism, quantum chromodynamics, and the string theory began to yield more and more concrete and successful results.

Seeds of all these developments in modern particle physics were sowed by prophetic papers of Nambu during 1960–70s.

The discovery of the Higgs-like scalar particle at the Large Hadron Collider in July of 2011 was a climax for the idea of spontaneous symmetry breaking Nambu has laid down at the beginning of 1960s.

In this present volume, we collect what we consider to be the most prophetic of Nambu's papers for the benefit of current and future generation of particle physicists of the world. It contains some papers that were already contained in the 1995 book, but also contains several papers which never appeared in a book form before.

We hope and believe that the readers would find this volume to be not only very useful but also cherish it as a window to the mind of a man many refer to as "Nambu the seer."

T. Eguchi and M. Y. Han
Editors

October 2013

FOREWORD

It is our great pleasure to publish a Collection of Papers of Professor Yoichiro Nambu who has been one of the most renowned masterminds among our contemporaries. We have chosen the title "Broken Symmetry" for this volume since we think it represents the best-known contributions of his repertoire.

Many of us are clients of his papers and subscribe to his messages which convey his foresights always leading to a new frontier of physics. It is not necessarily easy to follow his publications, however, since some of his ideas are beyond immediate comprehension. Indeed, on the occasion of his 65th birthday, one of the leading theorists made the following remark: "He is always 10 years ahead of us, so I tried to understand his works in order to contribute to a new area which will flourish 10 years later. Contrary to my expectation, however, it took me 10 years to understand them."

For those who try to grasp the entire flow of his ideas it is gratifying and helpful to have an access to a collection of 40 papers selected by Professor Nambu himself. Furthermore, it should be emphasized that some of his unpublished papers and lecture notes, unavailable otherwise, have been reproduced and included in this volume. They are the famous unpublished lecture notes on string theory and the first paper on spontaneous symmetry breaking read at the Midwest Conference on Theoretical Physics held in 1960.

In addition to his scientific publications, a memoir on his research career (*Research in Elementary Particle Theory*) prepared by himself and published first in "Butsuri", the Bulletin of the Physical Society of Japan, in 1977 (in Japanese) is also translated in English and included in this volume. This self-portrait of his offers a detailed description of how and why he was motivated to develop and pursue many of his pioneering works.

The chronology of this list of selected papers would eventually enable us to trace back some of his ideas to his earlier publications with the help of his memoir.

We hope that the readers would treasure this volume for years to come as a constant source of inspiration.

Editors, May 1995

CONTENTS

*Numbers in brackets refer to that in the List of Publications (pp. 179–194)

Professor Nambu was unable to attend the Nobel prize ceremony in Stockholm, Sweden and the Nobel committee arranged a special presentation ceremony to be held on the campus of the University of Chicago presided by the Swedish ambassador to the United States on behalf of the King of Sweden.

Photos — courtesy of the News Bureau of the University of Chicago

News Release by The University of Chicago

Yoichiro Nambu awarded 2008 Nobel Prize in Physics
December 10, 2008

UChicago News

Yoichiro Nambu, the Harry Pratt Judson Distinguished Service Professor Emeritus in Physics and the Enrico Fermi Institute, and one of the leading figures in the development of modern particle physics, on Tuesday was awarded the 2008 Nobel Prize in Physics.

In a citation posted after the award was announced shortly before 5:30 a.m., the Nobel Foundation credited Nambu with "the discovery of the mechanism of spontaneous broken symmetry in subatomic physics."

"It is a great honor for me to be awarded the prize this year — I did not expect it this year," Nambu said. "It is true that I've been told I had been on the list for many, many years — maybe 30 years. I did not expect it in any way this particular time and, I must say, I was very surprised when I got the news."

University of Chicago President Robert Zimmer said Nambu's work embodies the best traditions of Chicago scholarship, noting that Nambu is the 82nd person connected with the University as a faculty member, student or researcher to have received a Nobel Prize, including 28 Nobel Prizes in Physics.

"Professor Nambu really represents what it is we aspire to — the highest level of achievement that we aspire to. This is an institution that, from its beginning, was founded on the basis of rigorous inquiry. It has been viewed by many as the most serious and intense academic institution in the country, if not the world," Zimmer said.

"The environment that is created here, and the intensity that the faculty feel about their research and in their interaction with their colleagues, really is very distinctive," Zimmer said. "It is what fosters the work of so many of our faculty, and it is the environment that Professor Nambu has worked in and contributed to for so many years."

Nambu, 87, shares this year's prize with Makoto Kobayashi of the High Energy Accelerator Research Organization (KEK) in Tsukuba, Japan, and Toshihide Maskawa of Kyoto University in Kyoto, Japan.

"I know these people, and I have always appreciated their work very much, so I am very happy that I share the Nobel Prize with them," said Nambu, who has been a scholar at the University of Chicago since 1954.

University of Chicago Provost Thomas Rosenbaum, the John T. Wilson Distinguished Service Professor in Physics, said that Nambu's work cut a wide swath across the world of physics.

"Nambu's work is emblematic of what we try to accomplish here at the University of Chicago: a genuinely great scholar asking fundamental questions that then have impact across the discipline and across the disciplines," Rosenbaum said.

"What Nambu has done has, in fact, had implications, all the way from the structure of the early universe to the behavior of magnetic material," Rosenbaum added. "In addition — and this is not a criterion for winning the Nobel Prize — I can testify first hand that Nambu is a wonderful, modest man and a generous colleague."

Nambu's theories form an essential cornerstone of what physicists call the Standard Model, which explains in a unified way three of the four fundamental forces of nature: strong, weak and electromagnetic.

Nobel laureate James Cronin, University Professor Emeritus in Physics, said, "It has been clear for so many years that of all the people who have won the Nobel Prize, there has been one missing, and that is Yoichiro Nambu."

"Nambu was so far ahead of his time that it maybe was not completely realized the importance of his contributions, which are the cornerstone of the Standard Model of particle physics, which unites the strong interactions, the electromagnetic interactions and the weak interactions all in a clear, understandable whole," said Cronin.

Nambu was born in Tokyo in 1921. He said he developed his interest in science in high school when his father showed him science magazines. In college, one of his professors was Hideki Yukawa, one of the early figures in particle physics.

"I liked physics because it was a way of studying fundamental problems in natural science," he said.

He received his B.S. in 1942 from the University of Tokyo, and was then drafted into the Japanese military, where he worked at a radar laboratory. After the war, he studied with the famous Japanese physicist Sin-ichiro Tomonaga, receiving his D.Sc. in 1952 from the University of Tokyo.

Tomonaga recommended that Nambu study with J. Robert Oppenheimer at the Institute for Advanced Study in Princeton, where he went in 1952.

"The scientists were very warm to me," Nambu recalled. "Oppenheimer was brilliant, but everyone was a little afraid of him."

He joined the University of Chicago as a Research Associate in 1954 and became Associate Professor in 1956 and Professor in 1958.

Nambu found the concept of spontaneous symmetry breaking, or SSB, while studying superconductivity in the early 1960s. He said that it took him two years to determine that SSB contributes to the explanation of how superconductivity works, but that once he developed the theory, he quickly turned it around and applied it to particle physics.

"Physicists believed that fundamental models of physics should be simple and elegant with symmetry built into it," said Nambu. "Nevertheless, when it is realized in nature, some of the symmetries come out to be broken and that is a mechanism we have been seeing ever since."

He was named the Harry Pratt Judson Distinguished Service Professor in 1977 and served as Chairman of Physics from 1974 to 1977.

Nambu became a U.S. citizen in 1970. He and wife Chieko Nambu have a son, John Juichi Nambu.

Though he became a professor emeritus in 1991, Nambu noted how his theories are being tested today at some of the world's leading facilities, including the new particle accelerator at CERN in Geneva, where scientists hope to discover occurrences of symmetry breaking.

"Recent experiments or advances in cosmology can satisfy that people now can talk about the origins of the universe and see how the universe measures with that one which we know today," Nambu said. "Symmetry breaking is one of the fundamental ideas, which can be applied to many, many branches of physics."

Nambu's current research centers on why different quarks have different masses — a consequence, physicists believe, of SSB.

"Why quarks have different masses is one of the last unsolved puzzles of the Standard Model," he said.

And how does Nambu go about solving a problem like that? He once answered, laughing, "I think about it all the time." He said he hopes to keep thinking about it as long as he lives.

His previous honors include the Wolf Prize in Physics, the National Medal of Science, the Dannie Heineman Prize for Mathematical Physics, the J. Robert Oppenheimer Prize and the Order of Culture from the government of Japan.

Profiles of Nambu*

Madhusree Mukerjee

I first saw Yoichiro Nambu almost 10 years ago, from the back row of a graduate seminar in physics at the University of Chicago. A small man in a neat suit, he was sketching long, snaking tubes on the blackboard. Sometimes he said they were vortex lines, found in superconductors; other times he called them strings, connecting quarks. Mystified, and yet fascinated by a bridge between such disparate realms, I later asked him to be my thesis adviser.

Face to face, Nambu was still hard to understand. I was clearly not the first to try. Bruno Zumino of the University of California at Berkeley once recounted his own attempts: "I had the idea that if I can find out what Nambu is thinking about now, I'll be 10 years ahead in the game. So I talked to him for a long time. But by the time I figured out what he said, 10 years had passed." Edward Witten of the Institute for Advanced Study in Princeton, N.J., explains: "People don't understand him, because he is so farsighted."

Nambu was the first to see that when a physical system such as a superconductor — or an ocean of quarks — defies the symmetry imposed by physical laws, a new particle is born. Along with Moo-Young Han, then a graduate student at Syracuse University, he proposed the existence of gluons, the objects that hold quarks together. He also realized that quarks act as if they are connected by strings, an idea that became the foundation of string theory. "Over the years," remarks Murray Gell-Mann of the Santa Fe Institute, "you could rely on Yoichiro to provide deep and penetrating insights on very many questions."

The roots of Nambu's originality may lie in a singular childhood in pre-war Japan. Born in Tokyo in 1921, he was two when the city was destroyed by an earthquake. (He still has a vague recollection of flames.) Kichiro Nambu, his father, had run away from home to attend university and there had met his bride, Kimiko. The earthquake forced him to return to his hometown of Fukui, near Kyoto, with his wife and young son.

The prodigal was forgiven (although his wife never was). Retaining traces of defiance, Kichiro Nambu became a school teacher and built his house on the outskirts of town — an act that was later to save him from Allied bombs. From Tokyo he had brought back an

*Reprinted from *Scientific American*, February 1995

eclectic library. Browsing there, his growing boy learned of ideas that allowed him to flee, at least mentally, the excruciating regimen at the local school.

Fukui, in those days, prided itself on having the most militaristic school in Japan. The boys dressed in scratchy army uniforms and were taught to march, shoot and salute. "If you didn't see a senior boy and so didn't salute him, he would punch you out," Nambu recalls. "You had to keep one eye on every person." At 4:00 a.m. in midwinter, he would walk a mile to school to learn Samurai sword fighting, barefoot on bare floors in unheated halls. To the frail child, school proved as trying as, later, the real Imperial army.

Nor did the school neglect the mind. Heroic deeds — notably, that of a school teacher who died saving the emperor's picture from a fire — embellished the curriculum. Nambu was protected from such teachings by his father's anti-establishment diatribes. Yet they also prevented him from fitting in. "I had a longing to be like the other boys," he smiles ruefully. As he grew, he came to realize that his father's opinions were dangerous in an increasingly war-like Japan.

Thus, Nambu learned to keep his thoughts to himself. This trait served him well later, through years in the army — and perhaps even as a physicist. His originality might come from having to think everything through for himself, from being aware of, but ignoring, ideas in the world outside.

Moving on to a premier college in Tokyo in 1937, Nambu discovered a freer intellectual atmosphere and smart classmates who awed the country boy. Of his courses, physics caused him special trouble: "I couldn't understand entropy and flunked thermodynamics." Yet, possibly inspired by Hideki Yukawa, the pioneer who realized that particles transmit force, Nambu chose to aim for a master's in physics at Tokyo University.

Among his new classmates, he found some underground radicals. Japan was fighting China. "We were told of the victories," Nambu says, "but these communists somehow also knew about the massacres and defeats." The academic program turned out to be short: the class graduated six months early so that its members could be drafted.

In the army Nambu dug trenches and carried boats. "Physically it was hard," he shrugs, "but inside I was free. As long as you said, 'Yes, sir, yes, sir,' they left you alone." After a year he was assigned to help develop short wavelength radar. The navy already had such radar, but the army had no confidence in that equipment. Nor was Nambu's team especially successful: "To test our system, I set it up on a hilltop and hired a boat to take a metal rod out into the ocean. You could see it with your bare eyes — but not with our radar."

He was then ordered to steal a secret navy document, a paper on field theory by Sin-Itiro Tomonaga, who was applying his discoveries on particle waves to radar waveguides. (Werner Heisenberg's publications on field theory had arrived from Germany shortly before, after traveling by submarine for a year.) Obtaining these papers — simply by asking a professor — Nambu became acquainted with some of the newest ideas in physics.

Life was quite easy. The unit was housed in a golf club, and romance was budding between Nambu and his assistant, Chieko Hida. For the most part, the war seemed far away. Yet one night Nambu watched a fleet of B-29s fly over Osaka. For a change, they did not drop their bombs on the city but moved on to Fukui. Nambu lost his grandparents; his parents were spared.

After the war, Nambu and Hida married, whereupon he left for Tokyo to take up a long-promised research position. (Hida stayed on in Osaka to look after her mother.) Housing was scarce, and Nambu moved into his laboratory for three years. Gas and electricity were free, and he could bathe in the water basin intended for extinguishing air-raid fires. But his officemate, Ziro Koba, a diligent young man (he once shaved his head for missing a calculation), would come in early and often embarrassed Nambu, who was sleeping across both their desks.

"I was hungry all the time," Nambu says. Finding food took up most of the week. For the rest, he thought about physics, calculating on rolls of cash-register paper. Koba, a student of Tomonaga, kept Nambu informed about the latter's work. A group of solid-state physicists in a neighboring office also provided stimulating company.

All that these researchers knew of scientific developments in the West came from sporadic issues of Time magazine. Later, journals in a library set up by the Occupation forces helped to fill in the gaps. Yet much had to be reinvented by the Japanese physicists. Sometimes they got there first. After moving to Osaka City University in 1949, Nambu published a formula describing how two particles bind, now known as the Bethe–Salpeter equation. Along with others, he also predicted that strange particles should be created in pairs, a discovery usually attributed to Abraham Pais.

Describing Nambu's early work, Laurie M. Brown of Northwestern University writes of its "exuberant sense of play." As his student, I enjoyed Nambu's sheer pleasure in ideas and his ready laugh (even if I did not always get the joke). In the belief that too much work is harmful, he urged me to attend baseball games and to read the exploits of V. I. Warshawski, the fictional Chicago sleuth.

In 1952, Nambu was invited to visit the Institute for Advanced Study. There he found many brilliant and aggressive young men. "Everyone seemed smarter than I. I could not accomplish what I wanted to and had a nervous breakdown," Nambu wrote to me decades later, trying to bring comfort during my own travails as a postdoctoral fellow. In 1957, after having moved to Chicago, he proposed a new particle and met with ridicule. ("In a pig's eye!" Richard Feynman shouted at the conference, Brown recalls. The omega was discovered the next year, in an accelerator.) Meanwhile Nambu had heard J. Robert Schrieffer describe the theory of superconductivity that he had just devised with John Bardeen and Leon N. Cooper. The talk disturbed Nambu: the superconducting fluid did not conserve the number of particles, violating an essential symmetry of nature. It took him two years to crack the puzzle.

Imagine a dog faced with two bowls of equally enticing food. Being identical, the bowls present a twofold symmetry. Yet the dog arbitrarily picks one bowl. Unable to accept that the symmetry is entirely lost, Nambu discovered that the dog, in effect, cannot make up her mind and constantly switches from one bowl to the other. By the laws of quantum physics, the oscillation comes to life as a new particle, a boson.

Nambu points out that others, such as Bardeen, Philip W. Anderson, then at Bell Laboratories, and Gerald Rickayzen, then at the University of Illinois, also saw that a superconductor should have such a particle. It was Nambu, however, who detailed the circumstances and significance of its birth and realized that the pion, as well, was born in like manner (by breaking the chiral, or leftright, symmetry of quarks). While he searched for more of its siblings in nature, Nambu circulated his findings in a preprint.

Jeffrey Goldstone, then a postdoctoral fellow at CERN, the European laboratory for particle physics, realized the import of this work and soon published a simpler derivation, noting that the result was general. Thereafter the new particle was dubbed the Goldstone boson. ("At the very least, it should be called the Nambu–Goldstone boson," Goldstone comments.) When Nambu finally published his calculations in 1960, his paper also showed how the initially massless particle mixes with a magnetic field in a superconductor to become heavy. Recognized by Anderson, Peter Higgs, then at the Institute for Advanced Study, and others as a general phenomenon, it later became the Higgs mechanism of the Standard Model of particle physics.

In the years that followed, Nambu studied the dynamics of quarks, suggesting they were held together by gluons carrying a color quantum number to and fro. "He did this in 1965, while the rest of us were floundering about," Gell-Mann says. (Nambu, however, believed the quarks to be observable and assigned them integer electrical charges, an error that Gell-Mann and others corrected.) In 1970, perusing a complex mathematical formula on particle interactions, Nambu saw that it described strings. In the 1980s his "string action" became the backbone of string theory.

"He has an amazing power of coming up with pictures," says Peter G. O. Freund, a colleague at Chicago. While working with Nambu, I noticed that he would look at a problem from several different, yet simultaneous, points of view. It was as if instead of one mind's eye he had at least two, giving him stereoscopic vision into physical systems. Where anyone else saw a flat expanse of meaningless dots, he could perceive vivid, three-dimensional forms leaping out.

Over time, Nambu became known as a seer, albeit a shy one. "I can think of no one who gives such good advice," Witten says. Pierre M. Ramond of the University of Florida observes that the directions of particle physics were often predicted by Nambu's papers — encrypted in the footnotes.

These days Nambu puzzles over how quarks acquire their diverse masses. He suggests they might come from historical accident, such as the quarks being born at different stages of the early universe. His thoughts have also turned to biology and to an old bane, entropy. Nambu calculates that virus-size particles, when placed in a cusp-shaped container, seem to violate gravity and entropy. Perhaps they conceal a clue as to how life-forms defy entropy and become ever more organized. Prophecy or quixotic fancy? Ten years from now, we might know.

Reminiscences of the Youthful Years of Particle Physics[1]

Yoichiro Nambu
University of Chicago

1. I graduated from the Physics Department of the University of Tokyo in 1943. So I am one of the so-called "wartime guys". The Pacific War started in my second year. The third year was cut short and I was drafted into the Army, ended up being assigned to a radar laboratory near a city called Takarazuka, where I was able to gain experience in practical problems of physics. As the war ended, I was fortunate to be able to return to the University of Tokyo as a postdoctoral associate joining a group of repatriates and younger students.

Among the latter were Z. Koba, Y. Miyamoto and T. Kinoshita. Under the guidance of Prof. Tomonaga of the nearby Tokyo Bunrika University and the RIKEN Laboratory, they were collaborating with him to develop his so-called super-many-time theory. But progress was slow because of the difficult living condition. As for me, I was living in my office with a few others, and fortunately across my desk was Koba. Watching his work I was gradually able to learn the Tomonaga theory.

In 1947, the big news of the Lamb shift and of the discovery of the pion reached Japan. Immediately we got involved in a fierce competition with American physicists to develop quantum electrodynamics, or QED. So I also formally joined the Tomonaga group and listened to his lectures. The essence of the Tomonaga theory is relativistic invariance and renormalization. Tomonaga called renormalization the principle of "hohki", which means either "giving up" or "broom". I did not know which he meant, and did not like the word any way.

In time I became able to pursue my own research. In 1950, under the recommendation of Tomonaga, I joined S. Hayakawa, Y. Yamaguchi, K. Nishijima of University of Tokyo and T. Nakano of Osaka University to form a theory group at the newly created science department of Osaka City University.

First I would like to talk about the influence of the so-called Sakata–Taketani philosophy in the immediate postwar period. They were collaborators with Yukawa in creating the meson theory but also developed and advocated a unique methodology. Taketani often came to see his friend S. Nakamura near our office, and expounded his theory before us. I need not explain here his "three stages theory", but we youngsters were overwhelmed by his persuasive eloquence. He also spoke against our preoccupation with theoretical ideas, emphasized to pay attention to experimental physics. It was a good advice against the tendency resulting from the success of the Yukawa theory as well as the miserable postwar condition which made it difficult to do experimental physics. I believe that this advice has come to make a big influence on my attitude toward physics.

Still now I look back with nostalgia toward the three years at Osaka City University. It was the period when in America the technical and intellectual energies accumulated during the war got explosively released toward peaceful enterprises. In contrast, we in Japan were not yet able to foresee the future, feeling hungry every day in a barack office of the department. Nevertheless, I had never felt and enjoyed so much the sense of freedom. Being a new department, there were no senior professors to defer to, no need for formal lectures as there were only two students. We were completely free to engage in research by ourselves. Gradually news started to come in from abroad of discoveries found in accelerators and cosmic rays. As the international communication became easier, so did the internal one. Under these stimulations we started to make our own original work, which we published in our own tabloid magazine together with reports from people who were able to go overseas.

One of the successes of our collaborative efforts at Osaka City University was the pair production theory of the newly discovered "strange particles". It was surprise to us that we novices could do this ahead of the physicists of the world. Not only our group but those in Tokyo and Kyoto have also started to make original contributions. But we did not last. In a few years our very success made us disperse to more established institutions. Thus in 1952, I was invited to the Institute for Advanced Study (IAS) in Princeton together with Toichiro Kinoshita on the recommendation of Tomonaga.

2. Contrary to my expectation, however, my stay at IAS turned out to be a mixture of heaven and hell. Among the physicists, there were J. R. Oppenheimer the director, A. Einstein, and various kinds of members: W. Pauli, A. Pais, F. Dyson, C. N. Yang, T. D. Lee, G. C. Wick, G. Kallen, W. Thirring, etc. The short term members like me were housed in barracks on campus so we became close to each other right away. It was a heavenly environment compared to the miserable life we left behind. On the other hand, I could not help but feel of fierce competition among us. Moreover, my plan to pursue the saturation properties of nuclear forces and the origin of the spin-orbit force hit a wall and stalled. It made me recall the poem by Takuboku Ishikawa:

> "*On a day when all my friends look better than I,*
> *I would bring home flowers to enjoy with my wife*"

At the end of the first year, I was allowed to stay for another year, but did not receive support during the recess period from spring to summer. Oppenheimer kindly offered to use the fund left by Yukawa, arranged for Kinoshita and me to be able to spend the period at the California Institute of Technology. I decided to bring my wife and son from Japan and meet them there. So the Kinoshita couple and I started to drive two cars together and cross the country to the South. On the way we could witness the miserable status of the blacks. This was the time right after the famous ruling of the Supreme Court. We were also expecting some unpleasant incidents on the way, but did not experience them until we reached Pasadena and looked for apartments. (A similar incident also occurred later in Chicago.)

At CalTech there was a 500 GeV synchrocyclotron. I asked to join the experimental team but was politely declined. So I contented myself doing analysis of the $\gamma - \pi$ reaction. The famous Oppenheimer incident occurred in my second year at IAS. But I was not yet sufficiently informed of American politics to pay much attention to it. At the end of the two years, I wanted to stay a bit longer at a leading university in order to accomplish something. There was an offer but it was too distant. I searched for a research associate position but it was very scarce in those days. I almost gave up when an offer came from M. L. Goldberger of Chicago. I had already met him, and had also visited the university. But in those days the impression of the city was awful. During the war, America had also spent all its energies toward winning it, had no room to pay attention to domestic matters, I suppose. The commuter train from Hyde Park, the site of the university, to downtown was like that in Tokyo. I would never come here again, that was my impression. It is an irony of fate that we have come to live there. As our car approached Chicago, we had to pass the famous area of steel industry called Gary, Indiana. We felt as if driving into hell.

When I settled down at the Institute for Nuclear Study (INS) of the university, however, I began to feel that its atmosphere was like a paradise. The name has since changed, but it was formed together with its sister, the Institute for the Study of Metals (ISM, a wartime code name) in order to house people engaged in the Manhattan Project, and occupied two halves of a large building. I got to know all the people there right away, and was treated like a member of a big family. Every week Enrico Fermi held a discussion session with the theorists. I attended it a couple of times but soon he disappeared and passed away. The rest of INS members, besides Goldberger, included G. Wentzel, S. Allison, J. and M. Mayer, J. L. Marshall, H. Anderson, J. Simpson, H. Urey and S. Chandrasekhar. In the separate Physics Department were W. Zachariasen, M. Schein, and R. Mulliken. There was also a diverse group from Japan. Tadao Fujii had just got a PhD in nuclear physics,. H. Miyazawa was a research associate with Wentzel, R. Kubo and R. Kikuchi, my classmate at Tokyo, were members of ISM. M. Koshiba was a research associate in Schein's group. Y. Nakagawa was helping Chandrasekhar, doing experiments on magnetohydrodynamics. These people were all living within walking distance from the university. All the time we got invited to homes of senior professors. Especially famous was the grand year-end party at the Mayers, where an old barn house was filled up with people, virtually with standing room only. Regrettably this custom is gone now, presumably because it has become common for both husbands and wives to have a job. Following the tradition created by the universal physicist Fermi, we had a weekly institute-wide seminar. It was called Quaker meeting because there were no fixed agenda, everyone was encouraged to stand up and discuss any ideas one has at the moment, even if only half-baked. The topics ranged from all areas of the institute, from nuclear physics, particle physics, cosmic rays, solar phenomena to the physics and chemistry of the universe. This was a great stimulation for me.

INS had a 450 MeV synchrocyclotron built by Fermi and Anderson. It is here that the first hadron resonance called δ was discovered, and its interpretation was first given by K. Brueckner and H. Miyazawa ahead of Fermi. Also it is well known that the neutron–electron interaction was measured here. Frequently I would climb down a ladder to the control room in the pit, and chat with people there. I became especially close to Anderson.

On the theory side, Chicago was then a mecca of dispersion theory. It originated from the work of Gell-Mann and Goldberger, which was just the right tool for analyzing the experimental data coming in. Under Goldberger were H. Miyazawa, R. Oehme and myself. The mathematical structure of the dispersion theory fascinated me, and enabled me to forget the two years of bad dream. Besides collaborating with Golberger's friends F. Low and G. Chew at the University of Illinois, I made interesting discoveries about the representation of the Green's function, etc. These were just mathematical problems, but I was elated when L. Landau quoted my work. At the Kiev conference in 1959 he greeted me there as well as at his institute in Moscow.

Nevertheless, I had not forgotten the Sakata–Taketani philosophy. I was clearly conscious of it when I predicted the existence of a neutral vector meson in order to account for the property of the nucleon form factors. But the reaction I received was very cool. Resistance against new particles was still strong. The pion was the sole carrier of the strong interaction, the δ was only a π-nucleon resonance, they thought. Although I assumed the δ to be a new elementary particle, I postulated a low mass for it in order to make it stable against strong forces. But I was not clearly distinguishing between resonances and elementary particles.

According to Taketani's theory, the 1950s and 1960s were the time of transition from phenomenology to model building. It was generally recognized that Nature had, besides gravity, three forces, namely electromagnetic, strong, and weak, while the particles consisted of leptons (e and μ) and hadrons.[a] Regarding the weak forces, the Fermi theory modified with $V - A$ and Cabibbo mixing was satisfactory at least for the baryons. On the other hand, we were still in the dark about the true nature of the strong forces. Of course the QED was a complete success for electromagnetism. The concept of renormalization has introduced a new paradigm for quantum field theory. It was clear, though, that this was not enough for understanding the strong forces.

3. At this point I will stop for a moment to discuss the problems of methodology. There were two competing groups which I will tentatively call reductionists and generalists. The reductionists first try to find a symmetry or regularity in the phenomena, and trace it to the properties of the fundamental particles and interactions, Gell-Mann being their spokesman. The SU(2), SU(3) and SU(6) symmetries of flavor were established this way. But they were only approximate symmetries. Exact symmetries like the spacetime symmetry and the gauge invariance of electromagnetism were established by the first half of the 20th century.

[a]The name *lepton* is due to L. Rosenfeld; *hadron* is due to L. D. Okun.

On the other hand, approximate internal symmetries have since made their appearance in particle physics. They may be their inherent character, but may also reflect people's mode of thinking that is sensitive to similarities and analogies. In this respect I recall the following remark by A. Salam.[b]

"Classical physical theories are profound. Take the second law of thermodynamics, for instance. Heat cannot flow spontaneously from a colder body to a hotter body. Compare this to what you have been doing. You propose some symmetry, and ten seconds later you are already trying to figure out how to break it."

The gauge principle was generalized by Yang and Mills, and has come to be believed as an ideal theory for all fundamental forces. But it was not clear how it could be applied to the short range forces of the actual world having only approximate symmetries. J. J. Sakurai made a brave but tentative attempt to challenge the question with his vector dominance theory. He came to Chicago right after his PhD from Cornell University. His energetic efforts to develop his theory led to some successes, but what was lacking in completing his program was the recognition that the fundamental strong force were between hypothetical quarks, and the symmetries could break spontaneously.

The generalists, on the other hand, tried to clarify the general properties of quantum field theory, and use them in analyzing individual problems, in view of the fact that the fundamental laws are not completely known yet. It led to various theoretical tools such as axiomatics, S-matrix, dispersion relations, and Regge poles. The S-matrix was originally introduced in the 1930s by Heisenberg in connection with nuclear problems. Among the generalists, a radical group represented by G. Chew went beyond S-matrix and advocated the principle of particle democracy and bootstrap. This meant that there is no fundamental Lagrangian, all hadrons are bound states of each other. His charismatic character attracted many followers.

From the present point of view, it is clear that the royal road of physics was that taken by the reductionists. Yukawa's meson did what the S-matrix was not able to do. Thus the success of Yukawa's approach continued and eventually led to the present Standard Model. The road taken by the generalists, however, was not an idle one. Not only have the concepts like the CPT theorem been very useful for phenomenological analyses, but also today's superstring theory may be called its descendant. Thus the competition is still on, or rather they are moving toward unification, one might say.

Going back to the 1960s, I would first like to describe the social and political atmosphere of the time as I remember it. The 1960s were indeed an eventful decade. The assassination of the Kennedys was followed by the Vietnam war and the anti-war movement went hand-in-hand and gradually escalated. The anti-discrimination movement also followed the same path, leading to the assassination of Martin Luther King and the ensuing riots. At the same time, a rocket carrying man reached the moon. Hanging over all these events was the thickening dark cloud of the cold war and the atom bomb. The physicists had lost their purity because of their special role played in the War. This resulted in the big reward

[b]This is quoted by J. J. Sakurai in *Ann. Phys.* **11** (1960), p. 1.

they received from the government. As their research budget reached astronomical levels, the accelerator energy kept going up following the Livingstone curve. but they also had to collaborate in the cold war and accept constraints.

Under these circumstances the role played by R. Marshak in creating the so-called Rochester Conference merits special attention and appreciation. I first met him in Rochester on my way to Princeton. He was the chairman of the department, and had also started to recruit able students from Japan. It is well known that this program has produced several illustrious physicists. I attended the second Rochester Conference, with about 20 invitees. He made a special effort to make it easy for foreign physicists to attend it, particularly the Russians. The original small local meeting soon grew and became international. Going round from U.S. to Europe and Russia, its size increased to over a thousand. The Conference was now a symbol of particle physics. Several historical discoveries were announced here. I also saw before my eyes fierce altercations between people of strong ego.

4. My discovery of spontaneous symmetry breaking (SSB) was an accidental event. But I may also say that I owe this to my education at Tokyo which had induced in me an interest in condensed matter physics, as well as to the proximity of Chicago to the birth place of the BCS theory. In my days at Osaka City University, I was attracted by the plasma theory of Bohm and Pines, and tried to generalize the many-body problem as a quantum field theory. The paper I wrote with Kinoshita at IAS was its outcome. My boss, G. Wentzel, was also interested in it, and I already knew the people at the University of Illinois: J. Bardeen, D. Pines, F. Seitz, etc. One day in 1957, R. Schrieffer, a student of Bardeen's, gave a seminar at the invitation of Wentzel, and told us about the BCS theory. The theory of Cooper pairs by L. Cooper was already known, but the BCS theory was not yet published.

My impression of the seminar was a mixture of admiration and doubt. I admired the fact that their assumed Hamiltonian led to a condensate, and formed an energy gap. At the same time, I also realized that their wave function was not an eigenstate of charge, and violated gauge invariance. How can one discuss electromagnetic properties like the Meissner effect with this theory? Shouldn't one first check gauge invariance in calculating a process involving electromagnetism? The doubt grew deeper and deeper as the success of the BCS theory became more and more conspicuous.

It took me about two years before I became convinced of the correctness of the BCS theory. In essence, I realized that a collective mode satisfied the Ward–Takahashi identity, thereby restored charge-current conservation and gauge invariance. I also found that the SSB made the electromagnetic field acquire a mass and turn into a plasmon. It is a result of mixing the two zero-modes, the Coulomb field and the Nambu–Goldstone (NG) mode. Some condensed matter physicists gave a similar account, but I was quite satisfied that I was able to clarify the issue mathematically in terms of quantum field theory.

Another point that had attracted my attention in the BCS theory from the beginning was the similarity between the Bogoliubov–Valatin quasi-fermion and the Dirac electron. The former is a superposition of states of different charge, whereas the latter is a superposition of different chirality, and the mixing leads to the energy gap formation. As soon as I resolved the gauge invariance problem, I noticed that the Goldberger–Treiman relation between the pion decay constant and nucleon mass was nothing but the Ward–Takahashi identity for chiral invariance, except that I had to make a bold assumption to ignore the pion mass. It was nice that I could now understand at least the origin of the nucleon mass.

At the 1959 Rochester Conference in Kiev there was a talk by B. Toushek about the chirality of neutrinos. After the talk I made a brief comment about my ideas, but due to the miserable condition of the Soviet Union it took two years before the conference proceedings came out. In the meantime I continued my work with a research associate G. Jona-Lasinio. After a preliminary report at a conference at Purdue University, I presented my main paper (NJL) at the 1960 Rochester Conference in Geneva. There V. G. Vaks and A. I. Larkin also presented a brief paper, but did not mention the relation to pion physics.

As I have mentioned above, there are two approaches in the methodology of physics. But my work does not belong to either of them. Instead of establishing a symmetry, it gets broken. It also breaks the axiom of the uniqueness of the vacuum. I was most afraid of objections from the second group of people who values mathematical rigor above all. Being influenced by the Sakata school, however, I am always tempted to think of concrete objects behind mathematical expressions. Thus I was fond of Dirac's interpretation of the vacuum as a sea of negative energy states. In fact, my theory was more popular among condensed matter theorists than particle theorists, I felt. As regards the N-G boson, my intension was to write a paper after collecting examples in condensed matter physics. On this point I was conscious of Tomonaga's advice not to put too many things in one paper. Heisenberg's unified theory is a bad example of this. He may have chosen his nonlinear theory as an example to show that symmetry is the guiding principle of physics, but he then claims that it is "the theory of everything". But to do that he then has to introduce contraptions like a vacuum with indefinite metric and isospin, etc. If one of them turns out wrong, the whole theory will collapse even if the others remain right or useful. Put everything in one bag, and you can lose everything. I did not want to commit the same mistake. I wanted to show the possibility of SSB in the clearest way possible under the clearest assumptions so that it can be understood by everybody. First I chose a renormalizable example like QED, but I had trouble handling higher order effects. So I changed my mind and adopted a simpler Fermi–Heisenberg type model with simple cutoff. As an added remark, Heisenberg praised my work at the 1960 Rochester Conference. It was my fond memory that thereafter I visited him in Munich, had dinner at his home, and also spoke with his collaborator K. Yamazaki.

For a few years since then I was busy applying my theory to various hadronic problems, starting the so-called soft pion physics working with research associates Y. Hara and others. My paper on the loop expansion of the S-matrix is also related to it. In the meantime the SU(3) symmetry was discovered. The quark model made its appearance. The NJL model treats the nucleon as the fundamental field, but it can also be used as an effective theory of QCD by substituting nucleons with quarks. In the 1960s I was occupied with strong interactions as the first problem to resolve, and did not think of weak interactions yet. It was not clear either how to apply the theory of gauge fields to strong interactions.

I will add a story of what happened since then. The BCS theory can be rewritten as an effective theory of the Landau–Ginzburg–Higgs (LGH) type, so it also contains, besides the NG boson, a "Higgs" boson having twice the fermion mass. It is the sigma meson in the case of hadrons. But in the case of superconductivity, I did not realize this until told by C. M. Varma. I then studied this and discovered that the same relation holds in the collective mode of Helium superfluid. The approximate 1:2 ratio is a kind of supersymmetry which I called quasi-supersymmetry. As in the case of supersymmetry, one can also factorize the Hamiltonian. So I decided to investigate all BCS type phenomena systematically. In fact the BCS theory was already used to interpret the pairing energy. I directed my attention to the Interacting Boson Model (IBM) of Arima and Iachello. It is more complicated than superconductivity or superfluidity, but I found that one of the symmetry breaking chains can be interpreted as SSB. However, my proposal did not attract much attention by the nuclear physics community because I was not able to make precise predictions.

The next problem I took up was the more important one, namely the interpretation of the Higgs field. In the Standard Model it plays the role of giving mass to the W and Z bosons as well as to the fundamental fermions. But it cannot make predictions about the latter. They are mere adjustable parameters. I wanted to imagine some dynamics behind the Higgs field like the BCS theory. Naturally it gave rise to ideas like top condensation. The Higgs boson being a scalar, its exchange between fundamental fermions cause an attraction, so it plays the role of the phonon in superconductivity, causing SSB which in turn leads to the formation of a bound state, the Higgs boson. It is a kind of bootstrap mechanism Higgs $\rightarrow$ SSB $\rightarrow$ Higgs mediated by the top quark. This was my idea. W. Bardeen wrote it in the form of NJL and made more precise calculations. The Higgs mass turned out somewhat lower but did not agree with experimental indications. Besides, the problem of the masses of other fermions still remained.

5. Again I will return to the 1960s. Gauge invariance was my concern in the case of the BCS theory. On the other hand, it was the problem of statistics in the case of the quark model, where the quark is treated as a boson in hadronic phenomenology. On this point Gell-Mann was not very clear, saying that the quark may be a mere mathematical symbol. I was not satisfied with this. O. Greenberg proposed to solve the problem in terms of statistics of order three, but it was again too formal. So I thought of increasing quark's internal degrees of freedom. As a minimum I started with two, but the result was not very transparent. When I ventured to three, there emerged a beautiful symmetry; all hadronic states could be interpreted as SU(3) singlets. T. Takao, Y. Miyamoto and A. Tavkhelidze also published similar ideas but they were different in details.

As soon as my reprint was sent out, a modified one with some group theoretical improvements reached me from M. Y. Han, a post-doctoral research associate at Syracuse University. In this way our joint paper was born. There I postulated a superstrong force, to be represented by an SU(3) (or three-color) gauge field. But I was completely in the dark about how this could be realized. In those days the non-Abelian gauge theory had not yet been firmly established. Subsequently I noticed that the SU(3) degree of freedom could be used to assign integer charges to all quarks. But this assignment is not SU(3) invariant. To remedy this I used a QED analogy to show that the SU(3) singlet had the lowest energy. At any rate, the color degree was introduced for theoretical reasons. Then phenomenologically hadrons could also have color. As the number of new hadrons grew, however, there was no indication for the flavor SU(3) classification to break down. It was natural that my idea was ignored by the phenomenologists.

These efforts prompted me to study group representations in general. It was partly for reasons of general interest, but also because I wanted to find some order in hadronic spectroscopy. The linearly rising Regge trajectory seemed to suggest some dynamical symmetry like the SU(4) symmetry in hydrogen atom. In fact wave equations involving multiple spins or masses had been studied by many people. Dirac, for example, used an infinite dimensional representation of the Lorentz group. As a first exercise I found that its typical unitary representations can be constructed in terms of the familiar creation and annihilation operators, and found a spectrum of the inverse hydrogen type. Later I learned that this was a generalization of the equation discovered by E. Majorana in 1932. I also examined the nonrelativistic hydrogen atom, and was able to express the Schrödinger equation in terms of creation and annihilation operators. As it happens, similar programs were being pursued by A. O. Barut, C. Fronsdal, and T. Takabayashi independently, so I kept close touch with them for two years or so.

The program of unitary representations, however, soon collapsed. As it turned out, unitary representations of the Lorentz group are in general afflicted with various diseases such as violation of CPT, spin-statistics, causality, existence of tachyons; unnatural mass spectra and g-factors. Some of the troubles can be avoided by a judicious choice of representation, but a general principle is lacking. After all, the nature seems to have adopted the local field theory. It is not possible to pick only a part of the properties of nature and try to describe it in a simple way, it seems. This was also what I experienced in my days at Osaka City University in the case of the "Bethe–Salpeter equation" written in a differential form (in ladder approximation). Although it looked simple and elegant, I did not pursue it because of the existence of unphysical solution. So I decided to give up this time, too.

It is ironic that the next year, i.e. 1968, the Veneziano model made its appearance. It is not field theory, but a realization of the duality principle derived from Chew's idea of bootstrap Regge trajectory. I was immediately attracted by it although I had just given up on the mass spectrum problem. I wondered what physics lay behind his formula. The first thing to do was to decompose it into a sum of Breit–Wigner resonances, determine their multiplicity, and to check if the residues are positive. I started to work on this with P. Frampton, and soon found that the residues were positive at least asymptotically. The next task was to find a general algorithm for this decomposition. The Veneziano formula represents two-body scattering expressed by a beta function. While I was playing with its integral representation of Koba and Nielsen, one day I expanded the second factor into a power series:

$$B(-s+a,-t+b)=\int_0^1 x^{-s+a-1}(1-x)^{-t+b-1}dx,$$

$$(1-x)^{-t+b-1} = \exp\left[(-2p\cdot p' + 2m^2 - b + 1)\Sigma\frac{x^n}{n}\right].$$

If $2m^2-b+1=0$, each term is a function of p and p'. The factor $1/n$ is reminiscent of the energy denominator $1/E$ that appears in the Fourier expansion of Green's function. This means $E \approx n$, namely a one-dimensional harmonic oscillator, isn't it? Thus the idea that the hadron is like a string was born. In order to account for the constant terms, however, it was necessary to introduce extra dimensions like in the Kaluza–Klein theory. I thought this to be unnatural, and did not make further progress for a year or so.

In the summer of 1970, a Rochester Conference was held in Kiev. I also got an invitation to speak at a summer school in Copenhagen. I decided to leave my family with a friend in California before flying to Europe. But our car broke down while driving through the Great Salt Lake Desert in Utah. We had to call help from Salt Lake City to fix the engine, and stay for three days in a town called Wendover in the desert. (Years later J. Cronin set up an experimental project called CASA in a nearby desert to detect cosmic gamma rays.) I missed the flight to Europe, and had to spend the summer in California with my family. I had sent my lecture note to Europe already. Being only a note, I casually wrote down a worldsheet action (the Nambu–Goto action), and trusted that it will be published in the proceedings. But the proceedings never came out. I am grateful to the good will of people who kindly quoted the unpublished note.

6. The 1960s were the time of clarifying the properties of strong forces. We had reached Taketani's second stage, and a glimpse had appeared on the final stage of the Standard Model. The properties of quark dynamics could be gleaned from deep inelastic scattering, Feynman's parton model and the like. But there was no further decisive step. Toward the end of the decade, people were generally in a pessimistic mood. I was driven to this mood not only because of the unpopularity of the three-color theory and the failure of infinite component equations, but also I was greatly influenced by the social and political turmoil of the period. The latter affected my personal life, and I began to fear for my security. As the Vietnam war escalated, our research condition tightened. Budget cuts caused massive unemployment of scientists. Rumor was circulating of PhD cab drivers. R. Marshak, then president of American Physical Society, asked A. Wightman and me to do an unemployment survey among fresh PhDs. The result was a gloomy one, which affected me, too. That was the year of the accident in Utah.

Thereafter I made some progress. It was clear that strings can confine quarks. On the other hand, I also had a theory of observable integer charge quarks. Which of them should I favor? In the end, however, the string theory had hit a wall as the fundamental theory of hadrons. So reported K. Kikkawa to me as a conclusion reached at an Aspen meeting. That was because string theory was not a local field theory, as in the case of infinite component equations. Around just that time there occurred the discovery of asymptotic freedom which replaced string theory to resolve the problem of quark confinement by QCD, an orthodox quantum field theory. Thus string was an approximate description of one aspect of strong forces.

Particle physics began in the 1930s and took 40 years to reach maturity in the 1970s. QCD as a theory became firmly established, and together with the Salam–Weinberg theory formed the so-called Standard Model. The appearance of J/ψ and D meson were discovered and confirmed the existence of the second generation of quarks and leptons expected by the works of Cabibbo and the Sakata group, and immediately this was followed by the third generation as predicted by Kobayashi and Maskawa. The fundamental laws governing particle physics is essentially complete. From now on our task will be to check the Standard Model in more detail. If disagreements are found, it will then signal the beginning of the next cycle, as is expected or hoped for by everybody. These tests were already in progress.

More noteworthy, however, is the dramatic change of the character of particle physics that has taken place. It may be said that the 1970s were a period of phase transition. Throughout the history of physics, theory and experiment walked hand in hand. But now the Standard Model was immediately followed by supersymmetry and supergravity, and theory started to walk alone. Its character has become mathematical and formal. People trained at school in phenomenology cannot follow it and drop out. (A similar phenomenon must have occurred after relativity and quantum mechanics.) At the same time, theory and experiment joined hands to start a rapid growth together in cosmology and astronomy. Physics is going to cover the entire range from the ultimate small to the ultimate large in a unified way. Another change is that particle physics has become big science. Research is now a collective enterprise both in theory and experiment. A good example is superstring, as you can see in the great number of papers that appear on the internet everyday. All these circumstances also seem to have led to a slower pace of progress than in preceding years.

References

1. Translated from Bulletin of *Jap. Phys. Soc.* **57**, No. 1, 2003. For the history of particle physics, see, for example, *Prog. Theor. Phys. Suppl.* No. 105 (1991); V. L. Fitch and J. Rosner, *Twenty Century Physics* (Amer. Inst. Physics, 1995), p. 665.

Progress of Theoretical Physics, Vol. 5, No. 4, July~August, 1950

Force Potentials in Quantum Field Theory*

Yoichiro NAMBU
Osaka City University

(Received July 31, 1950)

1. Introduction and Summary

As early as 1935 Yukawa conjectured that the nuclear force, which ties together the component nucleons into a solid nucleus, could be attributed to an intermediate field with an intrinsic mass corresponding to the range of the nuclear force. The general success of the meson theory that followed the discovery of such particles in cosmic rays, has been so great that one cannot now discuss the nuclear and cosmic phenomena without the help of Yukawa's idea on the nature of mesons. One must admit, however, that the meson theory in the present stage is far from satisfactory almost in every detail regarding its quantitative predictions on various phenomena. This may be due to our insufficient knowledge about the nature of the real mesons as well as to a more profound crisis of the present quantum field theory which precludes us from drawing reasonable conclusions out of physical assumptions. Both difficulties are intrinsically connected with each other and their complete solution does not seem to be an easy one.

Here we will pick up one characteristic feature of the Yukawa theory, i.e. the theory of nuclear forces. Up to now, however, there is a wide gap between the nuclear potentials predicted on the basis of various meson theories and the one that is obtained empirically. No simple assumptions have ever succeeded in explaining the quantitative behavior of the internucleonic interactions. The theoretical side of the difficulty seems to consist of two facts. One is the high singularity of the nuclear potential which is more or less of a common origin in the present quantum theory, while the other is that we have no established formalism of the relativistic many-body problem, a peculiar situation encountered in the theory of the deuteron. At least formally, the S-matrix theory of Heisenberg affords a means to deal equally with scattering processes and bound systems. But the latter case is more helpless since the higher order terms in the coupling constant play an essential role in the determination of energy levels. In the non-relativistic limit, such higher order effects were properly represented in the interaction potential appearing in the Schrödinger eigenvalue

* Preliminary report: Prog. Theor. Phys. 5 (1950), 321. Some errors are corrected in the text.

problem, a typical case being that of the hydrogen atom. At the genesis of the meson theory, Yukawa borrowed this successful idea from the electromagnetic theory. There are, however, some remarkable distinctions between the electromagnetic and meson field. The latter has mass and charge, together with strong coupling constants. The Coulomb potential, being not quantized, was an exact theoretical consequence, while there is no such classical potential in the mesonic case. These situations exhibit themselves in the relativistic effect, or the retardation on one hand, and in the quantum effect, or the recoil and higher order forces on the other. Before deciding between the existing "korrespondenzmässigen" theories and any more revolutionary ones, we probably have to re-examine the old concept of potential and improve it so as to conform more closely to the rigorous field theoretical view.

In the following sections we apply the recent method of quantum electrodynamics to the analysis of these problems. At the present stage, however, we have to abandon the complete relativistic covariance, which is one of the most beautiful achievements of the new theory, but rather confine ourselves to more crude and approximate considerations. In Section 2 we begin with the derivation of the second order potential with the aid of the covariant field theory, and point out the ambiguities that naturally arise. The latter leads to the consideration of the fourth order potential in the next section. An extension of these methods to still higher order terms seems almost impracticable, but very simplified arguments are presented in the last section. The examination, however, of the existing individual meson models in the light of the obtained results will be carried out only on a later occasion.

2. The Second Order Potential

The derivation of the second order interaction potential between two Fermi particles has been carried out in various ways. Originally the Coulomb potential for charged particles was directly imported from the classical theory. Its relativistic correction was first derived by Breit[1] also on the basis of the correspondence principle. He afterwards developed his principle of the approximate relativistic covariance[2] to obtain some general expressions for electromagnetic as well as nuclear potentials with relativistic corrections. On the other hand, perturbation calculations based on the quantized field theory were carried out by many authors[3], with the results that were essentially the same as that expected from the correspondence arguments. In the meson theory an attempt was also made[4] to use canonical transformations instead of the ordinary perturbation method. Very recently this was replaced by the more elegant and perfect theory of Tomonaga and Schwinger[5], from which standpoint some papers have already been published on the meson potentials[6]. All these results seem to be in general, but not complete, agreement. The correct interpretation of them will only be given by a careful inspection and reflection on the meaning of the potential.

Here we start from the Tomonaga-Schwinger equation (in natural units):

$$i\frac{\delta}{\delta\sigma}\Psi[\sigma]=H(x)\Psi[\sigma], \tag{1}$$

$$H(x)=-\varepsilon j_\lambda \varphi_\lambda \tag{2}$$

for the interaction of a fermion field ψ with a boson field φ. ε is the coupling constant, and j_λ a quantity bilinear in $\bar{\psi}$ and ψ. λ represents the concurrent tensor and isotopic spin indices. By the routine transformation[11]

$$\Psi(\sigma)=\exp\left[-\frac{1}{2}\int_{-\infty}^{\infty}H(x')\varepsilon(\sigma\sigma')(dx')\right]\Psi_1[\sigma], \tag{3}$$

we obtain

$$i\frac{\delta}{\delta\sigma}\Psi_1[\sigma]=-\frac{1}{4}i[H(x),\int_{-\infty}^{\infty}H(x')\varepsilon(\sigma\sigma')(dx')]\Psi_1[\sigma] \tag{4}$$

to the second order in ε. For a system in which real bosons are absent, this gives

$$i\frac{\delta}{\delta\sigma}\Psi_1[\sigma]=\left[-\frac{1}{2}\varepsilon^2 j_\lambda(x)\int_{-\infty}^{\infty}j_\mu(x')\bar{\Delta}_{\lambda\mu}(x-x')(dx')\right]\Psi_1[\sigma]\equiv H^{(2)}(x)\Psi_1[\sigma]$$

using the commutation relation

$$\begin{aligned}&[\varphi_\lambda(x),\varphi_\mu(x')]=i\Delta_{\lambda\mu}(x-x')=iO_{\lambda\mu}\Delta(x-x'),\\&\bar{\Delta}_{\lambda\mu}(x)=-\frac{1}{2}\varepsilon(x)\Delta_{\lambda\mu}(x)\ \text{or}\ -\frac{1}{2}O_{\lambda\mu}(\varepsilon(x)\Delta(x)),\end{aligned} \tag{6}$$

where $\Delta_{\lambda\mu}(x)$ contains the essential factor $\Delta(x)$ as defined by Schwinger. In order to get the usual potential form we make use of the Fourier decomposition of $\bar{\Delta}$, or more conveniently the operational relation

$$\begin{aligned}&\int_{-\infty}^{\infty}\bar{\Delta}(x)f(x)dx_0=\frac{1}{4\pi}\frac{1}{r}\exp\left(-r\sqrt{\mu^2+(d/dx_0)^2}\right)f(x)|_{x_0=0},\\&r^2=\sum_{i=1}^{3}x_i^2,\qquad x_0=t,\end{aligned} \tag{7}$$

where μ characterizes the mass of the field. In case $\mu=0$,

$$\begin{aligned}\int\bar{D}(x)f(x)dx_0&=\frac{1}{4\pi}\frac{1}{2r}\left(\exp\left(-r\frac{d}{dx_0}\right)+\exp\left(r\frac{d}{dx_0}\right)\right)f(x)|_{x_0=0}\\&=\frac{1}{4\pi}\frac{1}{2r}(f(\boldsymbol{r},t-r)+f(\boldsymbol{r},t+r))\\&=\frac{1}{4\pi}\frac{1}{2r}(f(\boldsymbol{r})_{\text{ret}}+f(\boldsymbol{r})_{\text{adv}}).\end{aligned} \tag{8}$$

We should apply these formulas after making the space-like surface σ flat. Thus in Eq. (5) the interaction becomes

$$H^{(2)}(x)=-\frac{1}{2}\frac{\varepsilon^2}{4\pi}j_\lambda(\boldsymbol{r})\int O_{\lambda\mu}\,V(\boldsymbol{r}-\boldsymbol{r}')j_\mu(\boldsymbol{r}')d\boldsymbol{r}', \tag{9}$$

$$V(\boldsymbol{r}-\boldsymbol{r}')=\begin{cases}\dfrac{1}{|\boldsymbol{r}-\boldsymbol{r}'|}\exp(-|\boldsymbol{r}-\boldsymbol{r}'|\sqrt{\mu^2+(d/dt)^2}), & \text{or}\\ \dfrac{1}{|\boldsymbol{r}-\boldsymbol{r}'|}\dfrac{1}{2}\left\{\exp\left(-|\boldsymbol{r}-\boldsymbol{r}'|\dfrac{d}{dt}\right)+\exp\left(|\boldsymbol{r}-\boldsymbol{r}'|\dfrac{d}{dt}\right)\right\}.\end{cases} \tag{10}$$

$H^{(2)}(x)$ means the potential energy density for the particle at x due to other particles. In the two-particle system in which we are interested, the total potential energy will be written down in configuration space as

$$H^{(2)}(12)=-\frac{1}{2}\frac{\varepsilon^2}{4\pi}\{j_\lambda(1)\,O_{\lambda\mu}V(12)j_\mu(2)+j_\mu(2)O_{\mu\lambda}V(21)j_\lambda(1)\}, \tag{11}$$

where 1, 2 refer to the numbering of the particles, and j_λ means the quantity introduced before, but represented in the configuration space. V is an operator operating on the quantity that comes after it. We may regard (11) as the potential energy to be inserted in the ordinary Schrödinger equation, in which j_λ is a constant operator and d/dt is replaced by

$$\frac{d}{dt_i}=i\,[H_i^{(0)},\ \ ],\qquad i=1,2, \tag{12}$$

$H^{(0)}$ being the free state Hamiltonian for each particle. We note here that the r-dependent potential itself should always be thrown under the operation (12). Eq. (10) tells us that V corresponds to the attached field introduced by Dirac and others[8]. Its appearance is a consequence of the Schwinger transformation (3). When the real boson field is absent (no incoming and outgoing waves), the attached field is equivalent to the retarded field which follows from the transformation

$$\Psi[\sigma]=\exp[-i\int_{-\infty}^{\sigma}H(x')(dx')]\,\Psi_1[\sigma]. \tag{13}$$

Though they should be equivalent under the above condition, the form of the potential derived from these transformations is apparently not equivalent. Since such a condition is eliminated in the Schrödinger equation, we here meet with an ambiguity. On physical grounds, however, the symmetrical form may be preferred to the retarded form which is not invariant under the interchange of past and future. It is to be noted that the ambiguity does not arise in the non-relativistic limit for which the familiar potential $1/r$ or $\exp(-\mu r)/r$ results.

Instead of using the rather unesthetical one-time theory to derive the potential, we may also invoke the formula[11]

$$\psi(x)=\int_\sigma S(x-x')\,\gamma_\mu\,\psi(x')\,d\sigma'_\mu,$$

$$\bar\psi(x)=\int_\sigma d\sigma_\mu'\,\bar\psi(x')\,\gamma_\mu\,S(x'-x), \tag{14}$$

with
$$\frac{1}{i}\,S(x-x')=\{\psi(x),\,\bar\psi(x')\},$$

and express the retarded and advanced quantities $j_\lambda(x)$ in terms of ψ and $\bar\psi$ on the surface σ. The resulting interaction is non-local, or in other words, velocity dependent. The Schrödinger equation with such an interaction is then an integro-differential equation, and its solution will never be an easy task. The non-local effect, however, will set in only at very close distances for which large momentum transfers between the particles become important. Thus if we will take account of the retardation in an approximate manner, we may usually be allowed to expand (10) in powers of d/dt and retain only the first few terms. But if j_λ contains a Zitterbewegung (or non-static) part, we cannot do this, for it means the transition from positive to negative state or vice versa, and $d/dt\sim\pm 2mi$, where m is the mass of the particle. If $2m$ is much larger than μ, (10) gives a rapidly oscillating function of the form $\cos(2mr)/r$ which had rather better be substituted by a delta function:

$$\frac{1}{4\pi}\,\frac{1}{r}\cos 2mr\sim-\frac{1}{4m^2}\Big(1-\frac{1}{4m^2}\,\Delta+\cdots\Big)\,\delta(\boldsymbol{r}). \tag{15}$$

The problem of ambiguity mentioned above, however, is not yet settled at all. We show this with the example of the Breit interaction. In this case,

$$H(x)=-\varepsilon j_\lambda A_\lambda,\qquad j_\lambda=i\bar\psi\gamma_\lambda\psi, \tag{16}$$

and

$$H^{(2)}(x)=-\frac{1}{2}\,\frac{\varepsilon^2}{4\pi}\int j_\lambda(x)\,\frac{\cos\left(\frac{d}{dt'}\right)}{r}\,j_\lambda(x')\,d\boldsymbol{r}'. \tag{17}$$

This is the Møller formula.[3] The fourth component of the product $j_\lambda j_\lambda'$ corresponds to the Coulomb force with small retardation effect. An expansion in d/dt, which is allowed here, yields the potential in configuration space

$$H^{(2)}(12)=\frac{\varepsilon^2}{4\pi}\left[(\gamma_\lambda)_1\,\frac{1}{r}\,(\gamma_\lambda)_2+\frac{1}{4}\left(\frac{d^2}{dt_1^2}+\frac{d^2}{dt_2^2}\right)(\gamma_4)_1 r(\gamma_4)_2\right]+\cdots \tag{18}$$

When this is evaluated, we do not obtain the Breit formula, contrary to our expectation! However, (18) may also be written as

$$H^{(2)}(12)=\frac{\varepsilon^2}{4\pi}\left[(\gamma_\lambda)_1\,\frac{1}{r}(\gamma_\lambda)_2-\frac{1}{2}\,\frac{d}{dt_1}\,\frac{d}{dt_2}\,(\gamma_4)_1 r(\gamma_4)_2\right], \tag{19}$$

for the energy conservation means

$$\frac{d}{dt_1}=-\frac{d}{dt_2} \tag{20}$$

neglecting the interaction term which amounts to a higher order correction. From (19) now follows the desired formula :*

$$H^{(2)}(12)=\left[(\gamma_4)_1(\gamma_4)_2-\frac{1}{2}\left\{(\gamma_i)_1(\gamma_i)_2+\frac{((\gamma_i)_1,\boldsymbol{r})((\gamma_i)_2,\boldsymbol{r})}{r^2}\right\}\right]\frac{1}{r}. \tag{21}$$

The assumption (20) was also made by Bethe and Fermi[3], and more explicitly by Toyoda[9] in meson theory. It is true that an ambiguity does not influence the second order S-matrix corresponding to the elastic scattering of two particles, since in this case (20) holds exactly. But a potential in its proper sense should contain those matrix elements in which energy is not conserved and which brings the system into a virtual state. From the above result we see that the potential is indeterminate to within an arbitrary matrix whose diagonal elements with respect to energy are zero. Let us try to find out how this problem can be settled.

We make here a requirement that the two-body problem shall be rendered to a relativistic form as far as possible. Thus we assume a pair of the Dirac many-time equations for the wave function $\psi(x_1, x_2)$ describing a two-particle system under action at a distance :

$$\begin{aligned}\left\{i\frac{\partial}{\partial t_1}-H_1-V(12)\right\}\psi=0,\\ \left\{i\frac{\partial}{\partial t_2}-H_2-V(21)\right\}\psi=0.\end{aligned} \tag{22}$$

H_1 and H_2 being the Hamiltonians for free state. In general the potentials $V(12)$ and $V(21)$ may be different. By a standard transformation (22) can be brought to the interaction representation

$$\begin{aligned}\left\{i\frac{\partial}{\partial n_1}-V(12)\right\}\psi=0,\\ \left\{i\frac{\partial}{\partial n_2}-V(21)\right\}\psi=0.\end{aligned} \tag{23}$$

where the time-like parameters n may point to different directions, but for the time being assumed to be parallel. Now an integrability condition should be enounced :

$$i\left(\frac{\partial V(21)}{\partial n_1}-\frac{\partial V(12)}{\partial n_2}\right)=[V(12), V(21)]. \tag{24}$$

---

* An alternative method to derive (21) is to separate the pure Coulomb interaction beforehand and treat only the transversal part in the above way.

Suppose the potentials V to be defined in terms of the variables of the two particles which stand in a definite relation to each other with respect to their space-time position, thus taking account of the retardation effect. Then, since the relative position of the particles should remain invariant in V and n_1 is parallel to n_2, we have

$$\frac{\partial V}{\partial n_1}=\frac{\partial V}{\partial n_2}, \tag{25}$$

so that

$$i\frac{\partial}{\partial n}(V(21)-V(12))=[V(12),\ V(21)], \tag{26}$$

which in turn requires that

$$V(12)=V(12), \tag{27}$$

as will be seen by developing V in a power series in the coupling constant. (27) may be regarded as expressing Newton's third law of motion. Let us test this condition for the above mentioned example. The Møller interaction gives, in view of (17) and (18),

$$\begin{aligned} V(12)\,|_{t_1=t_2}&=\frac{\varepsilon^2}{4\pi}\cdot\frac{1}{2}\left\{(\gamma_\lambda)_1\frac{1}{r}(\gamma_\lambda)_2+\frac{1}{2}(\gamma_\lambda)_1\frac{d^2}{dt_2^2}r(\gamma_4)_2\right\},\\ V(21)\,|_{t_1=t_2}&=\frac{\varepsilon^2}{4\pi}\cdot\frac{1}{2}\left\{(\gamma_\lambda)_1\frac{1}{r}(\gamma_\lambda)_2+\frac{1}{2}(\gamma_4)_2\frac{d^2}{dt_1^2}r(\gamma_4)_1\right\}, \end{aligned} \tag{28}$$

hence clearly $V(12)\neq V(21)$! To make the equation of motion (23) integrable, we must therefore supplement the potentials by some additional terms. This can be performed by a transformation independent of past history:

$$\psi=\exp\left[\frac{i}{4}\frac{\varepsilon^2}{4\pi}\left(\frac{d}{dt_1}+\frac{d}{dt_2}\right)(\gamma_4)_1 r(\gamma_4)_2\right]\psi', \tag{29}$$

with the result

$$V'(12)=V'(21)=\frac{\varepsilon^2}{4\pi}\frac{1}{2}\left[(\gamma_\lambda)_1\frac{1}{r}(\gamma_\lambda)_2-\frac{1}{2}\frac{d}{dt_1}\frac{d}{dt_2}(\gamma_4)_1 r(\gamma_4)_2\right]+O(\varepsilon^4), \tag{30}$$

which just leads to (21).

The meaning of the above condition may be illustrated as follows. Let us describe the retarded interactions V by the Feynman diagram. So long as these interactions occur successively with sufficiently large intervals of time, there is nothing difficult. But when they come nearer and cross each other, there arises a difference in the order of emission and absorption of quanta represented by the process $V(21)\,V(12)$ or $V(12)\,V(21)$ and the one corresponding to the actual process, which always occurs in the chronological order. Such a situation is characteristic of the relativistic interaction, for the range of the time interval in which the crossing of the lines prevails will tend to zero as $c\to\infty$. In the expression of the S-matrix, however, these discrepancies are properly compensated, giving the correct result in the fourth order term. Thus we may suppose that in amending the second order potential so as to satisfy the condition (27) we are automatically taking account, at least partially, of the higher order interactions. The ambiguities in the second order potential will then be reflected in the higher order ones in such a way that the overall effect turns out identical. In other words, they will correspond simply to our freedom of canonical transformation of the whole representation.

Anticipating the justification of the above considerations in later sections, we shall next derive an alternative form of the potential which seems more convenient for our purpose. From the point of view of the S-matrix, the transformation (3) is a rather clumsy procedure. We had better write down the S-matrix after Dyson[10] as

$$S=1+\int H(x)\,dx+\frac{1}{2!}\iint P(H(x_1),\,H(x_2))\,(dx_1)\,(dx_2)+\cdots, \tag{31}$$

and identify its sub-matrix corresponding to the two-particle system with

$$1+\int V(x_1,x_2)\,(dX)+\frac{1}{2!}\iint P(V(x_1,x_2),\,V(x_1',x_2'))\,(dX)\,(dX)'+\cdots, \tag{32}$$

where X means a set of coordinates describing the motion of the system as a whole, and the symbol P is defined with respect to these coordinates. Indeed, if a relativistic two-body problem could actually be formulated, we could consider a. generalization of the center of gravity to which belongs the constants of motion, or the momentum-energy vector. In case the two particles are identical, it is natural to adopt for X

$$X_\mu=\frac{1}{2}\,((x_\mu)_1+(x_\mu)_2), \tag{33}$$

the remaining variables being the relative coordinates

$$x_\mu=(x_\mu)_1-(x_\mu)_2, \tag{34}$$

so that

$$x_1=X+\frac{x}{2},\qquad x_2=X-\frac{x}{2}. \tag{35}$$

In terms of these variables V becomes, to the second order,

$$H^{(2)}\equiv V(X)=\frac{1}{2}\int P\Big(H\Big(X+\frac{x}{2}\Big),H\Big(X-\frac{x}{2}\Big)\Big)(dx). \tag{36}$$

With the expression (2), it is

$$H^{(2)}=-\frac{i}{4}\varepsilon^2\int j_\lambda\left(X+\frac{x}{2}\right)\Delta_{F\lambda\mu}(x)j_\mu\left(X-\frac{x}{2}\right)(dx),$$
$$\Delta_F(x)=\Delta^{(1)}(x)-2i\bar{\Delta}(x). \tag{37}$$

The appearance of the function $\Delta_F(x)$ as the vacuum expectation value is a common feature of the Feynman-Dyson theory.[5][10] This makes, at first glance, the potential function non-Hermitic, but actually the term with $\Delta^{(1)}(x)$ does not contribute to the process and may be omitted. In case the retardation expansion is allowed, (37) gives

$$\begin{aligned}H^{(2)}=&-\frac{\varepsilon^2}{4\pi}\Big[j_\lambda(\boldsymbol{r}_1)O_{\lambda\mu}\frac{e^{-\mu r}}{r}j_\mu(\boldsymbol{r}_2)\\&-\frac{1}{2\mu}\left(\frac{d}{dx_0}\right)^2\left\{j_\lambda\left(\boldsymbol{r}_1,X_0+\frac{x_0}{2}\right)O_{\lambda\mu}e^{-\mu r}j_\mu\left(\boldsymbol{r}_2,X_0-\frac{x_0}{2}\right)\right\}_{x_0=0}\Big]\\=&-\frac{\varepsilon^2}{4\pi}\Big[j_\lambda(1)O_{\lambda\mu}\frac{e^{-\mu r}}{r}j_\mu(2)\\&-\frac{1}{2\mu}\left\{\frac{1}{4}\left(\frac{d}{dt_1}\right)^2-\frac{1}{2}\frac{d}{dt_1}\frac{d}{dt_2}+\frac{1}{4}\left(\frac{d}{dt_2}\right)^2\right\}j_\lambda(1)e^{-\mu r}j_\mu(2)O_{\lambda\mu}\Big].\end{aligned} \tag{38}$$

For $\mu=0$, replace $-\exp(-\mu r)/\mu$ by r. This formula applies, being an expansion in $d/dt_1-d/dt_2=i[H_1-H_2,\ ]$, to those transitions in which the two particles possess the same energy sign both in the initial and final state, a wider condition than the previous one. It is to be noted, however, that in this way the separate derivation of $V(12)$ and $V(21)$ in (23) is impossible, and the physical insight into the condition on V is accordingly obscured.

3. The Fourth Order Potential

As was shown in the preceding analysis, we cannot give in general an unambiguous second order interaction unless the fourth order one is specified. They are related to each other in such a way that the second and fourth order (and possibly higher order) terms in the S-matrix come out correctly as the combined effect of these potentials. In other words, the fourth order terms of the S-matrix are the sum of the first order effect of the fourth order potential and the second order effect of the second order potential. In the S-matrix theory a similar circumstance arises when S is expressed as[11]

$$S=\frac{1-iK}{1+iK}. \tag{39}$$

Expanding S and K:

$$S=1+S_2+S_4+\cdots,\qquad K=K_2+K_4+\cdots, \tag{40}$$

we see that

$$S_2^+=-S_2,\quad K_2=\frac{i}{2}S_2,\quad K_4=\frac{i}{2}S_4+i(K_2)^2=\frac{i}{2}S_4-i\left(\frac{S_2}{2}\right)^2. \tag{41}$$

Thus $2K_4$ is the difference of S_4 and a two-fold iteration of S_2, and is the real (Hermitic) part of iS_4. In this case, however, the term $(S_2)^2$ corresponds to the process in which the particles are scattered twice but the energy is conserved in the intermediate state (the damping part). In our case, on the other hand, a second order potential is meant to include the reactive effect as well and consequently the fourth order potential is of a nature different from K.

Now from the standpoint of the Feynman theory, the fourth order processes consist of two main graphs (a) and (b). Other graphs contain self-energy effects, and we shall neglect them for the present purpose, for it only gives a

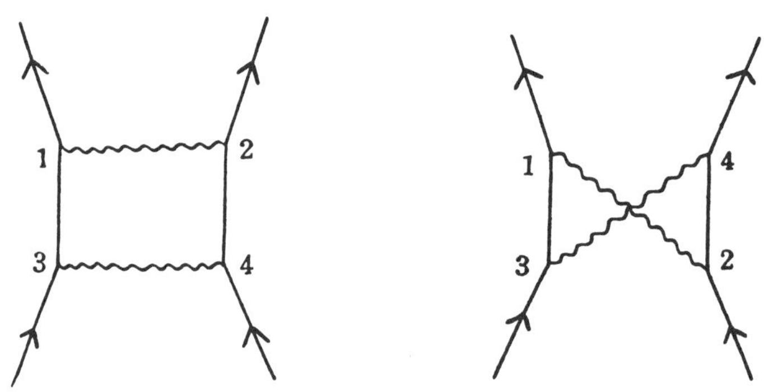

(a) (b)

small smearing out effect to the source if divergences can be avoided. On iterating the second order graph, we get a graph corresponding to (a) if the two boson lines do not cross, and a graph similar to (b) if they do. But the latter is not exactly (b), for the order of occurrence of the events at 1, 2, 3 and 4 is not chronological, and it does not give the correct matrix element corresponding to (b). Hence the difference between the true fourth order term and the one arizing from the iteration of the second order term can be written as

$$-\frac{1}{8}(-i\varepsilon^2)^2\int j_\mu(1)j_\nu(3)[j_\lambda(4),j_\rho(2)]\,\Delta_{F\mu\rho}(12)\Delta_{F\nu\lambda}(34)\times\frac{1+\varepsilon(13)}{2}\,\frac{1+\varepsilon(42)}{2}(dx_1)(dx_2)(dx_3)(dx_4), \tag{42}$$

where 1, 3 and 4, 2 are understood to refer to different particles. A more symmetrical form is provided by

$$\frac{1}{32}\varepsilon^4\int\Big\{\frac{1+\varepsilon(13)\varepsilon(42)}{2}[j_\mu(1)\cdot j_\nu(3)][j_\lambda(4),j_\rho(2)] + \frac{\varepsilon(13)+\varepsilon(42)}{4}\left(\{j_\mu(1),j_\nu(3)\}[j_\lambda(4),j_\rho(2)]+[j_\mu(1),j_\nu(3)]\{j_\lambda(4),j_\rho(2)\}\right)\Big\}\times\Delta_{F\mu\rho}(12)\,\Delta_{F\nu\lambda}(34)\,(dx_1)(dx_2)(dx_3)(dx_4). \tag{43}$$

(41) and (42) contain the sign function $\varepsilon(x)$ which depends on the surface parametrization. In fact, on writing down the matrix element explicitly, the second term of (43) is seen to contain a factor $\varepsilon(x)\ S^{(1)}(x)$. This shows that the potential energy and accompanying eigenvalue problem do not have in general the ordinary covariant character. As it will not be of much importance to obtain the rigorous form of the potential, we shall be content with rough estimations. For this purpose we specify the time axis and proceed with v/c expansions. Then two cases discriminate themselves. The first is the one in which the quantities j are static so that in an expression like $j_\mu(1)\,j_\nu(3)$ we can take account of the difference of t_1 and t_3 by a simple Taylor expansion in t_1-t_3. If, on the other hand, the j are non-static $(\sim v/c)$, then such an expansion fails because of the large Zitterbewegung frequency, and we consequently face the second case. The cross term, in which one j is static and the other non-static, will be considered afterwards.

a) The static case. We use the expression (43), not explicitly writing down the S-functions for the particles, but going with the j_λ themselves. We make beforehand a remark that the first term in the bracket in (43) gives a zeroth order $(\sim(v/c)^0)$ potential while the second is responsible for the retardation correction to it. Thus in the first term we put $j_\lambda(t)\sim 1$ with respect to the time dependence. The Δ-function is, in Fourier representation,

$$\Delta_F(x)=\frac{-2i}{(2\pi)^4}\int e^{i(kx)}\,\frac{(dk)}{k^2+\mu^2}, \tag{44}$$

with the understanding that μ^2 is furnished with a small negative imaginary part. The first term of Eq. (43), which we call (A), becomes then

$$\begin{aligned}&\frac{1}{32}\,\varepsilon^4\,\frac{(-2i)^2}{(2\pi)^8}\int\cdots\int\frac{1+\varepsilon(13)\,\varepsilon(42)}{2}\,[j_\mu(\boldsymbol{r}_1),j_\nu(\boldsymbol{r}_3)]\,[j_\lambda(\boldsymbol{r}_4),j_\rho(\boldsymbol{r}_2)]\,O_{\mu\rho}O_{\lambda\nu}\\ &\times\frac{1}{k^2+\mu^2}\cdot\frac{1}{k'^2+\mu^2}\cdot e^{-ik_0(t_1-t_2)}\,e^{-ik_0'(t_3-t_4)}\,e^{i(\boldsymbol{k},\,\boldsymbol{r}_1-\boldsymbol{r}_2)}\,e^{i(\boldsymbol{k}',\,\boldsymbol{r}_3-\boldsymbol{r}_4)}\\ &\times(d\boldsymbol{k})\,(d\boldsymbol{k}')\,dt_1\,dt_2\,dt_3\,dt_4\cdot d\boldsymbol{r}_1\,d\boldsymbol{r}_2\,d\boldsymbol{r}_3\,d\boldsymbol{r}_4.\end{aligned} \tag{45}$$

Making use of the fact that $j_\mu(r)$ and $j_\nu(r')$ commute (for equal t) if $r\neq r'$, we can identify r_1 and r_2 with r_3 and r_4 respectively, and put $r_1-r_2=r_3-r_4=r$, thereby reducing the quadruple integral $\int\Pi dr$ to a double one. As regards the integration in t, we invoke the centers of time

$$T=\frac{t_1+t_2+t_4+t_3}{4}=\frac{T'+T''}{2},\quad T'=\frac{t_1+t_2}{2},\quad T''=\frac{t_3+t_4}{2},$$

together with

$$t=T'-T'',\quad t'=t_1-t_2,\quad t''=t_3-t_4, \tag{46}$$

and leave out the integration in T to obtain the potential energy. Further putting

$$t'-t''=u, \qquad t=\frac{1}{2}uv, \tag{47}$$

we find

$$\int\cdots\int\frac{1+\varepsilon(13)\varepsilon(42)}{2}e^{-ik_0(t_1-t_2)-ik_0'(t_3-t_4)}dt_1\cdots dt_4$$

$$=\int dT\int_{-\infty}^{\infty}e^{-i(k_0-k_0')u}\frac{1}{2}|u|du\int_{-1}^{1}dv\cdot 2\pi\delta(k_0+k'_0)$$

$$=-\frac{8}{(k_0-k_0')^2}2\pi\delta(k_0+k_0')\int dT. \tag{48}$$

The potential (A) is given by

$$H_A^{(4)}=\left(\frac{\varepsilon^2}{4\pi}\right)^2\frac{1}{2}\int[j_\mu(\boldsymbol{r}_1),j_\nu(\boldsymbol{r}_1)][j_\lambda(\boldsymbol{r}_2),j_\rho(\boldsymbol{r}_2)]\,O_{\mu\rho}\,O_{\lambda\nu}\,V(\boldsymbol{r}_1-\boldsymbol{r}_2)\,d\boldsymbol{r}_1\,d\boldsymbol{r}_2,$$

$$V(\boldsymbol{r})=\frac{i}{16\pi^5}\int\frac{1}{k_0^2}\cdot\frac{e^{-i\boldsymbol{k}\boldsymbol{r}}}{k^2+\mu^2}\cdot\frac{e^{i\boldsymbol{k}'\boldsymbol{r}}}{k'^2+\mu^2}\delta(k_0+k'_0)(dk)(dk'). \tag{49}$$

V can be evaluated, with the above mentioned definition at the poles, as

$$V(\boldsymbol{r})=\frac{i}{16\pi^5}\iint\left[-\pi i\left\{\frac{1}{(\boldsymbol{k}^2+\mu^2)^{3/2}}\frac{1}{\boldsymbol{k}'^2-\boldsymbol{k}^2}+\frac{1}{(\boldsymbol{k}'^2+\mu^2)^{3/2}}\frac{1}{\boldsymbol{k}^2-\boldsymbol{k}'^2}\right\}e^{i\boldsymbol{k}\boldsymbol{r}}\,e^{i\boldsymbol{k}'\boldsymbol{r}}\right]d\boldsymbol{k}\,d\boldsymbol{k}'$$

$$=\frac{i}{16\pi^5}\int(-\pi i)\frac{e^{i\boldsymbol{k}\boldsymbol{r}}}{(\boldsymbol{k}^2+\mu^2)^{3/2}}(-2\pi^2)\frac{\cos kr}{r}\times 2\,d\boldsymbol{k}$$

$$=\frac{1}{2\pi r^2}\int\frac{\sin 2kr}{(\boldsymbol{k}^2+\mu^2)^{3/2}}k\,dk=\frac{1}{\pi r}K_0(2\mu r) \tag{50}$$

where K_ν is the modified Bessel function of order ν:

$$K_\nu(x)=\frac{1}{2}\pi i e^{-\frac{1}{2}\pi i\nu}H_{-\nu}^{(1)}(ix). \tag{51}$$

Asymptotically, V behaves as

$$V(r)\sim-(1/\pi r)\ln(\gamma\mu r),\ \ \gamma=1.781\ \text{ for } 2\mu r\ll 1$$
$$\sim(1/2r)(1/\pi\mu r)^{1/2}e^{-2\mu r}\ \text{ for } 2\mu r\gg 1. \tag{52}$$

The range of the force is thus half of that in the second order potential, corresponding to the fact that two mesons are interchanged between the particles.

The retardation correction has been neglected in the above calculation, but it can be shown that the $(v/c)^1$-effect is zero. In view of the approximate nature of the calcultion, we shall not investigate the $(v/c)^2$ -correction to the fourth order potential which is already small compared to the second order one.

On the other hand, the second term of Eq. (43), called (B), gives a $(v/c)^1$-effect. We accordingly put

$$j(1)\,j(3)\sim e^{i\Delta p_0(1)t_1}\cdot e^{i\Delta p_0(3)t_3}. \tag{53}$$

An integral of the form

$$\int\frac{\varepsilon(13)+\varepsilon(42)}{4}\,e^{-i(k_0-k'_0)\frac{t_1-t_3-t_2+t_4}{2}}\,e^{ip(t_1-t_3)}\,e^{iq(t_4-t_2)}\,dt_1\ldots dt_4 \tag{54}$$

is, by a similar procedure, found to be

$$=-\frac{\pi i}{2}\,\frac{1}{p-q}\,\{\delta(k_0-k_0'-2p)-\delta(k_0-k_0'-2q)\}. \tag{55}$$

Using this, the part (B) of the potential can be calculated, giving

$$H_B^{(4)}=\frac{i}{16}\left(\frac{\varepsilon^2}{4\pi}\right)^2\int\Big\{\Big[\frac{d}{dt}\,(j_\mu(\boldsymbol{r}_1)\,j_\lambda(\boldsymbol{r}_2)),j_\nu(\boldsymbol{r}_1)\,j_\rho(\boldsymbol{r}_2)\Big]-\Big[j_\mu(\boldsymbol{r}_1)j_\lambda(\boldsymbol{r}_2),$$

$$\frac{d}{dt}\,(j_\nu(\boldsymbol{r}_1)j_\rho(\boldsymbol{r}_2))\Big]\Big\}O_{\mu\rho}\,O_{\nu\lambda}\,\frac{e^{-\mu r}}{r}\,\frac{e^{-\mu r}}{2\mu}\,d\boldsymbol{r}_1\,d\boldsymbol{r}_2$$

$$=\frac{i}{8}\left(\frac{\varepsilon^2}{4\pi}\right)^2\int\Big[\frac{d}{dt}(j_\mu(\boldsymbol{r}_1)j_\lambda(\boldsymbol{r}_2))\,V_2(r),\,j_\nu(\boldsymbol{r}_1)j_\rho(\boldsymbol{r}_2)\,V_0(r)\Big]\,O_{\mu\rho}\,O_{\lambda\rho}\,d\boldsymbol{r}_1,\,d\boldsymbol{r}_2, \tag{56}$$

$$V_0=\frac{e^{-\mu r}}{r},\qquad V_2=\frac{e^{-\mu r}}{2\mu}=-\frac{1}{2\mu}\,\frac{\partial}{\partial\mu}\,V_0. \tag{57}$$

(b) The non-static case. We make the approximation

$$j(1)\,j(3)\sim e^{2mi(t_1-t_3)} \tag{58}$$

since the main contribution comes from processes in which the initial and final state is positive, while the intermediate is negative, in energy. Here we must be careful about the sign of m in the exponential. It is determined by the order of the operators $j(1)$ and $j(3)$, but not by the order of the times t_1 and t_3. The integral corresponding to (48) is then

$$-\left\{\frac{1}{\left(2m-\frac{1}{2}\,(k_0-k_0')\right)^2}+\frac{1}{\left(2m+\frac{1}{2}\,(k_0-k_0')\right)^2}\right\}\,2\pi\delta(k_0+k_0')\int dT$$

for the order 13·42,

and
$$\left\{-\frac{2\pi i}{m}\,\delta(k_0-k_0')+\frac{2}{(2m)^2-\left(\frac{k_0-k_0'}{2}\right)^2}\right\}\,2\pi\delta(k_0+k_0')\int dT \tag{59}$$

for the order 13·24.

We neglect k_0 and k_0' compared to m, since the important contribution comes from values of k_0 of the order μ, which is $\ll m$ for actual cases. The interaction (A) is thus given by

$$H_A^{(4)}=\left(\frac{\varepsilon^2}{4\pi}\right)^2\frac{1}{2}\int\{j_\mu(1),O_{\mu\rho}j_\rho(2)\}^2\cdot\frac{1}{4m}V_0(r)^2\,d\boldsymbol{r}_1\,d\boldsymbol{r}_2$$

$$+\left(\frac{\varepsilon^2}{4\pi}\right)^2\frac{1}{2}\int\{j_\mu(1),j_\nu(1)\}\{j_\lambda(2),j_\rho(2)\}O_{\mu\rho}\,O_{\nu\lambda}\frac{1}{8\pi}\frac{\mu}{m^2r^2}K_1(2\mu r)\,d\boldsymbol{r}_1\,d\boldsymbol{r}_2. \tag{60}$$

The interaction (B), on the other hand, becomes zero in the present approximation. We shall not, however, enter into more detailed treatments.

c) Cross terms. When some j are static and others non-static, there occur in the expression (43) cross terms of these quantities. They may be calculated in a similar manner. But the resulting fourth order potentials will be non-static with radial dependence essentially the same as in the case b). Consequently their order of magnitude will be smaller than the preceding ones. We shall not investigate these terms here.

In the above results, a finite field mass μ has been assumed. If $\mu=0$, they cannot be applied at once since some of the expressions diverge. In this case we can show that

in (57) $\qquad V_2\rightarrow\frac{r}{2}$,

and in (60) $\qquad \frac{1}{8\pi}\frac{\mu}{m^2r^2}K_1(2\mu r)\rightarrow\frac{1}{16\pi}\frac{1}{m^2r^3}. \qquad (61)$

A reasonable evaluation of (50) for $\mu=0$ is not known, but it may be left out of consideration since fortunately such a case does not occur actually.

4. The Relation between the Second and Fourth Order Potential

Now we have at hand the second and fourth order potential $H^{(2)}$ and $H^{(4)}$. Their sum should be taken as the potential energy for the system which is rigorous up to the fourth order in the coupling contant. $H^{(2)}$ is a familiar interaction which can be derived by various methods, except that the retardation effect should be taken account of in a manner specified above. In case j is static, a retardation expansion is allowed. But for general j the situation is different. In fact, for transitions in which either of the particles changes its energy sign but not both, (odd transition), the interaction is virtually of a delta-function type which will not fit in the category of the ordinary potential. In other words, we had better decompose every quantity j into a static part $\bar{j}$ and a non-static (Zitterbewegung) part* $\tilde{j}$, and apply the retarded potential (38) only for the

* The terms static and non-static are somewhat deviating from ordinary usage. They refer to the change of the energy sign.

even part $\bar{j}_\lambda \bar{j}_\mu + \tilde{j}_\lambda \tilde{j}_\mu$. In this sense the Pauli approximation plays a major role in the two-body problem.

The fourth order potential, on the other hand, exhibits several interesting features. It takes on different forms according to the v/c-dependence of j as above. Moreover, it has an additional dependence on the spin variables through commutator expressions. Suppose that all the j's commute with each other, then the term (49) or the first term of (43) vanishes, leaving only the retardation correction (56) in the static case. Thus there remain no fourth order forces except for a small retardation correction.

Let us see how these fourth order characteristics show up in the electromagnetic interaction of two electrons. Here we have one static and three non-static interactions, namely

$$j_4(\boldsymbol{r}_1)\, j_4(\boldsymbol{r}_2) = -(\bar{\psi}\gamma_4\psi)_1\, (\bar{\psi}\gamma_4\psi)_2 \tag{62a}$$

and

$$j_i(\boldsymbol{r}_1)\, j_i(\boldsymbol{r}_2) = -(\bar{\psi}\gamma_i\psi)_1\, (\bar{\psi}\gamma_i\psi)_2, \quad i=1,2,3. \tag{62b}$$

We apply the formula (49) and (50) for (62a) and (60) for (62b). The former, however, vanishes, showing that the Coulomb potential is exact even to the fourth order effect. The latter yields a relatively large spin-spin and a smaller spin-independent interaction of the electrons, but the spin-spin term is just cancelled by a second order perturbation of the second order potential with $j_i j_i$ *via* negative state. The retardation correction (56) can be applied only to (62a). We can, however, again eliminate it by a canonical transformation (which is equivalent to the previously considered one (29)):

$$\Psi = \exp\left[\frac{i}{8} \frac{d}{dt} \int j_4(\boldsymbol{r}_1)\, V_2 j_4(\boldsymbol{r}_2)\, d\boldsymbol{r}_1\, d\boldsymbol{r}_2\right] \Psi'. \tag{63}$$

In fact the transformed interaction becomes

$$-\frac{1}{2}\int j_\lambda(1) j_\lambda(2)\, V - \frac{i}{8}\iint\left[\frac{d}{dt}\,(j_4(1)j_4(2)),\, j_4(1)j_4(2)\right] V_0\, V_2$$

$$+\frac{i}{8}\iint\left[\frac{d}{dt}\,(j_4(1)j_4(2)),\, j_4(1)j_4(2)\right] V_0 V_2 + \frac{1}{8}\int \frac{d^2}{dt^2}\,(j_4(1)j_4(2))\, V_2 \tag{64}$$

$$+\text{higher order terms.}$$

The last term contributes

$$\frac{1}{8}\int\left\{\left(\frac{d}{dt_1}\right)^2 j_4(1)j_4(2) + 2\,\frac{d}{dt_1} j_4(1) \frac{d}{dt_2} j_4(2) + j_4(1)\left(\frac{d}{dt_2}\right)^2 j_4(2)\right\} V_2, \tag{65}$$

which combines with the first term with the expansion (38). Thus we get the corrected Breit interaction, with $j_\lambda j_\lambda$ replaced by $\bar{j}_\lambda \bar{j}_\lambda + \tilde{j}_\lambda \tilde{j}_\lambda \equiv \overline{j_\lambda j_\lambda}$,

$$H^{(2)\prime}=\frac{\varepsilon^2}{4\pi}\frac{1}{2}\int\left\{\overline{j_\lambda(1)\,j_\lambda(2)}\,V_0-\frac{1}{2}\,\overline{\frac{dj_4(1)}{dt}}\;\overline{\frac{dj_4(2)}{dt}}\,V_2\right\}d\boldsymbol{r}_1\,d\boldsymbol{r}_2$$

$$+\left(\frac{\varepsilon^2}{4\pi}\right)^2\frac{1}{2}\int\{\tilde{j}_i(1)\,\tilde{j}_i(2)\}^2\cdot\frac{1}{4m}\,V_0^2\,d\boldsymbol{r}_1\,d\boldsymbol{r}_2+\text{smaller terms.} \qquad (66)$$

There remains no net fourth order effect of detectable magnitude. On the other hand, if we try to perform a second order perturbation using the original Breit formula, a result is obtained which contradicts the experimental facts about the hyperfine structure of helium[12]. This is due to the lack of the fourth order corrections which cancel a part of the iteration of the second order potential, leaving no spin-dependent fourth order effects. The same circumstance is observed in the (symmetrical or charged) pseudoscalar meson theory with pseudoscalar coupling investigated by Watson and Lepore[18], in which the isotopic spin variables replace the ordinary ones.

5. Further Discussions

In the preceding analysis we have given a general form of the potential which is correct up to the fourth order in the coupling constant and takes an approximate account of the retardation effect. The result shows that the corrections, which should be made to the potential directly derivable from non-relativistic classical theory, is essentially dependent on the nature of the interaction. Those interactions which contain only classical variables give in the quantum theory also potentials not much different from classical ones. On the other hand, if the interaction contains quantum variables, such as the Dirac spin and isotopic spin matrices, the higher order corrections in general affect the classical potential. Although these corrections have been considered only in the fourth approximation, we can in principle obtain the higher order ones step by step, in such a manner that the resulting potential yield exactly the S-matrix derived from the field theory. The origin of these correction lies, as was seen above, in the fact that the order of the emission and absorption of quanta exchanged between the particles affect the matrix elements. As an illustration, let us consider a charged meson theory, which contains the non-commutative isotopic spin variables τ_λ. The physical meaning of the non-commutativity is that the charge must be conserved at any instant, so that a nucleon cannot emit or absorb successively more than one positive or negative mesons. In other words, there is a strong correlation between the partaking quanta, which produces deviations in the higher order S-matrix terms from those obtained by a mere iteration of the second order process. For scattering problems these higher order corrections may be neglected. But for the bound system they would give in general a large influence on the eigenvalues and eigenfunctions. It is very difficult, however, to give an explicit expression for these corrections. We give below only an idea about these circumstances on a simple model, which may rather be too rough an approximation.

Take a symmetrical meson theory with static interaction. The main nth order processes consist of those graphs in which n meson lines connect the two nucleon lines, the order of the joints on the particle lines being different for each graph. Every graph gives rise to a matrix element of the form

$$\sum_{\tau}(\tau_1\,\tau_2\ldots\tau_n)_1\,(\tau_{i1}\,\tau_{i2}\ldots\tau_{in})_2\,f_n(p_1\,q_1\,p_2\,q_2,\;(i_1\ldots i_n)),\tag{67}$$

where p_1, q_1, p_2, q_2 are the initial and final momenta of the particles. Now we *assume* that the function f virtually does not depend on the permutation symbol $(i_1\ldots i_n)$ for the *irreducible* graphs, in which every meson line crosses some other meson lines.

For these graphs we put

$$f_n=\langle f_n\rangle_{\mathrm{Av}}\sim-\frac{1}{C_n}v_0^{n-1}v=-kv_0^n,\tag{68}$$

$$v=ce^{-\mu r}/r,\qquad v_0\sim ce^{-\mu r}/\sqrt{\mu r},\quad k\sim\sqrt{\mu/r},$$

where C_n is the number of irreducible graphs. This form may be suggested by observing that

$$(-i)^2\,C_mf_m\cdot C_{n-m}f_{n-m}\cdot T\sim C_nf_n\tag{69}$$

and

$$\int\frac{e^{ikr}}{k^2+\mu^2}\,dk=4\pi\,K_0(\mu r)\sim\frac{1}{\sqrt{\mu r}}\,e^{-\mu r},\tag{70}$$

where $T\sim1/k$ is the time interval in which a mth and an $(n-m)$th order graph overlap, giving an nth order irreducible graph. Then we extend for simplicity the average over the irreducible graphs to all graphs, thereby using the formula

$$\frac{1}{n\,!}\sum_{(i_1\ldots i_n)}(\tau_i\ldots\tau_n)_1\,(\tau_{i_1}\ldots\tau_{i_n})_2=\begin{cases}n+1 & n=\text{even},\\ \dfrac{n+2}{3}\,\tau_1\cdot\tau_2 & n=\text{odd}.\end{cases}\tag{71}$$

Next, starting with an effective potential

$$\{(\tau_1\cdot\tau_2)+a\}\,V=-k\sum s_nv_0^n,\tag{72}$$

we get similar irreducible terms of S-matrix:

$$\{(\tau_1\cdot\tau_2)+a\}^nF_n,\quad F_n\sim-VV_0^{n-1}=-kV_0^n.\tag{73}$$

Equating both S-matrices, we obtain an equation determining the effective potential, which is solved as

$$1+\frac{1}{k}V=\frac{(A-2B)-B(\tau_1\cdot\tau_2)}{(A+B)(A-3B)},$$

$$A=\frac{1+v_0^2}{(1-v_0^2)^2},\quad B=\frac{1-v_0^2/3}{(1-v_0^2)^2}v_0. \tag{74}$$

This is a rather strange result. For large r and small coupling constant, V is nearly equal to $-v(\tau_1\cdot\tau_2)$. But for the contrary case v_0 will be large, and the isotopic dependence is of a linear combination of $\tau_1\cdot\tau_2$ and 1. Of course the above form of the functional dependence is not a reliable one, because of the nature of the present investigation. But it may be sufficient to make one modest in asserting the validity of lower order approximations.

Next we make a remark on the relation between the results obtained here and those which follows from the ordinary perturbation calculation. Writing the momentum-energy of the particle and emitted quantum as (p, p_0) and (k, k_0) respectively, the second order matrix element is

$$\left(\frac{1}{2\pi}\right)^3\left\{\frac{1}{-(\Delta p_0)_1-k_0}+\frac{1}{-(\Delta p_0)_2-k_0}\right\}O_1O_2\frac{1}{2k_0}. \tag{75}$$

The energy denominators can be rewritten as

$$\frac{1}{-(\Delta p_0)_1-k_0}+\frac{1}{-(\Delta p_0)_2-k_0}=\frac{1}{-(\Delta p_0)_1-k_0}+\frac{1}{(\Delta p_0)_1-k_0}$$
$$+\frac{1}{-(\Delta p_0)_2-k_0}-\frac{1}{(\Delta p_0)_1-k_0}$$
$$=\frac{-2k_0}{k_0^2-(\Delta p_0)_1^2}+\frac{(\Delta p_0)_1+(\Delta p_0)_2}{\{k_0-(\Delta p)_1\}\{k_0+(\Delta p_0)_2\}}. \tag{76}$$

The first term corresponds to the Eqs. (9) and (10), while the second is, being of the form $i(d/dt)V$ (non-Hermitic), contributes only to a higher order effect. The symmetrical form (37), however, does not follow from the perturbation theory. The fourth order interaction is therefore also different. In fact the perturbation theory will not give a fourth order potential as simple as the present one.

Thus far we have been confining ourselves to essentially non-relativistic treatments. Those regions where the distance of the particles is small and relativistic effects play a predominant role are out of consideration in the present investigation. Moreover, the difficulty of high singularities of the potential which haunts these regions is of a more profound origin and will only be solved with the general difficulties of the present quantum theory. A complete relativistic eigenvalue problem itself, if it could be formulated, would not give much help to these fundamental difficulties.

As for the relativistic formulation of the two-body problem, our results seem to present a slight progress over the older ones in that the many-time theory has been invoked as a guiding principle which requires a certain integrability condition for the interaction term. In view of the preceding analysis, we easily find that a pair of equations of the form (in the interaction representation)

$$\left\{i\frac{\partial}{\partial t_2}-V\left(\frac{t_1+t_2}{2}\right)\right\}\psi=0, \qquad \left\{i\frac{\partial}{\partial t_2}-V\left(\frac{t_1+t_2}{2}\right)\right\}\psi=0 \tag{77}$$

just meet our purpose. Transforming back to the Schrödinger representation, we have

$$\left\{i\frac{\partial}{\partial t_1}-H_1-V(t_1-t_2)\right\}\psi=0, \quad \left\{i\frac{\partial}{\partial t_2}-H_2-V(t_1-t_2)\right\}\psi=0\,; \tag{78}$$

or

$$\left\{i\frac{\partial}{\partial T}-(H_1+H_2)-2V(t)\right\}\psi=0, \qquad \left\{i\frac{\partial}{\partial t}-(H_1-H_2)\right\}\psi=0, \tag{79}$$

with

$$\begin{aligned} & i\frac{\partial V}{\partial t}-[H_1-H_2,\ V]=0, \\ & V=e^{-i(H_1-H_2)t}\,V(0)\,e^{i(H_1-H_2)t}, \end{aligned} \qquad T=\frac{t_1+t_2}{2}, \qquad t=t_1-t_2. \tag{80}$$

The space-time behavior of the wave function ψ is such that it is essentially determined at $t_1=t_2$, and extended to $t_1\neq t_2$ simply as free of interaction. The latter procedure means nothing but a canonical transformation of the whole representation by a function $\exp[i(H_1-H_2)t]$. Thus the above many-time equations do not really present a purely relativistic (four-dimensional) formulation. They are essentially the same as the ordinary one-time equation. However, we notice here a fact which seems to have been overlooked. That is the freedom of canonical transformation just mentioned. Indeed any equation of the form

$$\left(i\frac{\partial}{\partial t}-H-V\right)\psi=0 \tag{81}$$

allows a transformation $\exp[if(H)]\psi$ with the result that only the interaction potential suffers a modification. In this sense the potential is not uniquely determined though the ensuing S-matrix is unique[14].

Finally a remark may be added concerning the Eqs. (31) and (32). The S-matrix (31) is unitary, but its sub-matrix under consideration is not. For if two particles collide, the Bremsstrahlung will occur, which diminishes the norm of the final elastic state. This means that V is in general complex. In fact an atom in an excited state will "decay" by emitting photons, and this should be properly represented by the complex potential. We may, however, regard V approximately Hermitic, at least at low energies. The preceding results give it some justification.

In conclusion, we may say that the concept of potential energy cannot enjoy a wide and practical extension much beyond the classical and non-relativistic form. Consequently a more ambitious attempt at a relativistic eigenvalue problem for closed systems will have to deviate considerably from the present one. An example will be furnished by the equation

$$\left(\gamma_\lambda \frac{\partial}{\partial x_\lambda} - \kappa\right)_1 \left(\gamma_\mu \frac{\partial}{\partial x_\mu} - \kappa\right)_2 \psi = -\frac{\varepsilon^2}{2} (\gamma_\nu)_1 (\gamma_\nu)_2 D_F(12)\, \psi. \qquad (82)$$

This equation can be deduced, as a reasonable one, from the many-time equation with intermediate field at the cost of some assumption and alteration on the meaning of state vector. We might expect that it determines an eigen-momentum-energy vector of the system from some boundary conditions at (space-time) infinity. Eq. (81) only takes account of the second order effect. Inclusion of higher order effects is not possible in the above form and can only be achieved by an integral equation. If higher order effects play an important part on the interaction, we shall therefore have to resort to still different methods, such as the S-matrix theory, the strong coupling theory, direct tackling with integral equations,[15] and so on.

References

1) G. Breit, Phys. Rev. **34** (1929), 553.
2) G. Breit, Phys. Rev. **51** (1937), 248, 778; **53** (1938), 153.
3) G. Breit, Phys. Rev. **39** (1932), 616. C. Møller, ZS. f. Phys. **78** (1931), 786.
H. Bethe and E. Fermi, Zeit. Phys. **77** (1932), 296.
4) C. Møller and L. Rosenfeld, Det Kgl. Danske Wid. Selsk. Mat. Fys. Med. XVII, No. 11 (1940).
5) S. Tomonaga, Prog. Theor. Phys. **1** (1946), 27.
J. Schwinger, Phys. Rev. **74** (1948), 1439; **75** (1949), 651; **76** (1949), 790.
R. P. Feynman, Phys. Rev. **76** (1949), 749, 769.
6) L. Van Hove, Phys. Rev. **75** (1949), 1519.
Y. Nambu, Prog. Theor. Phys. **3** (1948), 444.
7) Y. Nambu, Ref. 6)
8) P. A. M. Dirac, Proc. Roy. Soc. **136** (1932), 453.
9) T. Toyoda, Phys. Rev. **77** (1950), 353; Soryûsiron-Kenkyû **1** (1949), No. 3, 129 (in Japanese).
10) F. J. Dyson, Phys. Rev. **75** (1949), 486.
11) J. Schwinger, Ref. 5).
12) G. Breit, Phys. Rev. **36** (1930), 303; **39** (1932), 616.
13) K. M. Watson and J. V. Lepore, Phys. Rev. **76** (1949), 1157.
14) C. Møller, Det Kgl. Danske Vid. Selsk. Mat. Fys. Med. XXIII, No. 1 (1945).
15) S. M. Dancoff, Phys. Rev. **78** (1950), 382.

On the Nature of V-Particles, I

Y. Nambu, K. Nishijima and
Y. Yamaguchi

*Department of Physics,
Osaka City University*

June 19, 1951

Recently Butler et al.[1] presented further evidences concerning the nature of the so-called V-particles which had been observed by Rochester and Butler[2] and by Anderson and collaborators[3]. According to these authors: i) The V-particles are found among penetrating showers with a rate of the order of 1%; ii) Their decay life is estimated to be about 10^{-10} sec; iii) There are two kinds of them, charged and neutral, the latter being about 5 to 10 times more abundant than the former, presumably due to their difference in lives; iv) According to Butler et al., moreover, they can be

Reprinted from Prog. Theor. Phys. 6 (1951) pp. 615–622.

classified in two groups of different masses. The heavier particles, with mass about 2200 to 2300 m_e, decay into a nucleon and a (π) meson, hence fermions. The lighter ones, with mass around 1000 m_e, decay into two (π) mesons, and hence bosons; v) One case is reported which shows a successive decay of a charged V through neutral V into three charged particles (three mesons or two mesons plus one proton), though no suggestion is given as to whether they belong to the heavier or the lighter group.

Then the arguments go on as follows. First of all, there is a possibility that the decay products of V and τ might include μ-mesons as well. This, however, seems rather improbable in view of the fairly large nuclear interaction of the decay mesons reported by Anderson et al.[3], and the relatively long life of the parents in contrast to their sizable production rate. A three particle decay, such as $V\rightarrow P+\mu+\nu$, is all the more unlikely according to Anderson et al.'s arguments about the coplanar character of the parent and daughter particles. It is also to be noted that soft showers are not associated with the V or τ events, which shows that decays giving rise to photon or electron must be rare compared to the main processes. The above mentioned contradiction between production and decay may be lifted if we postulate, as was just the case in the π-μ decay, that the observed V (though not necessarily including τ) are decay products of some unknown particles with sufficiently short life and strong nuclear interaction. But at the present stage we will try to solve this problem using only the observed particles, assuming the decay products to be π's and nucleons (of course excluding the case V)). This can be done as follows:

Now we face two alternative interpretations of the point v) above. One is to assume that the unknown particles involved in this reaction are V's, decaying according to the scheme:

$$V_{\pm}\rightarrow V_0+\pi_{\pm},\quad V_0\rightarrow N+\pi \quad (N\text{—nucleon});\qquad (1)$$

while the other is to assume them to be τ's:

$$\tau_{+}\rightarrow\tau_0+\pi_{\pm},\quad \tau_0\rightarrow 2\pi \qquad (2)$$

According to (1), the V's may be regarded as excited states of nucleons which make transitions to the lower states by emitting mesonic radiation. (τ-emission can be forbidden energetically.) Such a situation has long since been anticipated in the strong coupling theory of nucleon-meson interaction. This theory however, does not just seem to be very useful for our analysis. For it is too crude and incomplete to be relied upon quantitatively, especially in view of the extremely small width of the actual levels ($\sim 10^{-5}$ev). In the present stage we had rather better treat them as different elementary particles, obeying Fermi statistics and having half-odd spins, possibly higher than 1/2, and introduce formal interactions which cause the observed transitions. Thus we assume the following scheme

Interaction	$V_+V_0\pi$, $V_0N\pi$	$V_\pm N\tau$, $V_0N\tau$
Coupling const.	G_1	G_2

$VV\pi$	$NN\pi$	$VV\tau$	$NN\tau$	
G_3	g_3(known)	G_4	g_4	(3)

The conditions to be considered are: a) decay life roughly all of order 10^{-10} sec.; b) competition among various possible decay modes (especially for the τ decay process); and c) production mechanism, to give a yield of $\sim 10^{-2}$ times that of ordinary mesons.

For the calculation of $\tau\rightarrow\pi+x(x=\pi,\gamma,\cdots)$ processes, the results of covariant calculation by many authors[6] can directly be applied with only a few alterations. Unfortunately, however, most of the calculation involved are not free from the diverging ambiguities which have to be disposed of with the aid of the Pauli regulator. Accordingly, in the present order of magnitude consideration, we check them with more rough and intuitive estimation which only takes account of the coupling constants and the volume of phase space —in a manner more or less similar to Fermi's.[7] On the other hand, the various selection rules, such as described by Fukuda et al.,[6e] are more reliable and can be used to narrow down the possibilities.

In this way we get the following results:

$$G_1^2 \sim 10^{-11}-10^{-13},\ G_2^2 \sim 10^{-2}-10^{-3},$$
$$g_4^2 \sim 10^{-7}-10^{-9}$$
$$(G_3^2 \gtrsim g_3^2,\ G_4^2 \lesssim g_4^2, \text{ not necessary}). \quad (4)$$

The transformation property of τ(or at least τ_0) must be either scalar or vector, since it is very likely that the π mesons are pseudoscalar.[8] The range of values in (4) correspond to different assumptions as to Fermi's reaction volume (or alternatively the general trend of the covariant calculation), as well as the assignment of spin values for the V's*. Thus discrepancies with experiment of the order, say, 100, should be tolerated.

Next let us examine the assumption (2). In this case, $\tau_\pm$ must be pseudoscalar, while τ_0 scalar or vector.** This choice, which may also be adopted in the first model (1), excludes the process $\tau_\pm \to 2\pi$ in favor of $\tau_\pm \to \tau_0+\pi$ and $\tau_\pm \to 3\pi$ (Powell's case).[5e] There arise the following nine couplings:

$$\begin{array}{lll} VN\pi, & VN\tau_\pm, & VN\tau_0, \\ VV\pi, & VV\tau_\pm, & VV\tau_0, \\ NN\pi, & NN\tau_\pm, & NN\tau_0. \end{array} \quad (5)$$

Of these, $NN\pi$ is known, and $NN\pi_\pm$ and $NN\tau_0$ and $VN\tau_0$ turn out either unnecessary or harmful. Among many possible combinations of the remaining couplings few yield consistent results. Thus we take as a relatively reasonable choice (see Appendix):

$$\begin{array}{lll} VN\tau_0 & VV\tau_\pm & VV\tau_0 \\ G_1^2\sim10^{-2}, & G_2^2\sim10^{-1.5}, & G_3^2\sim10^{-6}, \\ VV\pi & VN\pi & \\ G_4^2\sim10^{-1.5}, & G_5^2\sim10^{-12}, & \text{others}=0. \end{array} \quad (6)$$

These estimations are of course susceptible to fluctuations, by as large a factor as $\sim10^2$, depending on different assumptions on the calculational procedure.

At present we cannot tell with confidence which of the above two alternatives (1) and (2) is the more preferable. The former is relatively free from ambiguities and conflictions in determining the coupling constants. The idea of the existence of a series of excited levels of nucleons also attracts us. On the other hand, however, the latter cannot be excluded and even seems natural in view of the other evidences concerning τ-mesons, e.g. Powell's $\tau_\pm \to 3\pi$ decay, which can well compete with $\tau_\pm \to \tau_0+\pi$ on this model. It is hoped that the forthcoming paper of Butler et al. will settle this question.

In conclusion, we should like to call attention to some effects which could be related to the V and τ mesons. First, the nuclear force would be modified. But the range being at most only $\sim1/m_\tau\sim2/m_N$, the effect would be too small to account for the existing N-N scattering data. Second, the conventional strong coupling theory, which allows the nucleon isobars of both charges to occur not as anti-particles, could be tested by the experiments, in which the stability of nuclei should also be taken account of. (According to our models, the anti-particles can only appear as pairs.)*** Third, the new particles would play some part in the anomalous magnetic moment of nucleons, though we do not know how and to what extent. Fourth, the production of these particles, though depending on the model, would occur mainly a V-τ pairs and V-V pairs, and to a lesser degree as single (or multiple) τ's. The corresponding threshold energy would be about 1.1 Bev $(\gamma+N\to V+\tau)$ and 0.8 Bev $(N+N\to V+V)$ respectively in the laboratory system. So far these predictions are not definitely at variance with experiments.[9]

A more detailed account of the present analysis will be given later.

Appendix

On the second model (Eq. (2)), the possible competing processes which need consideration are as follows**

$$\tau_0 \to 2\pi, \quad \tau_0 \to \pi_0+\gamma, \quad \tau_0 \to 2\gamma\,;$$

$$\tau_\pm \to \tau_0+\pi_\pm \quad \tau_\pm \to 3\pi, \quad \tau_\pm \to \pi_\pm+2\gamma\,;$$

$$V \to N+\pi.$$

From these processes we can first determine the relative magnitude of the coupling constants so as to fit the experimental facts, and then normalize them by some process such as the production rate. A crude life-time formula is, for example, furnished by

$$1/t \sim (G_3^2)(G_4^2)^2 I_2/m_\pi \sim (G_3^2)(G_4^2)^2 \cdot 10^{23}\text{sec}^{-1} \quad \text{for } \tau_0 \to 2\pi,$$

where I_2 means the availabe volume per unit energy of momentum space. An additional factor $\sim 10^{-3}(\sim(m_\pi/m_V)^3)$ may be introduced if we take account of the coordinate space volume corresponding to the third order process in which virtual V-pairs are created (see Figure). This estimation also agrees roughly with the results of covariant calculation using regulators. The numerical values given in (6) were obtained by the latter refined method assuming scalar and pseudoscalar coupling for the scalar τ_0 and pseudoscalar $\tau_\pm$ respectively.

The first model (1) may be treated analogously.

*) Spin 3/2 for V would make the decay $V \to N+\pi$ forbidden of first order if it were allowed for spin 1/2.

**) Actually pseudovector $\tau_\pm$ is to be discarded since they favor $\tau_\pm \to \pi_\pm+\gamma$ (assuming π to be pseudoscalar) For τ_0, either pure neutral scalar or symmetrical neutral vector must be taken in order to allow the process $\tau_\pm \to \tau_0+\pi_\pm$

***) For example, the process: anti-$V_0' \to P_- + \pi_+$ would be much rarer than $V_0 \to P_- + \pi_+$ in our case. This is consistent with experiments.

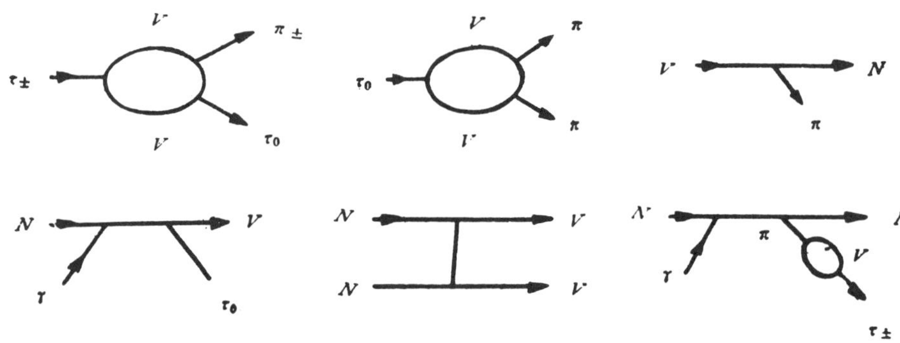

Feynman diagrams for decay and production on the second model (2). Those on the first model are analogous.

1) R. Armenteros, K. H. Barker, C. C. Butler, A. Cachon and A. H. Chapman, Nature **167** (1951), 501.

2) G. D. Rochester and C. C. Butler, Nature **160** (1947), 885.

3) A. J. Seriff, B. B. Leighton, C. Hsiao, E. W. Cowan and C. D. Anderson, Phys. Rev. **78** (1950), 290.
Also see: C. C. Butler, W. G. V. Rosser and K. H. Barker, Proc. Phys. Soc. A **63** (1950), 145; K. H. Barker and C. C. Butler, Proc. Phys. Soc. A **64** (1951), 4.

4) V. D. Hopper and S. Biswas, Phys. Rev. **80** (1950), 1099.
5) Cloud chamber data:
a) L. Leprince-Ringuet and M. L'heritier, C. R. **219** (1944), 616.
b) R. Brode, Rev. Mod. Phys. **21** (1949), 37.
c) H. S. Bridge and M. Annis, Phys. Rev. **81** (1951), 445.
Also see refs. 1), 2), 3) and 8).
Photographic emulsion data:
d) R. Brown, U. Camerini, P. H. Fowler, H. Muirhead, C. F. Powell and D. N. Ritson, Nature **163** (1949), 48 and 82.
e) A. I. Alikhanian, E. M. Samoilovich, I. I. Grevich and Kh. P. Babayan, J. Exp. Theor. Phys. **19** (1949), 667.
f) L. Leprince-Ringuet, Rev. Mod. Phys. **21** (1949), 42.
g) H. H. Forster, Phys. Rev. **77** (1950), 733.
h) N. Wagner and D. Cooper, Phys. Rev. **76** (1949), 449.
Also ref. 4).
6) Neutral mesons:
a) J. R. Finkelstein, Phys. Rev. **72** (1947), 414.
b) H. Fukuda and Y. Miyamoto, Prog. Theor. Phys. **4** (1949), 347, 394.
c) J. Steinberger, Phys. Rev. **76** (1949), 1180.
τ-mesons:
d) H. Fukuda, S. Hayakawa and Y. Miyamoto, Prog. Theor. Phys. **5** (1950), 283 and 352.
e) S. Ôneda, S. Sasaki and S. Ozaki, Prog. Theor. Phys. **4** (1949), 524; ibid. 5 (1950), 25 and 165.
f) H. Fukuda and S. Hayakawa, Prog. Theor. Phys. **5** (1950), 993.
7) E. Fermi, Prog. Theor. Phys. **5** (1950), 570.
8) E. g., see K. Aizu et al., Prog. Theor. Phys. **5** (1950), 931.
9) C. B. A. McCusker and D. D. Millar, Nuovo Cim. **8** (1951), 289.

On the Nature of V-Particles, II

Y. Nambu, K. Nishijima
and Y. Yamaguchi

Osaka City University

July 31, 1951

In a preceding letter[1] we have proposed two possible interpretations of the phenomena associated with V-particles, mainly on the basis of Butler et al.'s recent observations.[2] Of course these interpretations have not been claimed to be the only and definite, but merely possible and more or less natural ones. Also the whole argumentation itself was not complete and exhaustive enough to get full insight into the nature of the V-particles. In view of these points, we will here supplement the previous letter with some more general considerations, while adding to our knowledge the experimental results of American groups[3] in so far as they are accessible to us.

Of the various properties of the V-particles which are known to us, the most remarkable seems to be their large yield and long life, two apparently contradicting properties on the basis of simple detailed balance consideration. They suggest that production and decay are not inverse processes and/or some kind of selection rules (in a very general sense) are at work in the decay reaction. On the other hand, among what are not yet clarified experimentally, there are such important things as the identification of the decay products of the V-particles and their decay modes. Thus at least some of the decay mesons could be μ-mesons (e.g. $\tau\rightarrow\pi+\mu$), and some of the decays could or should involve three or more product particles[3d] (e.g. $\tau+\pi_0,\rightarrow 2\pi_\pm$ $V\rightarrow N+\mu+\nu$). Since no other crucial evidences are known (such as the mode of V production), we are led to a wide variety of possible interpretations if we will take all these points into account, and they can be recommended or disfavored only after a closer examination.

Following the above considerations, let us

first summarize as follows the various conceivable assumptions which shall serve to find out and characterize systematically the possible individual models: 1) Assume the V-particles as elementary entities obeying some basic field equations; or $\bar{1}$) Regard them as composite substance with substructure. 2) Regard V and τ as separate and independent events; or $\bar{2}$) Regard them as inherently related. 3) Assume the product mesons to be π's only; or $\bar{3}$) They may involve μ's as well. 4) Assume only two-particle decays; or $\bar{4}$) Allow of three or more particle decays. 5) In order to account for the large difference in production and decay rate: a) Use the relation in mass values (e.g. for production, $\tau+N\rightarrow V+\tau$, but for decay the inverse is forbidden energetically). b) Assume some unknown parents which, directly produced in nuclear events, decay instantly into the observed V's (analogy of π and μ mesons). c) Assume an interaction mechanism such that V's are generated in a nucleon-nucleon impact, but are hard to decay singly (e.g. $N+N\rightarrow V+V$). d) Take advantage of large internal angular momenta for V, which make the decay highly forbidden. e) Take advantage of parity and other selection rules related to transformation properties (including Furry's theorem). f) Other special devices. These various assumptions are not mutually exclusive, but may be appropriately combined. 6) Use the conventional perturbation theory (weak coupling); or $\bar{6}$) Exploit the strong interaction (like Fermi[4]) or do not depend on the details of coupling at all (e.g. selection rules which can be enunciated by observing only the initial and final states).

By suitable combinations of these characteristic assumptions, we shall be able to arrive at a large number of models (or interpretations) which can explain more or less consistently the essential features of V events. We give below some of the representative models that follow in this way from our present general consideration, and make comments on their merits and defects.

1) All observed processes are assumed to be direct ones. For example, we introduce the following couplings:

a) $NNNV\tau$, $VN\pi$, $\tau\pi\pi$ ($1\bar{2}345c6$)*.
b) $NV'\tau$, $V'V\pi$, $VN\pi$, $\tau\pi\pi$ ($m_{V'}>m_V+m_\pi$) ($1\bar{2}345b6$)[5],
c) $NNVV$, $NN\tau\tau$, $VN\pi$, $\tau\pi\pi$ or $\tau\pi\mu$ (12(3)45c6).

This standpoint is equivalent to determining all the relevant terms of the S-matrix independently in so far as they do not lead to inconsistencies. Though formally possible, it is not a very attractive procedure since we little understand the nature of the events by such a highly phenomenological approach. The introduction of a short-lived parent V's as in the above second model, seems to be an unnecessary complication unless some definite evidence on the existence of such a particle is presented.

2) Our two models proposed in I ($1\bar{2}3$(4)5ace6).[6] They were specially designed to account for the successive decays of V or τ. But the pairwise production which they predict does not seem to be favored by experiments, if not yet rejected.

3) A three particle decay is assumed for V: $V\rightarrow N+\mu+\nu$, while the production occurs, for example, as $V+\pi$ pairs ($12\bar{3}\bar{4}5a6$). Since some evidences show that the decay does not necessarily follow a two-body scheme, such a possibility will not be excluded as responsible for at least part of the decays. (But a universal Fermi-type coupling ($g\sim10^{-49}$) would lead to a rather long life ($>10^{-6}$ sec).

4) Gamow-type model. Turning to the structural theories, the most naive one may be to regard V as a bound $N+\pi$ system ($\bar{1}2345cd\bar{6}$). This, however, would require an unusually high potential barrier (or the order of several Bev) to assure the long life, and if this barrier were supplied by the centrifugal force, the angular momentum would have to be ~8, which seems too high to be easily attained by nucleon-nucleon collisions

$(N+N \to V+V)$.

5) We may modify the above model so that V is a bound $N+\tau$ system. The assumed interaction is, for example,

$\tau\tau NN$ (for τ and V production), $\tau\pi\pi$ (for τ and V decay) ($\bar{1}\bar{2}345ac6$),
or τNN, (for τ and V production), $\tau\pi\pi$ (for τ and V decay) („),

The decay of both V and τ are controlled by the same coupling $\tau\pi\pi$. The first model will predict the pairwise production, whereas on the second (τ must be a boson), the process $V \to N+\pi$ is threatened by $V \to N+\gamma$.

6) Isomeric transition. Alternatively we amend the defects of model **4** by assuming that in the decay of V the π meson is radiated from a small volume of the dimensions of nucleon Compton wave length ($\bar{1}2345cdf\bar{6}$). Using the well-known multiple radiation formula, the angular momentum L can be lowered to ~ 4, which is not unreasonable to be realized by nuclear collisions. But in order to suppress the γ-emission, the bound meson field which is responsible for the excitation energy would have to be extremely rigid. If we calculate ad hoc the moment of inertia of the bound meson field around a nucleon and quantize this rigid body we obtain the observed excitation of ~ 200 Mev for $L=4\frac{1}{2}$ with the coupling constant $g^2 \sim 10$, and only the transitions to $L=1/2$ or $3/2$ are energetically allowed for π emission. Although this model is interesting, it is hard to be founded upon orthodox field theory. The conventional strong coupling theory, on the other hand, which is similar to above in physical ideas, does not seem to give the required long life of the excited states.

Some remarks may be added on the spin (or transformation property) of V and τ particles. We did not take it into account in the above classification since it did not seem essential in the characterization of the models. In general both fermion and boson property may be allowed for these particles. But in any way the stability of protons and neutrons should be guaranteed. Thus, for instance, a coupling like $VN\mu$ (V=boson elementary particle), leading to $V \to \mu+N$, cannot be admitted.

In this way we have seen that at the present stage several different model can assert themselves just about as good. But, in general, we may say that the elementary particle theories (Alternative 1) are liable to predict pairwise production which remains to be confirmed by experiment, while the structural theories (Alternative $\bar{1}$) tend to suffer from high probabilities of radiative decay. It will be premature and useless to demand anything more definite from what we know at present about the V particles.

In concluding, we express deep gratitude to Prof. S. Hayakawa for informing us of the recent activities of American groups as his own opinions. The models mentioned here include those which have been proposed in America, but no published papers being available, we have refrained from explicit citation.[7] Thanks are also due to Messrs. G. Takeda, S. Takagi, H. Fukuda, K. Aizu, T. Kinoshita, H. Miyazawa and S. Ôneda for valuable discussions.

References

1) Y. Nambu, K. Nishijima and Y. Yamaguchi, Prog. Theor. Phys. **6** (1951), this number. Cited in the text as I. The notations used here follow those of I.
2) C. C. Butler et al., Nature **167** (1951), 501. See I.
3)
a) H. S. Bridge and M. Annis, Phys. Rev. **82** (1951), 445.
b) R. W. Thompson, H. O. Cohn and R. S. Flum, Bull. Am. Phys. Soc. **26** (1951), No. 3, p. 6.
c) W. B. Fretter, ibid. No. 5, p. 16.
d) R. B. Leighton, A. J. Seriff and C. D. Anderson, ibid.
d) R. B. Leighton, A. J. Seriff and C. D. Anderson, ibid.
e) E. J. Althaus, ibid.

4) E. Fermi, Prog. Theor. Phys. **5** (1950), 570.
5) This type of model was suggested by S. Takagi.
6) The first of these was also proposed by S. Ôneda.
7) According to Prof. Hayakawa, theories have been given by R. P. Feynman, E. Fermi, R. E. Marshak, R. F. Christy and others.

*) This notation signifies the nature of the model according to the above mentioned criteria.

POSSIBLE EXISTENCE OF A HEAVY NEUTRAL MESON*

YOICHIRO NAMBU
The Enrico Fermi Institute for Nuclear Studies, The University of Chicago, Chicago, Illinois

(Received April 25, 1957)

In an attempt to account for the charge distributions of the proton and the neutron as indicated by the electron scattering experiments,[1] we would like to consider the possibility that there may be a heavy neutral meson which can contribute to the form factor of the nucleon. We assume that this meson, ρ^0, is a vector field with isotopic spin zero and a mass two to three times that of the ordinary pion, coupled strongly to the nucleon field. An isolated ρ^0 would decay through virtual nucleon pair formation according to the following schemes:

$$\begin{aligned}
&(\mathrm{a}) \quad \rho^0 \to \pi^0 + \gamma, \qquad 2\pi^0 + \gamma, \qquad \pi^+ + \pi^- + \gamma;\\
&(\mathrm{b}) \quad \rho^0 \to e^+ + e^-, \qquad \mu^+ + \mu^-;\\
&(\mathrm{c}) \quad \rho^0 \to \pi^+ + \pi^-.
\end{aligned}$$

The process (a) would have a decay probability roughly of the order of $P_a \sim (\mu c^2/\hbar)(G^2/\hbar c)(e^2/\hbar c) \times(\mu/M)^2$, where G is the nuclear coupling constant, μ and M the ρ^0 and the nucleon masses, respectively. For the process (b), the probability would be $P_b \sim (\mu c^2/\hbar)(G^2/\hbar c)(e^2\hbar c)^2(\mu/M)^2$. The process (c) is a forbidden transition, so that it can take place only in violation of the isotopic spin conservation, with a decay rate comparable to that for (b).

Now the process (b) gives rise to a short-range interaction between a nucleon and an electron (or a muon) by exchange of a ρ^0. This will contribute a form factor $F'(k^2) \sim Gg/(\mu^2+k^2)$ to the electron-nucleon scattering, where g is the effective ρ^0-electron coupling, $g^2/\hbar c \sim (G^2/\hbar c)(e^2/\hbar c)^2(\mu/M)^2$. Since ρ^0 is an isotopic scalar, F' has the same sign for both proton and neutron, whereas the corresponding form factor F due to the pion cloud should change sign. Relativistic field theory shows on general grounds that $F(k^2)$ has the form

$$F(k^2) = \int_{2m_\pi}^{\infty} \frac{\rho(m)}{m^2 + k^2} dm,$$

where the lower limit of integration corresponds to the threshold for pion pair creation by an external electromagnetic field. With our assumptions about ρ^0, it is thus possible that the two form factors F and F' cancel approximately for the neutron but reinforce for the proton, in agreement with observation. If we equate tentatively the mean square radius of the proton with the one due to ρ^0:

$$Gg/\mu^2 \sim e^2\langle a^2\rangle/b,$$

we get

$$(G^2/\hbar c)(g^2/\hbar c) \sim [(e^2/\hbar c)/b]^2 \sim 10^{-6},$$

which checks with the previous estimate since $G^2/\hbar c$ would be of the order one. The decay lives become, very approximately,

$$\begin{aligned}
\tau_a &\sim 10^{-19}\text{–}10^{-20} \text{ sec},\\
\tau_b \sim \tau_c &\sim 10^{-17}\text{–}10^{-18} \text{ sec}.
\end{aligned}$$

Published in *Phys. Rev.* **106**, 1366, ©1957.

We can pursue further consequences of our assumption.

(1) ρ^0 could be produced by any strong nuclear reactions, but it would instantly decay mostly into a high-energy $\gamma(\gtrsim 140$ MeV) and a ρ^0. The ratio of charged to neutral components in high-energy reactions should accordingly be influenced.

(2) The second maximum of the pion-nucleon scattering around 1 BeV[2] could be attributed to the reaction

$$\pi^- + p \rightarrow n + \rho^0,$$

if a resonance should occur for such a system.

(3) ρ^0 would contribute a repulsive nuclear force of Wigner type and short range ($\lesssim 0.7 \times 10^{-13}$ cm), more or less similar to the phenomenological hard core.

(4) The anomalous moment of the nucleon[3] should be affected by ρ^0. The main effect seems to be that ρ^0 and the usual pion give opposite contributions to the isotopic scalar part of the core moment, thus tending to bring better agreement between theory and experiment.

(5) If it is energetically possible, we ought to expect that K mesons and hyperons would sometimes decay by emitting a ρ^0.

It should perhaps be added that the neutral meson considered here is similar in nature to the one introduced by Teller for quite different purposes.[4]

* This work was supported by the U. S. Atomic Energy Commission.

[1] R. Hofstadter, Revs. Modern Phys. **28**, 214 (1956). Other references are given in this paper. Also E. E. Chambers and R. Hofstadter, Phys. Rev. **103**, 1456 (1956); J. A. McIntyre, Phys. Rev. **103**, 1464 (1956); R. W. McAllister, Phys. Rev. **104**, 1494 (1956); Hughes, Harvey, Goldberg, and Stafne, Phys. Rev. **90**, 497 (1953); Melkonian, Rustad, and Havens, Bull. Am. Phys. Soc. Ser. II, **1**, 62 (1956). For theoretical interpretations, see Yennie, Lévy, and Ravenhall, Revs. Modern Phys. **29**, 144 (1957).

[2] Shapiro, Leavitt, and Chen, Phys. Rev. **92**, 1073 (1953); Cool, Madansky, and Piccioni, Phys. Rev. **93**, 637 (1954); O. Piccioni and other authors, *Proceedings of the Sixth Annual Rochester Conference on High-Energy Physics 1956* (Interscience Publishers, Inc., New York, 1956). See also F. Dyson, Phys. Rev. **99**, 1037 (1955); G. Takeda, Phys. Rev. **100**, 440 (1955).

[3] H. Miyazawa, Phys. Rev. **101**, 1564 (1956); see also G. Sandri, Phys. Rev. **101**, 1616 (1956).

[4] M. H. Johnson and E. Teller, Phys. Rev. **98**, 783 (1955); H. Duerr and E. Teller, Phys. Rev. **101**, 494 (1956); E. Teller, *Proceedings of the Sixth Annual Rochester Conference on High-Energy Physics 1956* (Interscience Publishers, Inc., New York, 1956); H. Duerr, Phys. Rev. **103**, 469 (1956).

PHYSICAL REVIEW VOLUME 117, NUMBER 3 FEBRUARY 1, 1960

Quasi-Particles and Gauge Invariance in the Theory of Superconductivity*

YOICHIRO NAMBU
The Enrico Fermi Institute for Nuclear Studies and the Department of Physics, The University of Chicago, Chicago, Illinois
(Received July 23, 1959)

Ideas and techniques known in quantum electrodynamics have been applied to the Bardeen-Cooper-Schrieffer theory of superconductivity. In an approximation which corresponds to a generalization of the Hartree-Fock fields, one can write down an integral equation defining the self-energy of an electron in an electron gas with phonon and Coulomb interaction. The form of the equation implies the existence of a particular solution which does not follow from perturbation theory, and which leads to the energy gap equation and the quasi-particle picture analogous to Bogoliubov's.

The gauge invariance, to the first order in the external electromagnetic field, can be maintained in the quasi-particle picture by taking into account a certain class of corrections to the charge-current operator due to the phonon and Coulomb interaction. In fact, generalized forms of the Ward identity are obtained between certain vertex parts and the self-energy. The Meissner effect calculation is thus rendered strictly gauge invariant, but essentially keeping the BCS result unaltered for transverse fields.

It is shown also that the integral equation for vertex parts allows homogeneous solutions which describe collective excitations of quasi-particle pairs, and the nature and effects of such collective states are discussed.

1. INTRODUCTION

A NUMBER of papers have appeared on various aspects of the Bardeen-Cooper-Schrieffer[1] theory of superconductivity. On the whole, the BCS theory, which leads to the existence of an energy gap, presents us with a remarkably good understanding of the general features of superconducivity. A mathematical formulation based on the BCS theory has been developed in a very elegant way by Bogoliubov,[2] who introduced coherent mixtures of particles and holes to describe a superconductor. Such "quasi-particles" are not eigenstates of charge and particle number, and reveal a very bold departure, inherent in the BCS theory, from the conventional approach to many-fermion problems. This, however, creates at the same time certain theoretical difficulties which are matters of principle. Thus the derivation of the Meissner effect in the original BCS theory is not gauge-invariant, as is obvious from the viewpoint of the quasi-particle picture, and poses a serious problem as to the correctness of the results obtained in such a theory.

This question of gauge invariance has been taken up by many people.[3] In the Meissner effect one deals with a linear relation between the Fourier components of the external vector potential A and the induced current J, which is given by the expression

$$J_i(q)=\sum_{j=1}^{3} K_{ij}(q)A_j(q),$$

with

$$K_{ij}(q)=-\frac{e^2}{m}\langle 0|\rho|0\rangle\delta_{ij}+\sum_n\left(\frac{\langle 0|j_i(q)|n\rangle\langle n|j_j(-q)|0\rangle}{E_n}+\frac{\langle 0|j_j(-q)|n\rangle\langle n|j_i(q)|0\rangle}{E_n}\right). \quad (1.1)$$

ρ and j are the charge-current density, and $|0\rangle$ refers to the superconductive ground state. In the BCS model, the second term vanishes in the limit $q\to 0$, leaving the first term alone to give a nongauge invariant result. It has been pointed out, however, that there is a significant difference between the transversal and longitudinal current operators in their matrix elements. Namely, there exist collective excited states of quasi-particle pairs, as was first derived by Bogoliubov,[2] which can be excited only by the longitudinal current.

As a result, the second term does not vanish for a longitudinal current, but cancels the first term (the longitudinal sum rule) to produce no physical effect; whereas for a transversal field, the original result will remain essentially correct.

If such collective states are essential to the gauge-invariant character of the theory, then one might argue that the former is a necessary consequence of the latter. But this point has not been clear so far.

Another way to understand the BCS theory and its problems is to recognize it as a generalized Hartree-Fock approximation.[4] We will develop this point a little further here since it is the starting point of what follows later as the main part of the paper.

* This work was supported by the U. S. Atomic Energy Commission.

[1] Bardeen, Cooper, and Schrieffer, Phys. Rev. **106**, 162 (1957); **108**, 1175 (1957).

[2] N. N. Bogoliubov, J. Exptl. Theoret. Phys. U.S.S.R. **34**, 58, 73 (1958) [translation: Soviet Phys. **34**, 41, 51 (1958)]; Bogoliubov, Tolmachev, and Shirkov, *A New Method in the Theory of Superconductivity* (Academy of Sciences of U.S.S.R., Moscow, 1958). See also J. G. Valatin, Nuovo cimento **7**, 843 (1958).

[3] M. J. Buckingam, Nuovo cimento **5**, 1763 (1957). J. Bardeen, Nuovo cimento **5**, 1765 (1957). M. R. Schafroth, Phys. Rev. **111**, 72 (1958). P. W. Anderson, Phys. Rev. **110**, 827 (1958); **112**, 1900 (1958). G. Rickayzen, Phys. Rev. **111**, 817 (1958); Phys. Rev. Letters **2**, 91 (1959). D. Pines and R. Schrieffer, Nuovo cimento **10**, 496 (1958); Phys. Rev. Letters **2**, 407 (1958). G. Wentzel, Phys. Rev. **111**, 1488 (1958); Phys. Rev. Letters **2**, 33 (1959). J. M. Blatt and T. Matsubara, Progr. Theoret. Phys. (Kyoto) **20**, 781 (1958). Blatt, Matsubara, and May, Progr. Theoret. Phys. (Kyoto) **21**, 745 (1959). K. Yosida, *ibid.* **731**.

[4] Recently N. N. Bogoliubov, Uspekhi Fiz. Nauk **67**, 549 (1959) [translation: Soviet Phys.—Uspekhi **67**, 236 (1959)], has also reformulated his theory as a Hartree-Fock approximation, and discussed the gauge invariance collective excitations from this viewpoint. The author is indebted to Prof. Bogoliubov for sending him a preprint.

Take the Hamiltonian in the second quantization form for electrons interacting through a potential V:

$$H=\int \sum_{i=1}^{2} \psi_i^+(x)K_i\psi_i(x)d^3x+\tfrac{1}{2}\int\int \sum_{i,k} \psi_i^+(x)$$
$$\times\psi_k^+(y)V(x,y)\psi_k(y)\psi_i(x)d^3xd^3y$$
$$\equiv H_0+H_{\text{int}}. \tag{1.2}$$

K is the kinetic energy plus any external field. $i=1$, 2 refers to the two spin states (e.g., spin up and down along the z axis).

The Hartree-Fock method is equivalent to linearizing the interaction H_{int} by replacing bilinear products like $\psi_i^+(x)\psi_k(y)$ with their expectation values with respect to an approximate wave function which, in turn, is determined by the linearized Hamiltonian. We may consider also expectation values $\langle\psi_i(x)\psi_k(y)\rangle$ and $\langle\psi_i^+(x)\psi_k^+(y)\rangle$ although they would certainly be zero if the trial wave function were to represent an eigenstate of the number of particles, as is the case for the true wave function.

We write thus a linearized Hamiltonian

$$H_0'=\int \sum_i \psi_i^+K_i\psi_i d^3x+\int\int \sum_{i,k} [\psi_i^+(x)\chi_{ik}(xy)\psi_k(y)$$
$$+\psi_i^+(x)\phi_{ik}(xy)\psi_k^+(y)$$
$$+\psi_k(x)\phi_{ki}^+(xy)\psi_i(y)]d^3xd^3y$$
$$\equiv H_0+H_s, \tag{1.3}$$

where

$$\chi_{ik}(xy)=\delta_{ik}\delta^3(x-y)\int V(xz)\sum_j\langle\psi_j^+(z)\psi_j(z)\rangle d^3z$$
$$-V(xy)\langle\psi_k^+(y)\psi_i(x)\rangle, \tag{1.4}$$
$$\phi_{ik}(xy)=\tfrac{1}{2}V(xy)\langle\psi_k(y)\psi_i(x)\rangle,$$
$$\phi_{ik}^+(xy)=\tfrac{1}{2}V(xy)\langle\psi_k^+(y)\psi_i^+(x)\rangle.$$

We diagonalize H_0' and take, for example, the ground-state eigenfunction which will be a Slater-Fock product of individual particle eigenfunctions. The defining equations (1.4) then represent just generalized forms of Hartree-Fock equations to be solved for the self-consistent fields χ and ϕ.

The justification of such a procedure may be given by writing the original Hamiltonian as

$$H=(H_0+H_s)+(H_{\text{int}}-H_s)\equiv H_0'+H_{\text{int}}',$$

and demanding that H_{int}' shall have no matrix elements which would cause single-particle transitions; i.e., no matrix elements which would effectively modify the starting H_0': to put it more precisely, we demand our approximate eigenstates to be such that

$$\langle n|H_{\text{int}}'|0\rangle=\langle n|H|0\rangle=0, \tag{1.5}$$

if in $|n\rangle$ more than one particle change their states from those in $|0\rangle$. This condition is contained in Eq. (1.4).[5] Since in many-body problems, as in relativistic field theory, we often take a picture in which particles and holes can be created and annihilated, the condition (1.5) should also be interpreted to include the case where $|n\rangle$ and $|0\rangle$ differ only by such pairs. The significance of the BCS theory lies in the recognition that with an essentially attractive interaction V, a non-vanishing ϕ is indeed a possible solution, and the corresponding ground state has a lower energy than the normal state. It is also separated from the excited states by an energy gap $\sim 2\phi$.

The condition (1.5) was first invoked by Bogoliubov[2] in order to determine the transformation from the ordinary electron to the quasi-particle representation. He derived this requirement from the observation that H_{int}' contains matrix elements which spontaneously create virtual pairs of particles with opposite momenta, and cause the breakdown of the perturbation theory as the energy denominators can become arbitrarily small. Equation (1.5), as applied to such pair creation processes, determines only the nondiagonal part (in quasi-particle energy) of H_s in the representation in which H_0+H_s is diagonal. The diagonal part of H_s is still arbitrary. We can fix it by requiring that

$$\langle 1'|H_{\text{int}}'|1\rangle=0, \tag{1.6}$$

namely, the vanishing of the diagonal part of H_{int} for the states where one more particle (or hole) having a Hamiltonian H_0' is added to the ground state. In this way we can interpret H_0' as describing single particles (or excitations) moving in the "vacuum," and the diagonal part of H_s represents the self-energy (or the Hartree potential) for such particles arising from its interaction with the vacuum.

The distinction between Eqs. (1.5) and (1.6) is not so clear when applied to normal states. On the one hand, particles and holes (negative energy particles) are not separated by an energy gap; on the other hand, there is little difference when one particle is added just above the ground state.

In the above formulation of the generalized Hartree fields, χ and ϕ will in general depend on the external field as well as the interaction between particles. There is a complication due to the fact that they are gauge dependent. This is because a phase transformation $\psi_i(x)\rightarrow e^{i\lambda(x)}\psi_i(x)$ applied on Eq. (1.3) will change χ and ϕ according to

$$\chi(xy)\rightarrow e^{-i\lambda(x)+i\lambda(y)}\chi(xy),$$
$$\phi(xy)\rightarrow e^{i\lambda(x)+i\lambda(y)}\phi(xy), \tag{1.6}$$
$$\phi^+(xy)\rightarrow e^{-i\lambda(x)-i\lambda(y)}\phi^+(xy).$$

[5] Equation (1.5) refers only to the transitions from occupied states to unoccupied states. Transitions between occupied states or unoccupied states are given by Eq. (1.6). These two together then are equivalent to Eq. (1.4). For the analysis of the Hartree approximation in terms of diagrams, see J. Goldstone, Proc.

It is especially serious for ϕ (and ϕ^+) since, even if $\phi(xy)=\delta^3(x-y)$ times a constant in some gauge, it is not so in other gauges. Therefore, unless we can show explicitly that physical quantities do not depend on the gauge, any calculation based on a particular ϕ is open to question. It would not be enough to say that a longitudinal electromagnetic potential produces no effect because it can be transformed away before making the Hartree approximation. A natural way to reconcile the existence of ϕ, which we want to keep, with gauge invariance would be to find the dependence of ϕ on the external field explicitly. If the gauge invariance can be maintained, the dependence must be such that for a longitudinal potential $A=-\text{grad}\lambda$, it reduces to Eq. (1.6). This should not be done in an arbitrary manner, but by studying the actual influence of H_{int} on the primary electromagnetic interaction when ϕ is first determined without the external field.

After these preliminaries, we are going to study the points raised here by means of the techniques developed in quantum electrodynamics. We will first develop the Feynman-Dyson formulation adapted to our problem, and write down an integral eauation for the self-energy part which corresponds to the Hartree approximation. It is observed that it can possess a nonperturbational solution, and the existence of an energy gap is immediately recognized.

Next we will introduce external fields. Guided by the well-known theorems about gauge invariance, we are led to consider the so-called vertex parts, which include the "radiative corrections" to the primary charge-current operator. When an integral equation for the general vertex part is written down, certain exact solutions are obtained in terms of the assumed self-energy part, leading to analogs of the Ward identity.[6] They are intimately related to inherent invariance properties of the theory. Among other things, the gauge invariance is thus strictly established insofar as effects linear in the external field are concerned, including the Meissner effect.

Later we look into the collective excitations. A very interesting result emerges when we observe that one of the exact solutions to the vertex part equations becomes a homogeneous solution if the external energy-momentum is zero, and expresses a bound state of a pair with zero energy-momentum. Then by perturbation, other bound states with nonzero energy-momentum are obtained, and their dispersion law determined. Thus the existence of the bound state is a logical consequence of the existence of the special self-energy ϕ and the gauge invariance, which are seemingly contradictory to each other.

When the Coulomb interaction is taken into account, the bound pair states are drastically modified, turning into the plasma modes due to the same mechanism as in the normal case. This situation will also be studied.

Roy. Soc. (London) **A239**, 267 (1957). Compare also T. Kinoshita and Y. Nambu, Phys. Rev. **94**, 598 (1953).

[6] J. C. Ward, Phys. Rev. **78**, 182 (1950).

2. FEYNMAN-DYSON FORMULATION

We start from the Lagrangian for the electron-phonon system, which is supposed to be uniform and isotropic.[7]

$$\mathcal{L}=\sum_p\sum_i\left[i\psi_i^+(p)\dot{\psi}_i(p)-\psi_i^+(p)\epsilon_p\psi_i(p)\right]+\sum_k\tfrac{1}{2}\left[\dot{\varphi}(k)\dot{\varphi}(-k)-c^2\varphi(k)\varphi(-k)\right]-g\frac{1}{\sqrt{\mathcal{V}}}\sum_{p,k}\psi_i^+(p+k)\psi_i(p)h(k)\varphi(k). \quad (2.1)$$

φ is the phonon field, with the momentum k (energy $\omega_k=ck$) running up to a cutoff value $k_m(\omega_m)$; c is the phonon velocity. ϵ_p is the electron kinetic energy relative to the Fermi energy; $gh(k)$ represents the strength of coupling.[8] ($\mathcal{V}$ is the volume of the system.)

The Coulomb interaction between the electrons is not included for the moment in order to avoid complication. Later we will make remarks whenever necessary about the modifications when the Coulomb interaction is taken into account.

It will turn out to be convenient to introduce a two-component notation[9] for the electrons

$$\Psi(x)=\begin{pmatrix}\psi_1(x)\\ \psi_2^+(x)\end{pmatrix} \quad\text{or}\quad \Psi(p)=\begin{pmatrix}\psi_1(p)\\ \psi_2^+(-p)\end{pmatrix}, \quad (2.2)$$

and the corresponding 2×2 Pauli matrices

$$\tau_1=\begin{pmatrix}0&1\\1&0\end{pmatrix},\quad \tau_2=\begin{pmatrix}0&-i\\i&0\end{pmatrix},\quad \tau_3=\begin{pmatrix}1&0\\0&-1\end{pmatrix}. \quad (2.3)$$

The Lagrangian then becomes:

$$\begin{aligned}\mathcal{L}&=\sum_p\Psi^+(p)\left(i\frac{\partial}{\partial t}-\epsilon_p\tau_3\right)\Psi(p)\\&\quad+\sum_k\tfrac{1}{2}\left[\dot{\varphi}(k)\dot{\varphi}(-k)-c^2\varphi(k)\varphi(-k)\right]\\&\quad-g\frac{1}{\sqrt{\mathcal{V}}}\sum_{p,k}\Psi^+(p+k)\tau_3\Psi(p)h(k)\varphi(k)+\sum_p\epsilon_p\\&=\mathcal{L}_0+\mathcal{L}_{\text{int}}+\text{const.}\end{aligned}$$

The last infinite c-number term comes from the rearrangement of the kinetic energy term. This is certainly uncomfortable, but will not be important except for the calculation of the total energy.

The fields obey the standard commutation relations.

[7] We use the units $\hbar=1$.

[8] For convenience, we have included in $h(k)$ the frequency factor: $h(k)=h_1(k)k_0$.

[9] P. W. Anderson [Phys. Rev. **112**, 1900 (1958)], has also introduced this two-component wave function.

Especially for Ψ, we have

$$\{\Psi_i(x),\Psi_j^+(y)\}\equiv\Psi_i(x)\Psi_j^+(y)+\Psi_j^+(y)\Psi_i(x)$$
$$=\delta_{ij}\delta^3(x-y), \quad (2.5)$$
$$\{\Psi_i(p),\Psi_j^+(p')\}=\delta_{ij}\delta_{pp'}.$$

We may now formally treat H_{int} as perturbation, using the formulation of Feynman and Dyson.[10] The unperturbed ground state (vacuum) is then the state where all individual electron states $\epsilon_p<0(>0)$ are occupied (unoccupied) in the representation where $\psi_i^+(p)\psi_i(p)$ is the occupation number.

Having defined the vacuum, the time-ordered Green's functions for free electrons and phonons

$$\langle T(\Psi_i(xt),\Psi_j^+(x't'))\rangle=[G_0(x-x',t-t')]_{ij},$$
$$\langle T(\varphi(xt),\varphi(x't'))\rangle=\Delta_0(x-x',t-t') \quad (2.6)$$

are easily determined. We get for their Fourier representation (in the limit $\mathcal{V}\to\infty$)[10a]

$$G_0(xt)=(1/(2\pi)^4)\int G_0(pp_0)e^{i\mathbf{p}\cdot\mathbf{x}-ip_0t}d^3pdp_0,$$

$$\Delta_0(xt)=\frac{1}{(2\pi)^4}\int_{|k|<k_m}\Delta_0(kk_0)e^{i\mathbf{k}\cdot\mathbf{x}-ik_0t}d^3kdk_0,$$

$$G_0(pp_0)=i\left[P\frac{1}{p_0-\epsilon_p\tau_3}-i\pi\,\text{sgn}(\tau_3\epsilon_p)\delta(p_0-\tau_3\epsilon_p)\right] \quad (2.7)$$
$$=i(p_0+\epsilon_p\tau_3)/(p_0^2-\epsilon_p^2+i\epsilon),$$
$$\Delta_0(kk_0)=i\left[P\frac{1}{k_0^2-c^2k^2}-i\pi\delta(k_0^2-c^2k^2)\right]$$
$$=i/(k_0^2-c^2k^2+i\epsilon).$$

With the aid of these Green's functions, we are able to calculate the S matrix and other quantities according to a well-defined set of rules in perturbation theory.

We will analyze in particular the self-energies of the electron and the phonon. In the many-particle system, these energies express (apart from the self-interaction of the electron) the average interaction of a single particle or phonon placed in the medium. Because the phonon spectrum is limited, there will be no ultraviolet divergences, unlike the case of quantum electrodynamics.

These self-energies may be obtained in a perturbation expansion with respect to H_{int}. We are, however, interested in the Hartree method which proposes to take account of them in an approximate but nonperturbational way. It is true that the self-energies are in general complex due to the instability of single particles. But to the extent that the single-particle picture makes physical sense, we will ignore the small imaginary part of the self-energies in the following considerations.

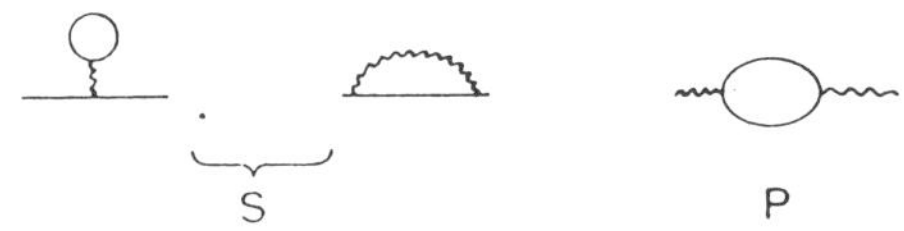

FIG. 1. Second order self-energy diagrams. Solid and curly lines represent electron and phonons, respectively, themselves being under the influence of the self-energies Σ and Π. All diagrams are to be interpreted in the sense of Feynman, lumping together all topologically equivalent processes.

Let us thus introduce the approximate self-energy Lagrangian $\mathcal{L}_s$, and write

$$\mathcal{L}=(\mathcal{L}_0+\mathcal{L}_s)+(\mathcal{L}_{\text{int}}-\mathcal{L}_s)$$
$$\equiv\mathcal{L}_0'+\mathcal{L}_{\text{int}}',$$
$$\mathcal{L}_0=\sum_p\Psi_p^+L_0\Psi_p+\sum_k\tfrac{1}{2}\varphi_kM_0\varphi_{-k}, \quad (2.8)$$
$$\mathcal{L}_s=-\sum_p\Psi_p^+\Sigma\Psi_p-\sum_k\tfrac{1}{2}\varphi_k\Pi\varphi_{-k},$$
$$L_0-\Sigma\equiv L,\quad M_0-\Pi\equiv M.$$

The free electrons with "spin" functions u and phonons obey the dispersion law

$$L_0(\mathbf{p},p_0=\epsilon_p)u_p=0,\quad M_0(\mathbf{k},k_0=\omega_k)=0, \quad (2.9)$$

whereas they obey in the medium

$$L(\mathbf{p},p_0=E_p)u_p=0,\quad M(\mathbf{k},k_0=\Omega_k)=0. \quad (2.9')$$

Σ will be a function of momentum p and "spin." Π will consist of two parts: $\Pi(k_0k)=\Pi_1(k)k_0^2+\Pi_2(k)$ in conformity with the second order character (in time) of the phonon wave equation.[11]

The propagators corresponding to these modified electrons and phonons are

$$G(pp_0)=i/(L(pp_0)+i\,\text{sgn}(p_0)\epsilon),$$
$$\Delta(kk_0)=i/(M(kk_0)+i\epsilon). \quad (2.10)$$

We now determine Σ and Π self-consistently to the second order in the coupling g. Namely the second order self-energies coming from the phonon-electron interaction have to be cancelled by the first order effect of $\mathcal{L}_{\text{int}}$.

These second order self-energies are represented by the nonlocal operators[12] (Fig. 1)

$$\mathcal{S}((t+t')/2)=\int\int\int\Psi^+(xt)S(x-x',t-t')$$
$$\times\Psi(x't')d^3xd^3x'd(t-t'), \quad (2.11)$$
$$\mathcal{P}((t+t')/2)=\tfrac{1}{2}\int\int\int\varphi(xt)P(x-x',t-t')$$
$$\times\varphi(x't')d^3xd^3x'd(t-t'),$$

[10] F. J. Dyson, Phys. Rev. **75**, 486, 1736 (1949); R. P. Feynman, Phys. Rev. **76**, 769 (1949); J. Schwinger, Phys. Rev. **74**, 1439 (1948). Although we followed here the perturbation theory of Dyson, there is no doubt that the relations obtained in this paper can be derived by a nonperturbational formulation such as J. Schwinger's: Proc. Natl. Acad. Sci. U. S. **37**, 452, 455 (1951).

[10a] P stands for the principal value; $i\epsilon$ in the denominator is a small positive imaginary quantity.

[11] In the same spirit Σ should actually be in the form $\Sigma_1(p)p_0+\Sigma_2(p)$. Here we neglect the renormalization term Σ_1 since the two conditions (2.13) can be met without it.

[12] We use the word nonlocal here for nonlocality in time.

where S and P have the Fourier representation

$$S(pp_0) = -ig^2\tau_3\delta^3(p)\delta(p_0)h^2(0)\Delta(0) \times \int \mathrm{Tr}[\tau_3 G(p'p_0)]d^3p dp_0 - ig^2\int \tau_3 G(p-k, p_0-k_0)\tau_3 h^2(kk_0) \times \Delta(kk_0)d^3k dk_0, \quad (2.12)$$

$$P(kk_0) = ig^2 h(kk_0)^2 \int \mathrm{Tr}[\tau_3 G(pp_0) \times G(p+k, p_0+k_0)]d^3p dp_0.$$

In Eq. (2.11) we have chosen more or less arbitrarily $(t+t')/2$ as the fixed time to which we refer the nonlocal operators $\mathcal{S}$ and $\mathcal{P}$. The self-consistency requirements (1.5) and (1.6) mean in the present case that Σ, Π must be identical with S, P (*a*): for the diagonal elements [on the energy shell, Eq. (2.9)], and (*b*): for the nondiagonal matrix elements for creating a pair out of the vacuum.

The pair creation of electrons is possible because Ψ, being a two-component wave function, can have in general two eigenfunctions u_{ps} $(s=1, 2)$ with different energies E_{ps} for a fixed momentum, p, only one of which is occupied in the ground state.

Thus taking particular plane waves $u_{ps}{}^*e^{-ip\cdot x+ip_0t}$, $u_{p's'}e^{ip'\cdot x'-ip_0't'}$ for Ψ^+ and Ψ in (2.11), we easily find that the diagonal matrix element of Σ corresponds to $u_{ps}{}^*S(p,E_{ps})u_{ps}$, while the nondiagonal part corresponds to $u_{ps}{}^*S(p,0)u_{ps}{}'$, $s\neq s'(p_0'=-p_0)$.

A similar situation holds also for the photon self-energy Π. Since Π consists of two parts, the diagonal and off-diagonal conditions will fix these.

With this understanding, the self-consistency relations may be written

$$\Sigma(\mathbf{p}E_p)_{\mathrm{D}} = S(\mathbf{p},E_p)_{\mathrm{D}}, \quad \Sigma(\mathbf{p}0)_{\mathrm{ND}} = S(\mathbf{p}0)_{\mathrm{ND}}, \quad \Pi(\mathbf{k}\Omega_k) = P(\mathbf{k}\Omega_k), \quad \Pi(\mathbf{k}0) = P(\mathbf{k}0), \quad (2.13)$$

where D, ND signify the diagonal and nondiagonal parts in the "spin" space. As stated before, we have agreed to omit possible imaginary parts in S and P. (The nondiagonal components, however, will turn out to be real.)

Before discussing the general solutions, let us consider the meaning of Eq. (2.13) in terms of perturbation theory. Suppose we expand G occurring in Eq. (2.12), with respect to Σ:

$$G = G_0 - iG_0\Sigma G_0 - G_0\Sigma G_0\Sigma G_0 + \cdots,$$

and expand Σ itself with respect to g^2, then we easily realize that Eq. (2.13) defines an infinite sum of a particular class of diagrams, which are illustrated in Fig. 2. The first term in S of Eq. (2.12) corresponds to the ordinary Hartree potential which is just a constant,

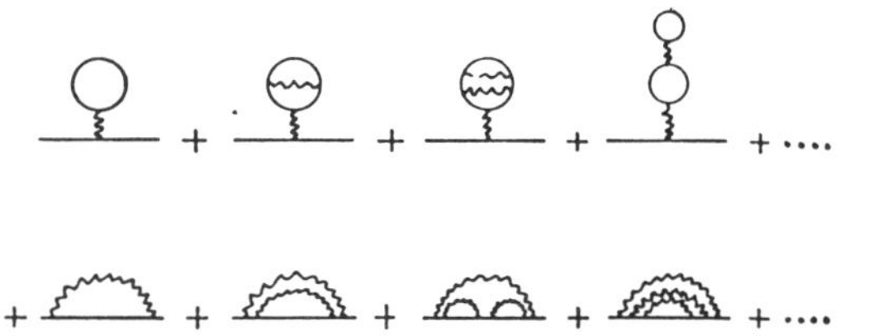

FIG. 2. Expansion of the self-consistent self-energy $\Sigma\sim S$ in terms of bare electron diagrams.

whereas the second term gives an exchange effect. In the latter, the approximation is characterized by the fact that no phonon lines cross each other.

It must be said that the Hartree approximation does not really sum the series of Fig. 2 completely since we equate in Eq. (2.13) only special matrix elements of both sides. For in the perturbation series the Σ obtained to any order is a function of p_0, whereas in Eq. (2.13) it is replaced by a p_0-independent quantity. Hence there will be a correction left out in each order (analogous to the radiative correction after mass renormalization in quantum electrodynamics).

In this perturbation expansion, S in Eq. (2.13) is always proportional to τ_3 on the energy shell since $H_0 \propto \tau_3$. Accordingly Σ will be $\propto \tau_3$ and commute with H_0, so that no off-diagonal part exists.[11]

It is important, however, to note the possibility of a nonperturbational solution by assuming that Σ contains also a term proportional to τ_1 or τ_2. Thus, take

$$\Sigma(p) = \chi(p)\tau_3 + \phi(p)\tau_1, \quad H_0' = (\epsilon+\chi)\tau_3 + \phi\tau_1 \equiv \bar{\epsilon}\tau_3 + \phi\tau_1. \quad (2.14)$$

This form bears a resemblance to the Dirac equation. Its eigenvalues are

$$E = \pm E_p \equiv \pm(\bar{\epsilon}_p{}^2 + \phi_p{}^2)^{\frac{1}{2}}. \quad (2.15)$$

Since H_0' describes by definition excited states, we have to adopt the hole picture and conclude that the ground state (vacuum) is the state where all negative energy "quasi-particles" ($E<0$) are occupied and no positive energy particles exist. If ϕ remains finite on the Fermi surface, the positive and negative states are separated by a gap $\sim 2|\phi|$. The corresponding Green's function G now has the representation

$$G(pp_0) = i\frac{p_0 + \bar{\epsilon}_p\tau_3 + \phi_p\tau_1}{p_0{}^2 - E_p{}^2 + i\epsilon}. \quad (2.16)$$

In order to extract the diagonal and nondiagonal parts in spin space, we will use the trick

$$O_{\mathrm{D}} = \tfrac{1}{2}\,\mathrm{Tr}\,(\Lambda O), \quad O_{\mathrm{ND}} = -(i/2)\,\mathrm{Tr}\,(\Lambda O\tau_2), \quad \Lambda = [E_p + H_0'(p)]/2E_p. \quad (2.17)$$

Applying this to Eq. (2.13a) with Eqs. (2.12), (2.14), and (2.15), we finally obtain the following equations for χ and ϕ

$$\frac{\epsilon_p\chi_p+\phi_p^2}{E_p}=\frac{g^2\pi}{(2\pi)^4}P\int\left[\frac{E_p}{\Omega_k}+\frac{E_{p-k}+\Omega_k}{E_pE_{p-k}\Omega_k}\right.$$
$$\left.\times(\bar{\epsilon}_p\bar{\epsilon}_{p-k}-\phi_p\phi_{p-k})\right]\frac{h(k)^2}{E_p^2-(E_{p-k}-\Omega_k)^2}d^3k,$$
$$\epsilon_p\phi_p=\frac{g^2\pi}{(2\pi)^4}\int(\bar{\epsilon}_{p-k}\phi_p+\bar{\epsilon}_p\phi_{p-k})$$
$$\times\frac{h(k)^2d^3k}{E_{p-k}\Omega_k(E_{p-k}+\Omega_k)}. \quad (2.18)$$

The second equation, coming from the nondiagonal condition, has a trivial solution $\phi=0$. If a finite solution ϕ exists, it cannot follow from perturbation treatment since there is no inhomogeneous term to start with.

Equation (2.18) is equivalent to, but slightly different from, the corresponding conditions of Bogoliubov because of a slightly different definition of the nondiagonal part of the self-energy operator, which is actually due to an inherent ambiguity in approximating nonlocal operators by local ones. (This is the same kind of ambiguity as one encounters in the derivation of a potential from field theory. The difference between the local operator Σ and the nonlocal one S shows up in a situation like that in Fig. 3, and the compensation between Σ and S is not complete.) We may avoid this unpleasant situation, by extending the Hartree self-consistency conditions to all virtual matrix elements, but this would mean that ϕ (and χ) must be treated as nonlocal. We will discuss this situation in a separate section since such a generalization brings simplification in dealing with the problem of gauge invariance and collective excitations.

For the moment we consider the second equation of (2.18) and rewrite it

$$\phi_p=A_p\frac{g^2\pi}{(2\pi)^4}\int\frac{\phi_{p-k}}{E_{p-k}}\frac{h(k)d^3k}{\Omega_k(E_{p-k}+\Omega_k)},$$
$$A_p=\bar{\epsilon}_p\Big/\left(\epsilon_p-\frac{g^2\pi}{(2\pi)^4}\int\frac{\bar{\epsilon}_ph(k)^2d^3k}{E_{p-k}\Omega_k(E_{p-k}+\Omega_k)}\right). \quad (2.19)$$

This is essentially the energy gap equation of BCS if $g^2A_ph(k)^2/\Omega_k(E_{p-k}+\Omega_k)$ is identified with the effective interaction potential V, and if $\bar{\epsilon}_p\sim\epsilon_p(\chi_p\sim0)$. It has a solution

$$\phi\sim\Omega_m\exp(-1/VN),$$

if $VN\ll1$, N being the density of states: $N=dn/d\epsilon_p$ on the Fermi surface.

The phonon self-energy Π may be studied similarly from Eq. (2.13), which should determine the renormalization of the phonon field. It does not play an essential role in superconductivity, though it gives rise to an important correction when the Coulomb effect is taken into account. (See the following section.)

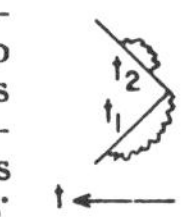

FIG. 3. An example of the situation where the cancellation of $\Sigma_{\rm ND}$ *versus* $S_{\rm ND}$ is not complete. The two self-energy parts overlap in time, and their centers of time t_1 and t_2 are such that $t_1>t_2$. If calculated according to the usual perturbation theory, this process will not be eliminated by the condition $\Sigma_{\rm ND}=(S_1)_{\rm ND}$.

From the nature of Eq. (2.12), it is clear that $\tau_1\phi$ can actually be pointed in any direction in the 1–2 plane of the τ space: $\tau_1\phi_1+\tau_2\phi_2$. It was thus sufficient to take $\phi_1\neq0$, $\phi_2=0$. Any other solution is obtained by a transformation

$$\Psi\to\exp(i\alpha\tau_3/2)\Psi,$$
$$(\phi,0)\to(\phi\cos\alpha,\phi\sin\alpha). \quad (2.20)$$

In view of the definition of Ψ, Eq. (2.20) is a gauge transformation with a constant phase. Thus the arbitrariness in the direction of ϕ is the 1–2 plane is a reflection of the gauge invariance.

For later use, we also mention here the particle-antiparticle conjugation C of the quasi-particle field Ψ. This is defined by

$$C:\quad \Psi\to\Psi^C=C\Psi^+=\tau_2\Psi^+,$$

or

$$\begin{pmatrix}\psi_1^C\\ \psi_2^{+C}\end{pmatrix}=\begin{pmatrix}-i\psi_2\\ i\psi_1^+\end{pmatrix}, \quad (2.21)$$

and changes quasi-particles of energy-momentum (p_0,p) into holes of energy-momentum $(-p_0,-p)$, or interchanges up-spin and down-spin electrons. Under C, the τ operators transform as

$$C:\quad \tau_i\to C^{-1}\tau_iC=-\tau_i^T,\quad i=1,2,3 \quad (2.22)$$

where T means transposition.

As a consequence, we have also

$$C:\quad L(p)\to L^C(-p)=-L(-p)^T. \quad (2.23)$$

Finally we make a remark about the Coulomb interaction. When this is taken into account, the phonon interaction factor $g^2h(k)^2\Delta(k,k_0)$ in Eq. (2.12a) has to be replaced by

$$[g^2h(k)^2\Delta(kk_0)+ie^2/k^2]/$$
$$\{1-i\Pi(kk_0)[\Delta(kk_0)+ie^2/g^2h(k)^2k^2]\}.$$

As is well known, the denominator represents the screening of the Coulomb interaction. Discussion about this point will be made later in connection with the plasma oscillations.

3. NONLOCAL (ENERGY-DEPENDENT) SELF-CONSISTENCY CONDITIONS

In the last section we remarked that the self-consistency conditions Eq. (2.13) may be extended to all virtual matrix elements, namely, not only on the energy shell (diagonal) and for the virtual pair creation out of the vacuum, but also for the self-energy effects which appear in intermediate states of any process.

This simply means that ϕ and π are now nonlocal; i.e., depend both on energy and momentum arbitrarily, and are to be completely equated with S and P, respectively,

$$\Sigma(pp_0)=S(pp_0),\quad \Pi(kk_0)=P(kk_0). \tag{3.1}$$

Actually, these self-energies can no more be incorporated in H_0' as the zeroth order Lagrangian since they contain infinite orders of time derivatives.[13] Nevertheless, Eq. (3.1) has a precise meaning in the bare particle perturbation theory. It defines the (proper) self-energy parts (in the sense of Dyson) as an infinite sum of the special class of diagrams illustrated in Fig. 2.

The earlier condition of Eq. (2.13) represented, as was noted there, only an approximation to this sum. In other words, Eqs. (2.13) and (3.1) are not exactly identical even on the energy shell.

The Hartree-Fock approximation based on Eq. (3.1) could be interpreted as a nonperturbation approximation to determine the "dressed" single particles (together with the "dressed vacuum") or the Green's function $\langle 0|T(\Psi(xt),\Psi^+(x't'))|0\rangle$ for the true interacting system. Such single particles will satisfy

$$L(p,p_0)u\cong 0,\quad M(k,k_0)\cong 0. \tag{3.2}$$

We use the approximate equality since a really stable single particle may not exist.

Let us assume that these determine the approximate renormalized dispersion law

$$p_0^2=E_r(p)^2,\quad k_0^2=\Omega_r(k)^2. \tag{3.3}$$

If we write for Σ

$$\Sigma(pp_0)=p_0\zeta(pp_0)+\chi(pp_0)\tau_3+\phi(pp_0)\tau_1, \tag{3.4}$$

where ζ, χ, ϕ are even functions of p_0, then

$$\begin{aligned}E_r^2(p)&=[\bar{\epsilon}(pp_0)^2+\phi(pp_0)^2]/[1-\zeta(pp_0)]^2|_{p_0^2=E_r(p)^2}\\ &\equiv E(pp_0)^2/Z(pp_0)^2|_{p_0^2=E_r(p)^2}.\end{aligned} \tag{3.5}$$

The Green's functions G and Δ will be given by

$$\begin{aligned}G(pp_0)&=i/L(pp_0)\\ &=i\int_0^\infty \frac{dx}{p_0^2-x+i\epsilon}\\ &\quad\times \mathrm{Im}\,\frac{p_0Z(px)+\bar{\epsilon}(px)\tau_3+\phi(px)\tau_1}{x^2Z(px)^2-E(px)^2},\end{aligned} \tag{3.6}$$

$$\begin{aligned}\Delta(kk_0)&=i/M(kk_0)\\ &=i\int_0^\infty \frac{dx}{k_0^2-x+i\epsilon}\,\mathrm{Im}\,\frac{1}{M(kx)}.\end{aligned}$$

This representation assumes that $G(p_0)[\Delta(k_0)]$ is analytic except for a branch cut on the real axis. The imaginary part in the integrand is expected to have a delta function or a sharp peak at $x=E_r^2(p)$ $[\Omega_r(k^2)]$. These properties are necessary in order that the vacuum is stable and the quasi-particles and phonons have a valid physical meaning as excitations.[14] In the following, we will generally consider this quasi-particle peak only, and write

$$G(pp_0)=i\frac{p_0Z(pp_0)+\bar{\epsilon}(pp_0)\tau_3+\phi(pp_0)\tau_1}{p_0Z(pp_0)^2-E(pp_0)^2+i\epsilon},\quad \text{etc.}$$

The Hartree equations now take the form

$$\Sigma(pp_0)=-i\frac{g^2}{(2\pi)^4}\int \tau_3G(p-k,\,p_0-k_0)\tau_3h(kk_0)^2d^3kdk_0,$$

$$\begin{aligned}\Pi(kk_0)=i\frac{g^2}{(2\pi)^4}\int \mathrm{Tr}\,[\tau_3G(k-p,\,k_0-p_0)\\ \times\tau_3G(pp_0)]d^3pdp_0.\end{aligned} \tag{3.7}$$

This equation for Σ is much simpler than the previous one (2.18) since we may just equate the coefficients of 1, τ_3, τ_1 on both sides. In particular, we get the energy gap equation

$$\begin{aligned}\phi(pp_0)=-\frac{ig^2}{(2\pi)^4}\int\frac{\phi(p'p_0')}{p_0^2Z(p'p_0')^2-E(p'p_0')^2+i\epsilon}\\ \times h(\mathbf{p}-\mathbf{p}',\,p_0-p_0')^2\Delta(\mathbf{p}-\mathbf{p}',\,p_0-p_0')d^3p'dp_0',\end{aligned} \tag{3.8}$$

which is to be compared with Eq. (2.19).

Although the existence of a solution to Eq. (3.6) may be difficult to establish, the solution, if it exists, should not be much different from the older solution to Eq. (2.19). At any rate, our assumption about the analyiticity of G and Π is consistent with Eq. (3.6) or (3.7) which implies that Σ and Π are also analytic except for a cut on the real axis.

In later calculations we shall encounter various integrals which we may classify into three types regarding their sensitivity to the energy gap. First, a normal self-energy part, for example, represents the effect of the bulk of the surrounding electrons on a particular electron, and is insensitive to the change of the small fraction $\sim\phi/E_F$ of the electrons near the Fermi surface in a superconductor. Such a quantity is

[13] It would seem then that we lose the advantage of the generalization since we cannot find the Bogoliubov transformation. However, we could still start from the older solution (2.13) as the zeroth approximation to Eq. (3.1), and then calculate the correction; namely, the "radiative" correction to the Bogoliubov vacuum and the Bogoliubov quasi-particle. These corrections would take account of the single-particle transitions which remain after the Bogoliubov condition (2.13) is imposed.

[14] This is a representation of the Lehmann type [H. Lehmann, Nuovo cimento **11**, 342 (1954)] which can be derived by defining the Green's functions in terms of Heisenberg operators. See also V. M. Galizkii and A. B. Migdal, J. Exptl. Theoret. Phys. U.S.S.R. **34**, 139 (1958) [translation: Soviet Phys. JETP **7**, 96 (1958)].

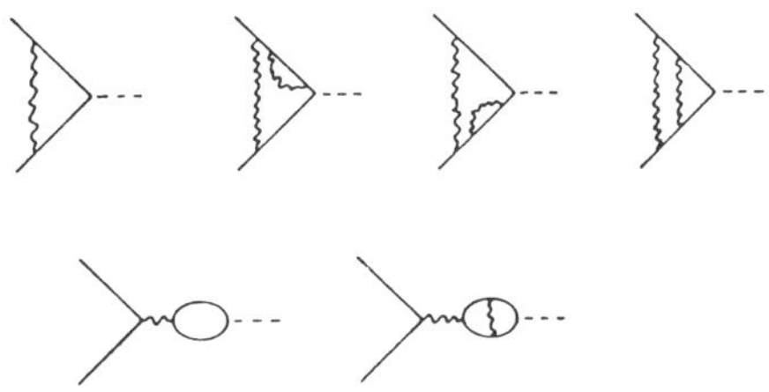

FIG. 4. Construction of the vertex part Γ in bare particle picture. The second line represents the polarization diagrams.

given by an integral like

$$g^2\int \frac{\epsilon_k}{E_k} f(\mathbf{p}-\mathbf{k})d^3k, \tag{3.9}$$

where the region $\epsilon_k \lesssim E_k = (\epsilon_k^2+\phi^2)^{\frac{1}{2}}$ makes little contribution if $f(p-k)$ is a smooth function.

Second, the energy gap itself is determined from an equation of the form

$$g^2\int \frac{d^3k}{E_k} f(\mathbf{p}-\mathbf{k}) \sim g^2 \int_{E_k \lesssim \omega_m} \frac{d^3k}{E_k} f(\mathbf{p}-\mathbf{k}) \sim 1, \tag{3.10}$$

which means that even if g^2 is small, such an expression is always of the order 1.

Finally we meet with integrals like

$$g^2\int \frac{\epsilon_k\phi}{E_k^3} f(\mathbf{p}-\mathbf{k})d^3k, \quad g^2\int \frac{\phi^2}{E_k^3} f(\mathbf{p}-\mathbf{k})d^3k, \quad \text{etc.} \tag{3.11}$$

They have an extra cutoff factor $\sim 1/E$, $1/E^2$, etc., in the integrand which restricts the contribution to an energy interval $\sim 2\phi$ near the Fermi surface. The integrals are thus of the order

$$g^2N\phi/\omega_m, \quad g^2N, \quad \text{etc.}$$

In the following, we will not be primarily concerned with the ordinary self-energy effects. We will assume that proper renormalization has been carried out, or else simply disregard it unless essential. When we carry out perturbation type calculations, we will arrange things so that quantities of the second type are taken into account rigorously, and treat quantities of the third type as small, and hence negligible ($g^2N \ll 1$).

4. INTEGRAL EQUATIONS FOR VERTEX PARTS[15]

In the presence of an electromagnetic potential, the original Lagrangian $\mathfrak{L}$ has to be modified according to the rule

$$i\frac{\partial}{\partial t} \to i\frac{\partial}{\partial t}+eA_0, \quad \mathbf{p}\to \mathbf{p}-\frac{e}{c}\mathbf{A}$$

for the electron. Going to the two-component representation, this corresponds to the prescription

$$i\frac{\partial}{\partial t} \to i\frac{\partial}{\partial t}+e\tau_3A_0, \quad \mathbf{p}\to \mathbf{p}-\frac{e}{c}\tau_3\mathbf{A} \tag{4.1}$$

acting on Ψ. It can also be inferred from the gauge transformation $\Psi \to \exp(i\alpha\tau_3)\Psi$ as was observed previously. So the ordinary charge-current operator turns out to be in our form given by

$$\rho(x) = \frac{e}{2}([\Psi^+(x),\tau_3\Psi(x)]+\{\Psi^+(x),\Psi(x)\}),$$

$$\begin{aligned}\mathbf{j}(x) = \frac{-ie}{4m}([\Psi^+(x), (\nabla-ie\tau_3\mathbf{A})\Psi(x)] \\ +[(-\nabla-ie\tau_3A)\Psi^+(x), \Psi(x)] \\ +\{\Psi^+(x), (\nabla\tau_3-ieA)\Psi(x)\} \\ -\{(\tau_3\nabla-ie\mathbf{A})\Psi^+(x), \Psi(x)\}).\end{aligned} \tag{4.2}$$

The second terms on the right-hand side, being infinite C numbers, arise from the rearrangement of ψ and ψ^+, and will actually be compensated for by the first terms.

This expression, however, has to be modified when we go to the quasi-particle picture.

For we have seen that the self-energy ϕ of a quasi-particle is a gauge-dependent quantity. If we want to have the quasi-particle picture and gauge invariance at the same time, then it is clear that the electromagnetic current of a quasi-particle must contain, in addition to the normal terms given by Eq. (4.2), terms which would cause a physically unobservable transformation of ϕ if the electromagnetic potential is replaced by the gradient of a scalar. In other words, the complete charge current of a quasi-particle has to satisfy the continuity equation, which Eq. (4.2) does not, since

$$\partial\rho/\partial t+\nabla\cdot\mathbf{j} = 2\Psi^+\phi\tau_2\Psi.$$

In order to find such a conserving expression for charge current, it is instructive to go back to the bare electron picture, in which the self-energy is represented by a particular class of diagrams discussed in the previous sections.

It is well known[16] in quantum electrodynamics that, in any process involving electromagnetic interaction, perturbation diagrams can be grouped into gauge-invariant subsets, such that the invariance is maintained by each subset taken as a whole. Such a subset can be constructed by letting each photon line in a diagram interact with a charge of all possible places along a chain of charge-carrying particle lines. The gauge-invariant interaction of a quasi-particle with an electromagnetic potential should then be obtained by attaching a photon line at all possible places in the diagrams of Fig. 2. The result is illustrated in Fig. 4,

[15] Hereafter we will often use the four-dimensional notation $x = (\mathbf{x},t)$, $p = (\mathbf{p},p_0)$, $d^4p = d^3pdp_0$.

[16] Z. Koba, Progr. Theoret. Phys. (Kyoto) 6, 322 (1951).

which consists of the "vertex" part Γ and the self-energy part Σ.

In this way we are led to consider the modification of the vertex due to the phonon interaction in the same approximation as the self-energy effect is included in the quasi-particle. It is not difficult to see that it corresponds to a "ladder approximation" for the vertex part, and we get an integral equation[17]

$$\Gamma_i(p',p)=\gamma_i(p',p)-g^2\int \tau_3 G(p'-k)\Gamma_i(p'-k,\, p-k) \times G(p-k)\tau_3 h(k)^2\Delta(k)d^4k, \quad (4.3)$$

where γ_i, $i=0, 1, 2, 3$ stand for the free particle charge current $[\tau_3, (1/2m)(\mathbf{p}+\mathbf{p}')]$ which follows from Eq. (4.2). Similar equations may be set up for any type of vertex interactions.

Equation (4.3) is the basis of the rest of this paper. It expresses a clear-cut approximation procedure in which the "free" charge-current operator γ_i of a quasi-particle is modified by a special class of "radiative corrections" due to H_{int}'.

As the next important step, we observe that there exist exact solutions to Eq. (4.3) for the following four types of vertex interactions

$$\begin{aligned}
(a)\quad & \gamma^{(a)}(p',p)=L_0(p')-L_0(p)\\
&\qquad =(p_0'-p_0)-\tau_3(\epsilon_{p'}-\epsilon_p),\\
& \Gamma^{(a)}(p',p)=L(p')-L(p)\\
&\qquad =\gamma_a(p')-[\Sigma(p')-\Sigma(p)],\\
(b)\quad & \gamma^{(b)}(p',p)=L_0(p')\tau_3-\tau_3L_0(p)\\
&\qquad =(p_0'-p_0)\tau_3-(\epsilon_{p'}-\epsilon_p),\\
& \Gamma^{(b)}(p',p)=L(p')\tau_3-\tau_3L(p),\\
(c)\quad & \gamma^{(c)}(p',p)=L_0(p')\tau_1+\tau_1L_0(p'),\\
& \Gamma^{(c)}(p',p)=L(p')\tau_1+\tau_1L(p),\\
(d)\quad & \gamma^{(d)}(p',p)=L_0(p')\tau_2+\tau_2L(p),\\
& \Gamma^{(d)}(p',p)=L(p')\tau_2+\tau_2L(p).
\end{aligned} \quad (4.4)$$

The verification is straightforward by noting that $G(p)=i/L(p)$, and making use of Eq. (3.7).

The fact that there are simple solutions is not accidental. These solutions express continuity equations and other relations following from the four types of operations, which do not depend on the presence or absence of the interaction:

$$\begin{aligned}
(a)\quad & \Psi(x)\to e^{i\alpha(x)}\Psi(x), && \Psi^+(x)\to\Psi^+(x)e^{-i\alpha(x)},\\
(b)\quad & \Psi\to e^{i\tau_3\alpha}\Psi, && \Psi^+\to\Psi^+e^{-i\tau_3\alpha},\\
(c)\quad & \Psi\to e^{\tau_1\alpha}\Psi, && \Psi^+\to\Psi^+e^{\tau_1\alpha},\\
(d)\quad & \Psi\to e^{\tau_2\alpha}\Psi, && \Psi^+\to\Psi^+e^{\tau_2\alpha},
\end{aligned} \quad (4.5)$$

where $\alpha(x)$ is an arbitrary real function.

(*a*) and (*b*) correspond, respectively, to the spin rotation around the *z* axis, and the gauge transformation. The entire Lagrangian is invariant under them, so that we obtain continuity equations for the *z* component of spin and charge, respectively:

$$\begin{aligned}
&\frac{\partial}{\partial t}\boldsymbol{\Psi}^+\boldsymbol{\Psi}+\nabla\cdot\boldsymbol{\Psi}^+\frac{\mathbf{p}}{m}\tau_3\boldsymbol{\Psi}=0,\\
&\frac{\partial}{\partial t}\boldsymbol{\Psi}^+\tau_3\boldsymbol{\Psi}+\nabla\cdot\boldsymbol{\Psi}^+\frac{\mathbf{p}}{m}\boldsymbol{\Psi}=0,
\end{aligned} \quad (4.6)$$

where $\boldsymbol{\Psi}$ is the true Heisenberg operator.

These equations are identical with

$$\boldsymbol{\Psi}^+\gamma^{(a)}\boldsymbol{\Psi}=0, \quad \boldsymbol{\Psi}^+\gamma^{(b)}\boldsymbol{\Psi}=0. \quad (4.7)$$

Taken between two "dressed" quasi-particle states, the left-hand side of Eq. (4.7) will become

$$\begin{aligned}
e^{-i(p'-p)\cdot x}\langle p'|\,\boldsymbol{\Psi}^+(x)\gamma^{(n)}\boldsymbol{\Psi}(x)\,|p\rangle
&=u_{p'}^*\Gamma^{(n)}(p',p)u_p\\
&=0, \quad (n=a,b)
\end{aligned} \quad (4.7')$$

where u_p, $u_{p'}$ are single-particle wave functions satisfying $L(p)u_p=u_{p'}^*L(p')=0$.

In this way we have shown the existence of spin and charge currents $\Gamma_i^{(a)}(p',p)$ and $\Gamma_i^{(b)}(p',p)$ for a quasi-particle, for which the continuity equations

$$(p_0'-p_0)\cdot\Gamma_0^{(n)}-\sum_{i=1}^{3}(p'-p)_i\cdot\Gamma_i^{(n)}=\Gamma^{(n)}(p',p)=0$$

will hold.

The last two transformations of Eq. (4.4) are not unitary, but mix ψ_1 and ψ_2^+ in such a way as to keep $\Psi^+\tau_3\Psi$ invariant. From infinitesimal transformations of these kinds we get

$$\begin{aligned}
&i\boldsymbol{\Psi}^+\tau_1\left(\frac{\overrightarrow{\partial}}{\partial t}-\frac{\overleftarrow{\partial}}{\partial t}\right)\boldsymbol{\Psi}+\nabla\cdot\boldsymbol{\Psi}^+\tau_2\left(\frac{\overrightarrow{\mathbf{p}}}{m}+\frac{\overleftarrow{\mathbf{p}}}{m}\right)\boldsymbol{\Psi}=0,\\
&-i\boldsymbol{\Psi}^+\tau_2\left(\frac{\overrightarrow{\partial}}{\partial t}-\frac{\overleftarrow{\partial}}{\partial t}\right)\boldsymbol{\Psi}+\nabla\cdot\Psi^+\tau_1\left(\frac{\overrightarrow{\mathbf{p}}}{m}+\frac{\overleftarrow{\mathbf{p}}}{m}\right)\boldsymbol{\Psi}=0,
\end{aligned} \quad (4.8)$$

which bear the same relations to $\gamma^{(c),(d)}$ and $\Gamma^{(c),(d)}$ as Eq. (4.6) did to $\gamma^{(a),(b)}$ and $\Gamma^{(a),(b)}$. Note that the above equations are unaffected by the presence of the phonon interaction.

The fact that we can find a conserved charge-current

[17] This equation may also be derived simply by considering the self-energy equation (3.7) in the presence of an external field, and expanding Σ in *A*. Σ should be now a function of initial and final momenta, and we define

$$\Sigma^{(A)}(p',p)\equiv\Sigma(p)\delta^4(p'-p)+\sum_{i=0}^{3}(\Gamma_i(p',p)-\gamma_i(p',p)) \times A^i(p'-p)+O(A^2).$$

In the limit $p'-p=0$, $\Gamma_i-\gamma_i=\partial\Sigma/\partial A^i$, which is the content of the Ward identity.[6] Investigation of the higher order terms in *A* is beyond the scope of this paper.

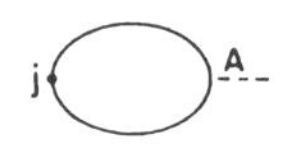

FIG. 5. The diagram for the kernel $K^{(2)}$.

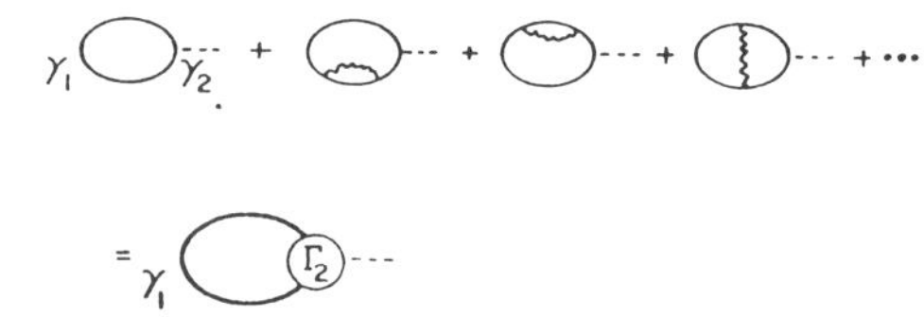

FIG. 6. Graphical derivation of Eq. (5.5). The thick lines represent quasi-particles.

for a quasi-particle is rather surprising. A quasi-particle cannot be an eigenstate of charge since it is a linear combination of an electron and a hole, tending to an electron well above the Fermi surface, and to a hole well below. We must conclude than that an accelerated wave packet of quasi-particles, whose energy is confined to a finite region of space, continuously picks up charge from, or deposits it with, the surrounding medium which extends to infinity. This situation will be studied in Sec. 7, where we will derive the charge current operators Γ_i explicitly.

5. GAUGE INVARIANCE IN THE MEISSNER EFFECT

We will next discuss how the gauge invariance is maintained in the problem of the Meissner effect when the external magnetic field is static. We calculate the Fourier component of the current $J(q)$ induced in the superconducting ground state by an external vector potential $A(q)$:

$$J_i(q)=\sum_{j=1}^{3} K_{ij}(q)A_j(q), \tag{5.1}$$

where q is kept finite.

For free electrons, K is represented by

$$K_{ij}=K_{ij}^{(1)}+K_{ij}^{(2)}, \quad K_{ij}^{(1)}=-\delta_{ij}ne^2/m, \tag{5.2}$$

where n is the number of electrons inside the Fermi sphere. $K^{(1)}$ comes from the expectation value of the current operator Eq. (4.2), whereas $K^{(2)}$ corresponds to the diagram in Fig. 5. [Compare also Eq. (1.1).] It is well known that in this case K_{ij} is of the form

$$K_{ij}(q)=(\delta_{ij}q^2-q_iq_j)K(q^2), \tag{5.3}$$

so that for a longitudinal vector potential $A_i(q)\sim q_i\lambda(q)$, we have

$$J_i(q)=K_{ij}q_j\lambda(q)=0, \tag{5.4}$$

establishing the unphysical nature of such a potential.

In the case of a superconducting state, the free electron lines in Fig. 5 will be replaced by quasi-particle lines. But then we have $K^{(2)}(q)\to 0$ as $q\to 0$ since the intermediate pair formation is suppressed due to the finite energy gap, whereas $K^{(1)}$ is essentially unaltered. Thus Eq. (5.2) takes the form of the London equation, except that even a longitudinal field creates a current.

According to our previous argument, this lack of gauge invariance should be remedied by taking account of the vertex corrections. Starting again from the free electron picture, and inserting the phonon interaction effects, as indicated in Fig. 6, we arrive at the conclusion that either one of the vertices γ in Fig. 5 has to be replaced by the full Γ^{10}. In addition, there is the polarization correction represented by a string of bubbles. Let us, however, first neglect this correction. $K_{ij}^{(2)}$ is then

$$K_{ij}^{(2)}(q)=\frac{-ie^2}{(2\pi)^4}\int \mathrm{Tr}[\gamma_i(p-q/2,\,p+q/2)G(p+q/2)$$
$$\times\Gamma_j(p+q/2,\,p-q/2)G(p-q/2)]d^4p. \tag{5.5}$$

Although we do not know $\Gamma_j(p,p)$ explicitly, we can establish Eq. (5.4) easily. For

$$-\sum_{j=1}^{3}\Gamma_j(p+q,\,q)q_j$$

is exactly the solution $\Gamma^{(b)}(p+q,\,p)$ of Eq. (4.4) where q_0 is equal to zero. Substituting this solution in Eq. (4.5) we find

$$K_{ij}^{(2)}(q)q_j$$
$$=\frac{-1}{(2\pi)^4}\int \mathrm{Tr}\{\gamma_i(p-q/2,\,p+q/2)$$
$$\times[\tau_3G(p+q/2)-G(p-q/2)\tau_3]\}d^4p$$
$$=\frac{-1}{(2\pi)^4}\int \mathrm{Tr}\{[\gamma_i(p-q,\,p)-\gamma_i(p,\,p+q)]$$
$$\times\tau_3G(p)\}d^4p$$
$$=\frac{1}{(2\pi)^4}\frac{q_i}{m}\int \mathrm{Tr}[\tau_3G(p)]d^4p, \tag{5.6}$$

where the properties of γ_i and G under particle conjugation and a translation in p space were utilized in going from the first to the second line.

On the other hand, the part $K^{(1)}$ is, according to Eq. (4.2) given by

$$K_{ij}^{(1)}=-\delta_{ij}\frac{e^2}{2m}(\langle 0|[\Psi^+(x),\tau_3\Psi(x)]|0\rangle+\{\Psi^+(x),\Psi(x)\})$$
$$=K_{ij}^{(1a)}+K_{ij}^{(1b)}. \tag{5.7}$$

The first term becomes further

$$-\delta_{ij}\frac{e^2}{2m}\langle 0|[\Psi^+(x),\tau_3\Psi(x)]|0\rangle$$

$$=-\delta_{ij}\frac{e^2}{m}\tfrac{1}{2}\lim_{\epsilon\to 0}\sum_{\pm}\langle 0|T(\Psi^+(x,t\pm\epsilon)\tau_3\Psi(x,t))|0\rangle$$

$$=-\delta_{ij}\frac{e}{m}\,\mathrm{Tr}[\tau_3 G(xt=0)]$$

$$=-\delta_{ij}\frac{e}{m}\frac{1}{(2\pi)^4}\int \mathrm{Tr}[\tau_3 G(p)]x^4p. \quad (5.8)$$

Thus

$$[K_{ij}^{(1a)}(q)+K_{ij}^{(2)}(q)]q_j=0. \quad (5.9)$$

The second term $K^{(1b)}$ comes from the c-number term of the current operator (4.2), and is just the anticommutator of the electron field, which does not depend on the quasi-particle picture, nor on the presence of interaction. Therefore we may write for this contribution

$$K_{ij}^{(1b)}(q)A_j(q)=\frac{-ie}{2m}\frac{1}{(2\pi)^3}\int e^{-iqx}d^3x$$
$$\times\{\Psi^+(x),(\tau_3\nabla-ieA(x))_i\Psi(x)\} \quad (5.10)$$

to show its formal gauge invariance since $\tau_3\nabla-ieA(x)$ is certainly a gauge-invariant combination for free electron field.

As for the polarization correction, we can easily show in a similar way that it vanishes for the static case ($q_0=0$) because

$$\int \mathrm{Tr}\,\Gamma_i(p-q/2,p+q/2)G(p+q/2)$$
$$\times\gamma_0(p+q/2,p-q/2)G(p-q/2)d^4p=0.$$

Thus the above proof is complete and independent of the Coulomb interaction which profoundly influences the polarization effect. Although the proof is thus rigorous, it is still somewhat disturbing since $K^{(1a)}$, $K^{(1b)}$ and $K^{(2)}$ are all infinite. Actually there is a certain ambiguity in the evaluation of $K^{(2)}$, Eq. (5.6), which is again similar to the one encountered in quantum electrodynamics.[18] An alternative way would be to expand quantities in q without making translations in p space. In this case we may write

$$-\Gamma^{(b)}(p+q/2,p-q/2)=\bar{\epsilon}(p+q/2)-\bar{\epsilon}(p-q/2)$$
$$-i\tau_2[\phi(p+q/2)+\phi(p-q/2)]$$
$$\approx p\cdot q/m-2i\tau_2\phi. \quad (5.11)$$

The first term then gives

$$\frac{e^2}{(2\pi)^3}\int\frac{\phi^2}{4E_p{}^5}\left(\frac{\mathbf{p}\cdot\mathbf{q}}{m}\right)^2 p_i\mathbf{p}\cdot\mathbf{q}d^3p\propto q^2q_i, \quad (5.12)$$

which is convergent and the same as the one obtained from Eq. (1.1) using the bare quasi-particle states. The second term also is finite and equal to

$$\frac{e^2}{(2\pi)^3}\int\frac{\phi^2}{E_p{}^3}\frac{p^2}{3m^2}q_id^3p+O(q^2)q_i\approx N\alpha^2q_i=n(e^2/m)q_i. \quad (5.13)$$

The last line follows from Eqs. (6.11) and (6.11′) below.

The calculation of $K^{(1)}$ from Eqs. (5.7) and (5.10), gives, on the other hand, the same value as Eq. (5.2), so that we get $(K_{ij}^{(1)}+K_{ij}^{(2)})q_j=0$ in the limit of small q. (The polarization correction is again zero.)

Since Eq. (5.13) is a contribution from the collective intermediate state (see Secs. 6 and 7), we may say that the collective state saves gauge invariance, as has been claimed by several people.[3,19]

It goes without saying that the effect of the vertex correction on K_{ij} will be felt also for real magnetic field. But as we shall see later, it is a small correction of order g^2N (except for the renormalization effects), and not as drastic as for the longitudinal case.

6. THE COLLECTIVE EXCITATIONS

In order to understand the mechanism by which gauge invariance was restored in the calculation of the Meissner effect, and also to solve the integral equations for general vertex interactions, it is necessary to examine the collective excitations of the quasi-particles. In fact, people[3] have shown already that the essential difference between the transversal and longitudinal vector potentials in inducing a current is due to the fact that the latter can excite collective motions of quasi-particle pairs.

We see that the existence of such collective excitations follows naturally from our vertex solutions Eq. (4.4). For taking $p=p'$, the second solution $\Gamma^{(b)}(p',p)$ becomes

$$\Gamma^{(b)}(p,p)=L(p)\tau_3-\tau_3L(p)$$
$$=2i\tau_2\phi, \quad (6.1)$$
$$\gamma^{(b)}=0.$$

In other words $\tau_2\phi(p)\equiv\Phi_0(p)$ satisfies a homogeneous integral equation:

$$\Phi_0(p)=-\frac{g^2}{(2\pi)^4}\int\tau_3G(p')\Phi(p')G(p')$$
$$\times\tau_3h(p-p')^2\Delta(p-p')d^4p. \quad (6.2)$$

We interpret this as describing a pair of a particle and an antiparticle interacting with each other to form a bound state with zero energy and momentum $q=p'-p=0$.

[18] H. Fukuda and T. Kinoshita, Progr. Theoret. Phys. (Kyoto) 5, 1024 (1950).

[19] On the other hand, the way in which the collective mode accomplishes this end seems to differ from one paper to another. We will not attempt to analyze this situation here.

In fact, by defining

$$F(p,-p)\equiv -G(p)\Phi_0(p)G(p), \tag{6.3}$$

Eq. (6.2) becomes

$$L(p)F(p,-p)L(p)=-\frac{g^2}{(2\pi)^4}\int \tau_3F(p',-p')\times\tau_3h(p-p')^2\Delta(p-p')d^4p',$$

or

$$\sum_{j,l=1}^{2}L(p)_{ij}L^C(-p)_{kl}F(p,-p)_{jl}=\frac{-g^2}{(2\pi)^4}\int\sum_{j,l}(\tau_3)_{ij}(\tau_3)_{kl}F(p',-p')_{jl}\times h(p-p')^2\Delta(p-p')d^4p. \tag{6.4}$$

The particle-conjugate quantity L^C was defined in Eq. (2.23).

Equations (6.2) and (6.4) are the analog of the so-called Bethe-Salpeter equation[20] for the bound pair of quasi-particles with zero total energy-momentum. $F_{ij}(p,-p)$ is the four-dimensional wave function with the spin variables i, j and the relative energy-momentum $(p_0,\mathbf{p})$.

Since there, thus, exists a bound pair of zero momentum, there will also be pairs moving with finite momentum and kinetic energy. In other words, there will be a continuum of pair states with energies going up from zero. We have to determine their dispersion law.

For a finite total energy-momentum q, the homogeneous integral equation takes the form

$$\Phi_q(p)\equiv L(\tfrac{1}{2}q+p)F(\tfrac{1}{2}q+p,\tfrac{1}{2}q-p)L(p-\tfrac{1}{2}q)=-g^2\frac{1}{(2\pi)^4}\int\tau_3F(\tfrac{1}{2}q+p',\tfrac{1}{2}q-p')\times\tau_3h(p-p')^2\Delta(p-p')d^4p. \tag{6.5}$$

From here on we carry out perturbation calculation. Let us expand F and L in terms of the small change $L(p\pm q/2)-L(p)$, thus

$$\begin{aligned}F(\tfrac{1}{2}q+p,\tfrac{1}{2}q-p)&=F^{(0)}(p)+F^{(1)}(p,q/2)+\cdots,\\ L(p\pm q/2)&=L(p)+\Delta L(p,\pm q/2).\end{aligned} \tag{6.6}$$

Collecting terms of the first order, we get

$$\begin{aligned}L(p)F^{(1)}(p,q/2)L(p)+U^{(1)}(p,q/2)&=-g^2\frac{1}{(2\pi)^4}\int\tau_3F^{(1)}(p',q/2)\times\tau_3h(p-p')^2\Delta(p-p')d^4p',\\ U^{(1)}(p,q/2)&=\Delta L(p,q/2)F^{(0)}(p)L(p)+L(p)F^{(0)}(p)\Delta L(p,-q/2).\end{aligned} \tag{6.7}$$

[20] E. E. Salpeter and H. A. Bethe, Phys. Rev. **84**, 1232 (1951).

This is an inhomogeneous integral equation for $F^{(1)}$. In order that it has a solution, the inhomogeneous term $U(p)$ must be orthogonal to the solution $\Phi_0(p)$ of the homogeneous equation. This condition can be derived as follows:

We multiply Eq. (6.7) by $F^{(0)}(p)=-G(p)\Phi_0(p)G(p)$, and integrate thus:

$$\int \mathrm{Tr}\,F^{(0)}(p)L(p)F^{(1)}(p,q/2)L(p)d^4p+\int \mathrm{Tr}\,F^{(0)}(p)U^{(1)}(p,q/2)d^4p$$
$$=-g^2\frac{1}{(2\pi)^4}\int\int \mathrm{Tr}\,F^{(0)}(p)\tau_3F^{(1)}(p',q/2)\times\tau_3h(p-p')^2\Delta(p-p')d^4pd^4p'.$$

In view of Eq. (6.5) the last line is

$$=\int \mathrm{Tr}\,L(p')F^{(0)}(p')L(p')F^{(1)}(p',q/2)d^4p',$$

so that

$$(F^{(0)},U^{(1)})\equiv\int \mathrm{Tr}\,F^{(0)}(p)U^{(1)}(p,q/2)d^4p=0. \tag{6.8}$$

This is the desired condition.

For the evaluation of Eq. (6.8), we will neglect the p dependence of the self-energy terms. Thus

$$\begin{aligned}F^{(0)}(p)&=\tau_2\phi/(p_0^2-E_p^2+i\epsilon),\quad E_p^2=\epsilon_p^2+\phi^2,\\ \Delta L(p,q/2)&=q_0/2-\tau_3(\mathbf{p}\cdot\mathbf{q}/2m+(q/2)^2/2m).\end{aligned} \tag{6.9}$$

We then obtain

$$(F^{(0)},U^{(1)})=2\pi i\int\frac{\phi^2}{E_p^3}\left[\left(\frac{q_0}{2}\right)^2-\left(\frac{\mathbf{p}\cdot\mathbf{q}}{2m}\right)^2-\frac{\epsilon_p}{m^2}\left(\frac{q}{2}\right)^2\right]d^3p=0,$$

or

$$\left(\frac{q_0}{2}\right)^2-\left(\frac{q}{2}\right)^2\left[\frac{1}{3}\frac{\bar{p}^2}{m^2}-\frac{\bar{\epsilon}_p}{m}\right]=0, \tag{6.10}$$

where the average $\bar{f}$ is defined

$$\bar{f}=\int f(p)\frac{\phi^2}{E^3}d^3p\Big/\int\frac{\phi^2}{E^3}d^3p. \tag{6.10'}$$

The weight function $\phi^2/E_p^3=\phi^2/(\epsilon_p^2+\phi^2)^{\frac{3}{2}}$ peaks around the Fermi momentum, so that $p^2\sim p_F^2$, $\epsilon_p\sim 0$. Thus

$$q_0^2\approx q^2\frac{1}{3}\frac{\bar{p}^2}{m^2}\equiv\alpha^2q^2,\quad \alpha^2\approx p_F^2/3m^2, \tag{6.11}$$

which is the dispersion law for the collective excitations.[2,3] We also note, incidentally, that

$$\frac{1}{(2\pi)^3}\int \frac{\phi^2}{E^3}d^3p \approx N = mp_F/\pi^2, \quad (6.11')$$
$$\alpha^2 N \approx p_F^3/3\pi^2 m = n.$$

We would like to emphasize here that these collective excitations are based on Eq. (6.2), which takes account of the phonon-Coulomb scattering of the quasi-particle pairs, but does not take into account the annihilation-creation process of the pair due to the same interaction.

It is well known that this annihilation-creation process is very important in the case of the Coulomb interaction, and plays the role of creating the plasma mode of collective oscillations. We will consider it in a later section.

As for the wave function $F^{(1)}$ itself, we have still to solve the integral equation (6.7). But this can be done by perturbation because on substituting $U^{(1)}$ in the integrand, we find that all the terms are of the type (3.11). In other words, to the zeroth order we may neglect the integral entirely and so

$$F^{(1)}(p,q/2) = -G(p)U^{(1)}(p,q/2)G(p). \quad (6.12)$$

The original function

$$\Phi_q(p) = -L(p+q/2)F(p,q/2)L(p-q/2)$$

is even simpler. We get

$$\Phi_q(p) \approx \Phi_0(p) \quad (6.13)$$

to this order.

7. CALCULATION OF THE CHARGE-CURRENT VERTEX FUNCTIONS

In this section we determine explicitly the charge-current vertex functions Γ_i, $(i=0, 1, 2, 3)$ from their integral equations. Only the particular combination $\Gamma^{(b)}$ of these was given before.

Let us first go back to the integral equation for Γ_0 generated by τ_3:

$$\Gamma_0(p+q/2, p-q/2)$$
$$= \tau_3 - g^2 \int \tau_3 G(p'+q/2)\Gamma_0(p'+q/2, p'-q/2) \times G(p'-q/2)\tau_3 h(p-p')^2\Delta(p-p')d^4p',$$

or

$$L(p+q/2)F_0(p+q/2, p-q/2)L(p-q/2)$$
$$= \tau_3 + g^2 \int \tau_3 F_0(p'+q/2, p'-q/2) \times \tau_3 h(p-p')^2 \Delta(p-p')d^4p'. \quad (7.1)$$

For small g^2, the standard approach to solve the equation would be the perturbation expansion in powers of g^2. We know, however, that there are low-lying collective excitations, discussed before, to which τ_3 can be coupled, and these excitations do not follow from perturbation.[21]

Fortunately, if we assume $q=0$, $q_0 \neq 0$, then we have an exact solution to Eq. (7.1) in terms of $\Gamma^{(b)}$ of Eq. (4.4). Namely,

$$\Gamma_0(p+q/2, p-q/2) = \Gamma^{(b)}(p+q/2, p-q/2)/q_0$$
$$= \tau_3\{[Z(p+q/2)+Z(p-q/2)\}/2 + (p_0/q_0)[Z(p+q/2)-Z(p-q/2)]\}$$
$$-[\chi(p+q/2)-\chi(p-q/2)]/q_0 + i\tau_2[\phi(p+q/2)+\phi(p-q/2)]/q_0, \quad (7.2)$$

which can readily be verified.

The second term is the result of the coupling of τ_3 to the collective mode. This can be understood in the following way. Γ_0 contains matrix elements for creation or annihilation of a pair out of the vacuum. These processes can go through the collective intermediate state with the dispersion law (6.11), so that Γ will contain terms of the form

$$R_\pm/(q_0 \pm \alpha q).$$

The residues $R_\pm$ can be obtained by taking the limit

$$R_\pm = \lim_{q_0 \pm \alpha q \to 0} \Gamma_0(p+q/2, p-q/2)(q_0 \pm \alpha q). \quad (7.3)$$

Applying this procedure to the integral equation (7.1) for Γ_0, we find that $R_\pm$ must be a solution of the homogeneous equation; namely,

$$R_\pm = C_\pm \Phi_q(p), \quad (7.4)$$

under the condition $q_0 \pm \alpha q = 0$.

For the particular case $q=0$, $\Phi_q(p)$ reduces to $\tau_2\phi(p)$, which in fact agrees with Eq. (7.2) if

$$C_\pm = -2i. \quad (7.5)$$

This observation enables us to write down Γ_0 for $q \neq 0$. According to the results of Sec. 6, $\Phi_q(p) = \Phi_0(p)$ in the zeroth order in g^2N. Since corrections to the non-collective part of Γ_0 also turn out to be calculable by perturbation, we may now put

$$\Gamma_0(p+q/2, p-q/2) \approx \tau_3\bar{Z} + 2i\tau_2\bar{\phi}q_0/(q_0^2-\alpha^2q^2),$$
$$\bar{\phi} \equiv [\phi(p+q/2)+\phi(p-q/2)]/2,$$
$$\bar{Z} \equiv [Z(p+q/2)+Z(p-q/2)\}/2$$

to the extent that terms of order g^2N and/or the p-dependence of the renormalization constants are neglected.

In quite a similar way the current vertex $\mathbf{\Gamma}$ may be constructed. This time we start from the longitudinal

[21] If we proceeded by perturbation theory, we would find in each order terms of order 1.

component for $q_0=0$, $q\neq 0$, which has the exact solution

$$\mathbf{\Gamma}(p+q/2,\, p-q/2)\cdot \mathbf{q}/q=-\Gamma^{(b)}(p+q/2,\, p+q/2)/q$$

$$=\frac{\mathbf{p}\cdot\mathbf{q}}{mq}\left\{1+\frac{\chi(p+q/2)-\chi(p-q/2)}{\mathbf{p}\cdot\mathbf{q}/m}\right\}$$

$$-\tau_3 p_0[\zeta(p+q/2)-\zeta(p-q/2)]/q$$

$$-2i\tau_2\frac{\phi(p+q/2)+\phi(p-q/2)}{2q}. \quad (7.7)$$

For $q_0\neq 0$, then, we get

$$\mathbf{\Gamma}(p+q/2,\, p-q/2)\cdot\mathbf{q}/q$$
$$\approx(\mathbf{p}\cdot\mathbf{q}/q)\bar{Y}+2i\tau_2\phi\alpha^2 q/(q_0^2-\alpha^2q^2), \quad (7.8)$$
$$\bar{Y}\equiv 1+[\chi(p+q/2)-\chi(p-q/2)]/(\mathbf{p}\cdot\mathbf{q}/m).$$

Combining (7.6) and (7.8), the continuity equation takes the form

$$q_0\Gamma_0-\mathbf{q}\cdot\mathbf{\Gamma}=q_0\tau_3\bar{Z}+(\mathbf{p}\cdot\mathbf{q}/m)\bar{Y}+2i\tau_2\bar{\phi}$$
$$\approx\Gamma^{(b)},$$

which is indeed zero on the energy shell.

The transversal part of $\mathbf{\Gamma}$, on the other hand, is not coupled with the collective mode because the latter is a scalar wave.[22] We may, therefore, write instead of Eq. (7.8)

$$\mathbf{\Gamma}(p+q/2,\, p-q/2)\approx(\mathbf{p}/m)\bar{Y}$$
$$+2i\tau_2\bar{\phi}\alpha^2 q/(q_0^2-\alpha^2q^2). \quad (7.10)$$

Equations (7.6) and (7.10) for Γ_i have a very interesting structure. The noncollective part is essentially the same as the charge current for a free quasi-particle except for the renormalization $\bar{Z}$ and $\bar{Y}$, whereas the collective part is spread out both in space and time. Neglecting the momentum dependence of $\bar{Z}$, $\bar{Y}$, and ϕ, we may thus write the charge-current density (ρ,j) as

$$\rho(x,t)\cong e\Psi^+\tau_3 Z\Psi(x,t)+\frac{1}{\alpha^2}\frac{\partial f(x,t)}{\partial t}\equiv\rho_0+\frac{1}{\alpha^2}\frac{\partial f}{\partial t}, \quad (7.11)$$
$$\mathbf{j}(x,t)\cong e\Psi^+(\mathbf{p}/m)Y\Psi(x,t)-\nabla f(x,t)\equiv\mathbf{j}_0-\nabla f,$$

where f satisfies the wave equation

$$\left(\Delta-\frac{1}{\alpha^2}\frac{\partial^2}{\partial t^2}\right)f\approx-2e\Psi^+\tau_2\phi\Psi. \quad (7.12)$$

$(\rho_0,\mathbf{j}_0)$ is the charge-current residing in the "core" of a quasi-particle. The latter is surrounded by a cloud of the excitation field f. In a static situation, for example, f will fall off like $1/r$ from the core. When the particle is accelerated, a fraction of the charge is exchanged between the core and the cloud.

The total charge residing in a finite volume around a core is not constant because the current $-\nabla f$ reaches out to infinity.

[22] There may be transverse collective excitations (Bogoliubov, reference 2), but they do not automatically follow from the self-energy equation nor affect the energy gap structure.

8. THE PLASMA OSCILLATIONS

The inclusion of the annihilation-creation processes in the equations of the previous sections means that the vertex parts get multiplied by a string of closed loops, which represent the polarization (or shielding effect) of the surrounding medium. We will call the new quantities Λ, which now satisfy the following type of integral equations

$$\Lambda(p',p)=\gamma-i\int\tau_3G(p'-k)\Lambda(p'-k,\,p-k)$$
$$\times G(p-k)\tau_3D(k)d^4k$$
$$+iD(p'-p)\tau_3\int\mathrm{Tr}[\tau_3G(p'-k) \quad (8.1)$$
$$\times\Lambda(p'-k,\,p-k)G(p-k)]d^4k,$$
$$D(q)\equiv-ig^2h(q)^2\Delta(q)+e^2/q^2.$$

$\hat{D}(q)$ includes the effect of the Coulomb interaction [see Eq. (2.24)]. Putting

$$\bar{X}(p'-p)\equiv i\int\mathrm{Tr}[\tau_3G(p'-k)\Lambda(p'-k,\,p'-k)$$
$$\times G(p-k)]d^4k, \quad (8.2)$$

Eq. (8.1) takes the same form as Eq. (4.3) for Γ with the inhomogeneous term replaced by $\gamma+\tau_3D\bar{X}$, so that Λ is a linear combination of the Γ corresponding to γ and Γ_0:

$$\Lambda=\Gamma+\Gamma_0D\bar{X}. \quad (8.3)$$

Substitution in Eq. (8.2) then yields

$$\bar{X}(p'-p)=i\int\mathrm{Tr}[\tau_3G(p'-k)$$
$$\times\Gamma(p'-k,\,p-k)G(p-k)]d^4k$$
$$+iD(p'-p)\bar{X}(p'-p)\int\mathrm{Tr}[\tau_3G(p'-k)$$
$$\times\Gamma_0(p'-k,\,p-k)G(p-k)]d^4k,$$

or

$$\bar{X}(p'-p)=i\int\mathrm{Tr}[\tau_3G(p'-k)$$
$$\times\Gamma(p'-k,\,p-k)G(p-k)]d^4k$$
$$\times\left\{1-iD(p'-p)\int\mathrm{Tr}[\tau_3G(p'-k)\right.$$
$$\left.\times\Gamma_0(p'-k,\,p-k)G(p-k)]d^4k\right\}^{-1}$$
$$\equiv X(p'-p)/[1-D(p'-p)X_0(p'-p)]. \quad (8.4)$$

Especially for $\gamma=\tau_3$, we get

$$\bar{X}_0(p'-p)=X_0(p'-p)/[1-D(p'-p)X_0(p'-p)],$$
$$\Lambda_0(p',p)=\Gamma_0(p',p)/[1-D(p'-p)X_0(p',p)]. \quad (8.5)$$

To obtain the collective excitations, let us next write down the homogeneous integral equation:

$$\Theta_q(p)=-i\int \tau_3 G(p'+q/2)\Theta_q(p')$$
$$\times G(p'-q/2)\tau_3 D(p-p')d^4p'$$
$$+i\tau_3 D(q)\int \mathrm{Tr}\,[\tau_3 G(p'+q/2)\Theta_q(p')$$
$$\times G(p'-q/2)]d^4p', \quad (8.6)$$

which means

$$\Theta_q(p)=\Gamma_0(p+q/2,\,p-q/2)D(q)\chi(q),$$

$$\chi(q)\equiv i\int \mathrm{Tr}[\tau_3 G(p'+q/2)$$
$$\times\Theta_q(p')G(p'-q/2)\tau_3]d^4p'. \quad (8.7)$$

Substituting Θ_q in the second equation from the first, we get

$$1=D(q)X_0(q), \quad (8.8)$$

where $X_0(q)$ is defined in Eq. (8.4).

The solutions to Eq. (8.8) determine the new dispersion law $q_0=f(q)$ for the collective excitations.

With the solution (7.6), the quantity X_0 in Eq. (8.8) can be calculated. After some simplifications using Eq. (6.11), we obtain

$$X_0=\frac{1}{(2\pi)^3}\left[\frac{\alpha^2q^2}{q_0^2-\alpha^2q^2}\int\frac{\phi^2d^3p}{E_p(E_p^2-\alpha^2q^2/4)}\right.$$
$$+\frac{q^2}{4}\int\frac{\phi^2d^3p}{E_p(q_0^2/4-E_p^2)}\left(\frac{p^2}{3m^2E_p^2}\right.$$
$$\left.\left.-\frac{\alpha^2}{E_p^2-\alpha^2q^2/4}\right)\right]+O(q^4). \quad (8.9)$$

For $\alpha q\ll\phi$, and $q_0\gg\phi$ or $\ll\phi$, the second integral may be dropped and

$$X_0\cong\alpha^2q^2N/(q_0^2-\alpha^2q^2). \quad (8.10)$$

For small q^2, the dominant part of $D(q)$ in Eq. (8.8) is the Coulomb interaction e^2/q^2. Equation (8.8) then becomes

$$q_0^2=e^2\alpha^2N=e^2n \quad (q^2\to 0), \quad (8.11)$$

where n is the number of electrons per unit volume. This agrees with the ordinary plasma frequency for free electron gas.

We see thus that the previous collective state with $q_0^2=\alpha^2q^2$ has shifted its energy to the plasma energy as a result of the Coulomb interaction.

On the other hand, if Coulomb interaction is neglected, Eq. (8.8) leads to[23]

$$q_0^2=\alpha^2q^2[1-ig^2\Delta(q,q_0)h(q,q_0)^2N]. \quad (8.12)$$

The correction term, however, is of the order g^2N, hence should be neglected to be consistent with our approximation.

We can also study the behavior of X_0 in the limit $q_0\to 0$ for small but finite q^2:

$$X_0\approx\frac{1}{(2\pi)^3}\int\frac{\phi^2}{E^3}d^3p\approx N, \quad (8.13)$$

which comes entirely from the noncollective part of Γ_0, but again agrees with the free electron value.

Another observation we can make regarding $\bar{X}_0(q,q_0)$ is the following. $\bar{X}_0$ represents the charge density correlation in the ground state:

$$\bar{X}_0(q,q_0)=\int\langle 0|T(\rho(xt),\rho(0))|0\rangle e^{-i\mathbf{q}\cdot\mathbf{x}+iq_0t}d^3xdt.$$

If $|0\rangle$ is an eigenstate of charge, $\bar{X}_0$ should vanish for $q\to 0$, $q_0\neq 0$ since the right-hand side then consists of the nondiagonal matrix elements of the total charge operator Q:

$$\bar{X}_0(0,q_0)\propto\sum_n\left(\frac{1}{q_0-E_n}-\frac{1}{q_0+E_n}\right)|\langle n|Q|0\rangle|^2.$$

The converse is also true if $E_n>|q_0|$, $n\neq 0$ for some $q_0\neq 0$. Our result for $\bar{X}_0$, as is clear from Eqs. (8.5) and (8.9), has indeed the correct property in spite of the fact that the "bare" vacuum, from which we started, is not an eigenstate of charge.

9. CONCLUDING REMARKS

We have discussed here formal mathematical structure of the BCS-Bogoliubov theory. The nature of the approximation is characterized essentially as the Hartree-Fock method, and can be given a simple interpretation in terms of perturbation expansion. In the presence of external fields, the corresponding approximation insures, if treated properly, that the gauge invariance is maintained. It is interesting that the quasi-particle picture and charge conservation (or gauge invariance) can be reconciled at all. This is possible because we are taking account of the "radiative corrections" to the bare quasi-particles which are not eigenstates of charge. These corrections manifest themselves primarily through the existence of collective excitations.

There are some questions which have been left out. We would like to know, for one thing, what will happen if we seek corrections to our Hartree-Fock approximation by including processes (or diagrams) which have not been considered here. Even within our ap-

[23] Compare Anderson, reference 7.

proximation, there is an additional assumption of the weak coupling ($g^2N \ll 1$), and the importance of the neglected terms (of order g^2N and higher) is not known.

Experimentally, there has been some evidence[24] regarding the presence of spin paramagnetism in superconductors. This effect has to do with the spin density induced by a magnetic field and can be derived by means of an appropriate vertex solution. However, this does not seem to give a finite spin paramagnetism at 0°K.[25] The collective excitations do not play an important role here as they are not excited by spin density. [$\Gamma^{(a)}$, Eq. (4.4), does not have the characteristic pole.]

It is desirable that both experiment and theory about spin paramagnetism be developed further since this may be a crucial test of the fundamental ideas underlying the BCS theory.

[24] Knight, Androes, and Hammond, Phys. Rev. **104**, 852 (1956); F. Reif, Phys. Rev. **106**, 208 (1957); G. M. Androes and W. D. Knight, Phys. Rev. Letters **2**, 386 (1959).
[25] K. Yosida, Phys. Rev. **110**, 769 (1958).

ACKNOWLEDGMENT

We wish to thank Dr. R. Schrieffer for extremely helpful discussions throughout the entire course of the work.

VOLUME 4, NUMBER 7 PHYSICAL REVIEW LETTERS APRIL 1, 1960

AXIAL VECTOR CURRENT CONSERVATION IN WEAK INTERACTIONS*

Yoichiro Nambu
Enrico Fermi Institute for Nuclear Studies and Department of Physics
University of Chicago, Chicago, Illinois
(Received February 23, 1960)

In analogy to the conserved vector current interaction in the beta decay suggested by Feynman and Gell-Mann, some speculations have been made about a possible conserved axial vector current.[1-3] One can formally construct an axial vector nucleon current, which satisfies a continuity equation,

$$\Gamma_\mu^A(p',p) = i\gamma_5\gamma_\mu - 2M\gamma_5 q_\mu/q^2, \quad q = p'-p, \qquad (1)$$

where p and p' are the initial and final nucleon momenta. Such an attempt has some appeal in view of the apparently modest renormalization effect on the axial vector beta decay constant ($g_A/g_V \approx 1.25$), although the second appealing point,[1] namely, the possible forbidding of $\pi \to e + \nu$, has now lost its relevance.

The expression (1), unfortunately, can be easily ruled out experimentally, as was pointed out by Goldberger and Treiman,[3] since it introduces a large admixture of pseudoscalar interaction.

VOLUME 4, NUMBER 7 PHYSICAL REVIEW LETTERS APRIL 1, 1960

On the other hand, Eq. (1) arouses theoretical curiosity as to the origin of the second term if it really exists; according to our conventional field theory, we would have to interpret the denominator q^2 as implying a massless, pseudoscalar, and charged quantum bridging the nucleon and lepton currents.

We would like to suggest that there may not be a strict pseudovector current conservation, but that we may have an approximate conservation which becomes rigorous in the limit $q^2 \gg m_\pi^2$, m_π being the pion mass. Specifically, we propose that the axial vector part of the nucleon beta decay vertex has the following form and properties:

$$g_A \Gamma_\mu^A(p', p) = g_V \left[i\gamma_5 \gamma_\mu F_1(q^2) - \frac{2M\gamma_5 q_\mu}{q^2 + m_\pi^2} F_2(q^2) \right],$$

$$F_1(0) = g_A / g_V \approx F_2(0),$$

$$F_1(q^2) \sim F_2(q^2) \quad \text{for} \quad q^2 \gg m_\pi^2. \tag{2}$$

The pion is then the analog of the massless quantum mentioned above. This is consistent with the dispersion relations expected for Γ_μ^A. Namely, F_1 and F_2 should have in general the form

$$F_i(q^2) = F_i(-m_\pi^2) - (q^2 + m_\pi^2) \int_{m_0^2}^{\infty} \frac{\rho_i(m^2) dm^2}{(q^2 + m^2)(m^2 - m_\pi^2)} \quad (i = 1, 2), \tag{3}$$

where $m_0 = 3m_\pi$ unless there are new particles of low mass. Thus the F's will be slowly varying for $|q^2| \ll m_0^2$. The conditions in Eq. (2) imply that $F_1/F_2 \approx 1$ for all q^2. If $m_\pi = 0$ and $F_1/F_2 \equiv 1$, then we restore exact current conservation,[2] and we also expect $F_1(0) = g_A/g_V = 1$.

If we adopt Eq. (2), the second term of Γ_μ^A immediately gives a relation between g_A, the pion decay (pseudovector) constant g_π, and the pion-nucleon (pseudoscalar) coupling G_π:

$$2Mg_A \approx 2Mg_V F_2(-m_\pi^2) = \sqrt{2}\, G_\pi g_\pi. \tag{4}$$

With $g_A = 1.25\, g_V = 1.75 \times 10^{-49}$ erg cm^3,[4] $G_\pi^2/4\pi = 13.5$, this gives a π-μ decay life of 2.7×10^{-8} sec as compared with the observed value 2.56×10^{-8} sec.

Goldberger and Treiman[5] have arrived at the same relation Eq. (4) (in the limit of their self-energy integral $J \to \infty$) from an entirely different approach. In our opinion, this is not a coincidence, as will be explained elsewhere.

We are tempted to extend this approximate conservation of the axial vector (and naturally also the vector current) to the strangeness-nonconserving beta decays. We take, for example, the ΛN axial vector in the form

$$\Gamma_\mu^A(p_N', P_\Lambda) \approx i\gamma_5\gamma_\mu - \frac{(M_\Lambda + M_N)\gamma_5 q_\mu}{q^2 + m_K^2}, \tag{5}$$

and attribute the second term to the pseudoscalar K meson.[6] The degree of accuracy of the relation (5) will be poorer than in the previous case in view of the Λ-N mass difference (which destroys vector conservation) and the large K-meson mass. At any rate, we obtain an analog of Eq. (4):

$$(M_\Lambda + M_N) g_A' \approx G_K g_K, \tag{6}$$

which relates the Λ beta decay axial vector coupling g_A', the ΛNK coupling G_K, and the K_μ decay coupling g_K.

With the observed $K_{\mu 2}$ lifetime 2.1×10^{-8} sec and a tentative value $G_K^2/4\pi = \frac{1}{4} G_\pi^2/4\pi$, we get

$$g_A'/g_A \approx 1/10. \tag{7}$$

This is not inconsistent with the observed beta decay of Λ which seems an order of magnitude less than predicted from a universal coupling scheme $g_V' = g_A' = g_V$.[7]

We can still go further, though the argument becomes more arbitrary. Let us assume that a fundamental weak coupling $(\bar{N}N\bar{N}\Lambda)$ gives rise to an effective V-A interaction (or at least part of it) of the form

$$g''(\Gamma_\mu^V - \Gamma_\mu^A)_{\bar{N}N} (\Gamma_\mu^V - \Gamma_\mu^A)_{\bar{N}\Lambda}. \tag{8}$$

Here $\Gamma_\mu^V = i\gamma_\mu$ which is approximately conserved by itself, and Γ_μ^A stands for Eq. (2) or (5). We see easily that Eq. (8) contains information about the $\Lambda \to N + \pi$ decay matrix element:

$$(2M_N g''/\sqrt{2}\, G_\pi) q_\mu (\Gamma_\mu^V - \Gamma_\mu^A)_{\bar{N}\Lambda}. \tag{9}$$

Combined with the assumption of $\Delta T = \frac{1}{2}$ selection rule, this gives a lifetime of 2.5×10^{-10} sec for $g'' = g_V$ as compared with the observed value 2.8×10^{-10} sec.

It is possible to apply this kind of consideration to other hyperons. Moreover, if the Feynman–Gell-Mann coupling scheme such as $(\pi\pi e\nu)$ is formally extended to $(K\pi e\nu)$, etc. as has been tried by some people, all the observed decay processes may be covered. Here we would like to point out that if all baryons should satisfy Eqs. (4) and (6), the ratios g_A/G_π and g_A'/G_K must be approximately common constants.

Our final remark concerns the theoretical basis for the assumptions made here. If the baryons are derived from some fundamental field ψ which possesses an invariance under a transformation of the type $\psi \rightarrow \exp(i\vec{\alpha}\cdot\vec{\tau}\gamma_5)\psi$,[8] then there will be a conservation of the pseudovector charge-current. A finite observed mass can be compatible with the conservation if the particle is coupled with a boson as was noted in Eq. (1).

This situation may be understood by making an analogy to the theory of superconductivity originated by Bardeen, Cooper, and Schrieffer,[9] and refined by Bogoliubov.[10] There gauge invariance, the energy gap, and the collective excitations are logically related to each other as was shown by the author.[11] In the present case we have only to replace them by γ_5 invariance, baryon mass, and the mesons. In fact, the mathematical method used in superconductivity may be taken over to study the self-energy problem of elementary particles. It is interesting that pseudoscalar mesons automatically emerge in this theory as bound states of baryon pairs. The nonzero meson masses and baryon mass splitting would indicate that the γ_5 invariance of the bare baryon field is not rigorous, possibly because of a small bare mass of the order of the pion mass.

The above-mentioned model of elementary particles will be studied in a separate paper.

---

*This work was supported by the U. S. Atomic Energy Commission.

[1]J. C. Taylor, Phys. Rev. 110, 1216 (1958).

[2]J. C. Polkinghorne, Nuovo cimento 8, 179 and 781 (1958).

[3]M. L. Goldberger and S. B. Treiman, Phys. Rev. 110, 1478 (1958).

[4]A. I. Alikhanov, Ninth Annual International Conference on High-Energy Physics, Kiev, 1959 (unpublished).

[5]M. L. Goldberger and S. B. Treiman, Phys. Rev. 110, 1178 (1958); M. L. Goldberger, Revs. Modern Phys. 31, 797 (1959).

[6]It is also possible to associate a scalar K meson with the ΛN vector current conservation, while leaving the axial vector unaccounted for.

[7]Again Eq. (5) and the subsequent conclusions are essentially the same as those of C. H. Albright, Phys. Rev. 114, 1648 (1959) and B. Sakita, Phys. Rev. 114, 1650 (1959), which are based on the Goldberger-Treiman method. For the Λ-decay case below, see L. Tenaglia, Nuovo cimento 14, 499 (1959).

[8]F. Gürsey (private communication) has recently obtained similar results on the π decay based on this γ_5 invariance. We do not here specify the interaction of the ψ field, which may be of the nonlinear Heisenberg type, or due to an intermediate boson (different from π or K).

[9]J. Bardeen, L. N. Cooper, and J. R. Schrieffer, Phys. Rev. 106, 162 (1957).

[10]N. N. Bogoliubov, V. V. Tolmachev, and D. V. Shirkov, A New Method in the Theory of Superconductivity (Academy of Sciences of USSR, Moscow, 1958).

[11]Y. Nambu, Phys. Rev. 117, 648 (1960).

PHYSICAL REVIEW VOLUME 122, NUMBER 1 APRIL 1, 1961

Dynamical Model of Elementary Particles Based on an Analogy with Superconductivity. I*

Y. NAMBU AND G. JONA-LASINIO†
The Enrico Fermi Institute for Nuclear Studies and the Department of Physics, The University of Chicago, Chicago, Illinois

(Received October 27, 1960)

It is suggested that the nucleon mass arises largely as a self-energy of some primary fermion field through the same mechanism as the appearance of energy gap in the theory of superconductivity. The idea can be put into a mathematical formulation utilizing a generalized Hartree-Fock approximation which regards real nucleons as quasi-particle excitations. We consider a simplified model of nonlinear four-fermion interaction which allows a γ_5-gauge group. An interesting consequence of the symmetry is that there arise automatically pseudoscalar zero-mass bound states of nucleon-antinucleon pair which may be regarded as an idealized pion. In addition, massive bound states of nucleon number zero and two are predicted in a simple approximation.
The theory contains two parameters which can be explicitly related to observed nucleon mass and the pion-nucleon coupling constant. Some paradoxical aspects of the theory in connection with the γ_5 transformation are discussed in detail.

I. INTRODUCTION

IN this paper we are going to develop a dynamical theory of elementary particles in which nucleons and mesons are derived in a unified way from a fundamental spinor field.[1] In basic physical ideas, it has thus the characteristic features of a compound-particle model, but unlike most of the existing theories, dynamical treatment of the interaction makes up an essential part of the theory. Strange particles are not yet considered.

The scheme is motivated by the observation of an interesting analogy between the properties of Dirac particles and the quasi-particle excitations that appear in the theory of superconductivity, which was originated with great success by Bardeen, Cooper, and Schrieffer,[2] and subsequently given an elegant mathematical formulation by Bogoliubov.[3] The characteristic feature of the BCS theory is that it produces an energy gap between the ground state and the excited states of a superconductor, a fact which has been confirmed experimentally. The gap is caused due to the fact that the attractive phonon-mediated interaction between electrons produces correlated pairs of electrons with opposite momenta and spin near the Fermi surface, and it takes a finite amount of energy to break this correlation.

Elementary excitations in a superconductor can be conveniently described by means of a coherent mixture of electrons and holes, which obeys the following equations[3,4]:

$$E\psi_{p+}=\epsilon_p\psi_{p+}+\phi\psi_{-p-}^*,\qquad E\psi_{-p-}^*=-\epsilon_p\psi_{-p-}^*+\phi\psi_{p+}, \tag{1.1}$$

near the Fermi surface. ψ_{p+} is the component of the excitation corresponding to an electron state of momentum p and spin + (up), and ψ_{-p-}^* corresponding to a hole state of momentum p and spin +, which means an absence of an electron of momentum $-p$ and spin − (down). ϵ_p is the kinetic energy measured from the Fermi surface; ϕ is a constant. There will also be an equation complex conjugate to Eq. (1), describing another type of excitation.

Equation (1) gives the eigenvalues

$$E_p=\pm(\epsilon_p^2+\phi^2)^{\frac{1}{2}}. \tag{1.2}$$

The two states of this quasi-particle are separated in energy by $2|E_p|$. In the ground state of the system all the quasi-particles should be in the lower (negative) energy states of Eq. (2), and it would take a finite energy $2|E_p|\geqslant 2|\phi|$ to excite a particle to the upper state. The situation bears a remarkable resemblance to the case of a Dirac particle. The four-component Dirac equation can be split into two sets to read

$$E\psi_1=\sigma\cdot p\psi_1+m\psi_2,\qquad E\psi_2=-\sigma\cdot p\psi_2+m\psi_1,\qquad E_p=\pm(p^2+m^2)^{\frac{1}{2}}, \tag{1.3}$$

where ψ_1 and ψ_2 are the two eigenstates of the chirality operator $\gamma_5=\gamma_1\gamma_2\gamma_3\gamma_4$.

According to Dirac's original interpretation, the ground state (vacuum) of the world has all the electrons in the negative energy states, and to create excited states (with zero particle number) we have to supply an energy $\geqslant 2m$.

In the BCS-Bogoliubov theory, the gap parameter ϕ, which is absent for free electrons, is determined essentially as a self-consistent (Hartree-Fock) representation of the electron-electron interaction effect.

---

* Supported by the U. S. Atomic Energy Commission.

† Fulbright Fellow, on leave of absence from Instituto di Fisica dell' Universita, Roma, Italy and Istituto Nazionale di Fisica Nucleare, Sezione di Roma, Italy.

[1] A preliminary version of the work was presented at the Midwestern Conference on Theoretical Physics, April, 1960 (unpublished). See also Y. Nambu, Phys. Rev. Letters **4**, 380 (1960); and Proceedings of the Tenth Annual Rochester Conference on High-Energy Nuclear Physics, 1960 (to be published).

[2] J. Bardeen, L. N. Cooper, and J. R. Schrieffer, Phys. Rev. **106**, 162 (1957).

[3] N. N. Bogoliubov, J. Exptl. Theoret. Phys. (U.S.S.R.) **34**, 58, 73 (1958) [translation: Soviet Phys.-JETP **34**, 41, 51 (1958)]; N. N. Bogoliubov, V. V. Tolmachev, and D. V. Shirkov, *A New Method in the Theory of Superconductivity* (Academy of Sciences of U.S.S.R., Moscow, 1958).

[4] J. G. Valatin, Nuovo cimento **7**, 843 (1958).

One finds that

$$\phi \approx \omega \exp[-1/\rho], \tag{1.4}$$

where ω is the energy bandwidth ($\approx$ the Debye frequency) around the Fermi surface within which the interaction is important; ρ is the average interaction energy of an electron interacting with unit energy shell of electrons on the Fermi surface. It is significant that ϕ depends on the strength of the interaction (coupling constant) in a nonanalytic way.

We would like to pursue this analogy mathematically. As the energy gap ϕ in a superconductor is created by the interaction, let us assume that the mass of a Dirac particle is also due to some interaction between massless bare fermions. A quasi-particle in a superconductor is a mixture of bare electrons with opposite electric charges (a particle and a hole) but with the same spin; correspondingly a massive Dirac particle is a mixture of bare fermions with opposite chiralities, but with the same charge or fermion number. Without the gap ϕ or the mass m, the respective particle would become an eigenstate of electric charge or chirality.

Once we make this analogy, we immediately notice further consequences of special interest. It has been pointed out by several people[3,5–8] that in a refined theory of superconductivity there emerge, in addition to the individual quasi-particle excitations, collective excitations of quasi-particle pairs. (These can alternatively be interpreted as moving states of bare electron pairs which are originally precipitated into the ground state of the system.) In the absence of Coulomb interaction, these excitations are phonon-like, filling the gap of the quasi-particle spectrum.

In general, they are excited when a quasi-particle is accelerated in the medium, and play the role of a backflow around the particle, compensating the change of charge localized on the quasi-particle wave packet. Thus these excitations are necessary consequences of the fact that individual quasi-particles are not eigenstates of electric charge, and hence their equations are not gauge invariant; whereas a complete description of the system must be gauge invariant. The logical connection between gauge invariance and the existence of collective states has been particularly emphasized by one of the authors.[8]

This observation leads to the conclusion that if a Dirac particle is actually a quasi-particle, which is only an approximate description of an entire system where chirality is conserved, then there must also exist collective excitations of bound quasi-particle pairs. The chirality conservation implies the invariance of the theory under the so-called γ_5 gauge group, and from its nature one can show that the collective state must be a pseudoscalar quantity.

It is perhaps not a coincidence that there exists such an entity in the form of the pion. For this reason, we would like to regard our theory as dealing with nucleons and mesons. The implication would be that the nucleon mass is a manifestation of some unknown primary interaction between originally massless fermions, the same interaction also being responsible for the binding of nucleon pairs into pions.

An additional support of the idea can be found in the weak decay processes of nucleons and pions which indicate that the γ_5 invariance is at least approximately conserved, as will be discussed in Part II. There are some difficulties, however, that naturally arise on further examination.

Comparison between a relativistic theory and a nonrelativistic, intuitive picture is often dangerous, because the former is severely restricted by the requirement of relativistic invariance. In our case, the energy-gap equation (4) depends on the energy density on the Fermi surface; for zero Fermi radius, the gap vanishes. The Fermi sphere, however, is not a relativistically invariant object, so that in the theory of nucleons it is not clear whether a formula like Eq. (4) could be obtained for the mass. This is not surprising, since there is a well known counterpart in classical electron theory that a finite electron radius is incompatible with relativistic invariance.

We avoid this difficulty by simply introducing a relativistic cutoff which takes the place of the Fermi sphere. Our framework does not yet resolve the divergence difficulty of self-energy, and the origin of such an effective cutoff has to be left as an open question.

The second difficulty concerns the mass of the pion. If pion is to be identified with the phonon-like excitations associated with a gauge group, its mass must necessarily be zero. It is true that in real superconductors the collective charge fluctuation is screened by Coulomb interaction to turn into the plasma mode, which has a finite "rest mass." A similar mechanism may be operating in the meson case too. It is possible, however, that the finite meson mass means that chirality conservation is only approximate in a real theory. From the evidence in weak interactions, we are inclined toward the second view.

The observation made so far does not yet give us a clue as to the exact mechanism of the primary interaction. Neither do we have a fundamental understanding of the isospin and strangeness quantum numbers, although it is easy to incorporate at least the isospin degree of freedom into the theory from the beginning. The best we can do here is to examine the various existing models for their logical simplicity and experimental support, if any. We will do this in Sec. 2, and settle for the moment on a nonlinear four-fermion interaction of the Heisenberg type. For reasons of simplicity in presentation, we adopt a model without isospin and strangeness degrees of freedom, and possessing complete γ_5 invariance. Once the choice is made,

[5] D. Pines and J. R. Schrieffer, Nuovo cimento **10**, 496 (1958).
[6] P. W. Anderson, Phys. Rev. **110**, 827, 1900 (1958); **114**, 1002 (1959).
[7] G. Rickayzen, Phys. Rev. **115**, 795 (1959).
[8] Y. Nambu, Phys. Rev. **117**, 648 (1960).

we can explore the whole idea mathematically, using essentially the formulation developed in reference 8. It is gratifying that the various field-theoretical techniques can be fully utilized. Section 3 will be devoted to introduction of the Hartree-Fock equation for nucleon self-energy, which will make the starting point of the theory. Then we go on to discuss in Sec. 4 the collective modes. In addition to the expected pseudoscalar "pion" states, we find other massive mesons of scalar and vector variety, as well as a scalar "deuteron." The coupling constants of these mesons can be easily determined. The relation of the pion to the γ_5 gauge group will be discussed in Secs. 5 and 6.

The theory promises many practical consequences. For this purpose, however, it is necessary to make our model more realistic by incorporating the isospin, and allowing for a violation of γ_5 invariance. But in doing so, there arise at the same time new problems concerning the mass splitting and instability. This refined model will be elaborated in Part II of this work, where we shall also find predictions about strong and weak interactions. Thus the general structure of the weak interaction currents modified by strong interactions can be treated to some degree, enabling one to derive the decay processes of various particles under simple assumptions. The calculation of the pion decay rate gives perhaps one of the most interesting supports of the theory. Results about strong interactions themselves are equally interesting. We shall find specific predictions about heavier mesons, which are in line with the recent theoretical expectations.

II. THE PRIMARY INTERACTION

We briefly discuss the possible nature of the primary interaction between fermions. Lacking any radically new concepts, the interaction could be either mediated by some fundamental Bose field or due to an inherent nonlinearity in the fermion field. According to our postulate, these interactions must allow chirality conservation in addition to the conservation of nucleon number. The chirality X here is defined as the eigenvalue of γ_5, or in terms of quantized fields,

$$X=\int\bar{\psi}\gamma_4\gamma_5\psi d^3x. \tag{2.1}$$

The nucleon number is, on the other hand

$$N=\int\bar{\psi}\gamma_4\psi d^3x. \tag{2.2}$$

These are, respectively, generators of the γ_5- and ordinary-gauge groups

$$\psi\rightarrow\exp[i\alpha\gamma_5]\psi,\quad \bar{\psi}\rightarrow\bar{\psi}\exp[i\alpha\gamma_5], \tag{2.3}$$

$$\psi\rightarrow\exp[i\alpha]\psi,\quad \bar{\psi}\rightarrow\bar{\psi}\exp[-i\alpha], \tag{2.4}$$

where α is an arbitrary constant phase.

Furthermore, the dynamics of our theory would require that the interaction be attractive between particle and antiparticle in order to make bound-state formation possible. Under the transformation (2.3), various tensors transform as follows:

$$\begin{aligned}
&\text{Vector:} && i\bar{\psi}\gamma_\mu\psi\rightarrow i\bar{\psi}\gamma_\mu\psi,\\
&\text{Axial vector:} && i\bar{\psi}\gamma_\mu\gamma_5\psi\rightarrow i\bar{\psi}\gamma_\mu\gamma_5\psi,\\
&\text{Scalar:} && \bar{\psi}\psi\rightarrow\bar{\psi}\psi\cos2\alpha+i\bar{\psi}\gamma_5\psi\sin2\alpha,\\
&\text{Pseudoscalar:} && i\bar{\psi}\gamma_5\psi\rightarrow i\bar{\psi}\gamma_5\psi\cos2\alpha-\bar{\psi}\psi\sin2\alpha,\\
&\text{Tensor:} && \bar{\psi}\sigma_{\mu\nu}\psi\rightarrow\bar{\psi}\sigma_{\mu\nu}\psi\cos2\alpha+i\bar{\psi}\gamma_5\sigma_{\mu\nu}\psi\sin2\alpha.
\end{aligned} \tag{2.5}$$

It is obvious that a vector or pseudovector Bose field coupled to the fermion field satisfies the invariance. The vector case would also satisfy the dynamical requirement since, as in the electromagnetic interaction, the forces would be attractive between opposite nucleon charges. The pseudovector field, on the other hand, does not meet the requirement as can be seen by studying the self-consistent mass equation discussed later.

The vector field looks particularly attractive since it can be associated with the nucleon number gauge group. This idea has been explored by Lee and Yang,[9] and recently by Sakurai.[10] But since we are dealing with strong interactions, such a field would have to have a finite observed mass in a realistic theory. Whether this is compatible with the invariance requirement is not yet clear. (Besides, if the bare mass of both spinor and vector field were zero, the theory would not contain any parameter with the dimensions of mass.)

The nonlinear fermion interaction seems to offer another possibility. Heisenberg and his co-workers[11] have been developing a comprehensive theory of elementary particles along this line. It is not easy, however, to gain a clear physical insight into their results obtained by means of highly complicated mathematical machinery.

We would like to choose the nonlinear interaction in this paper. Although this looks similar to Heisenberg's theory, the dynamical treatment will be quite different and more amenable to qualitative understanding.

The following Lagrangian density will be assumed ($\hbar=c=1$):

$$L=-\bar{\psi}\gamma_\mu\partial_\mu\psi+g_0[(\bar{\psi}\psi)^2-(\bar{\psi}\gamma_5\psi)^2]. \tag{2.6}$$

The coupling parameter g_0 is positive, and has dimensions $[\text{mass}]^{-2}$. The γ_5 invariance property of the interaction is evident from Eq. (2.5). According to the Fierz theorem, it is also equivalent to

$$-\tfrac{1}{2}g_0[(\bar{\psi}\gamma_\mu\psi)^2-(\bar{\psi}\gamma_\mu\gamma_5\psi)^2]. \tag{2.7}$$

This particular choice of γ_5-invariant form was taken without a compelling reason, but has the advantage

[9] T. D. Lee and C. N. Yang, Phys. Rev. 98, 1501 (1955).
[10] J. J. Sakurai, Ann. Phys. 11, 1 (1960).
[11] W. Heisenberg, Z. Naturforsch. 14, 441 (1959). Earlier papers are quoted there.

that it can be naturally extended to incorporate isotopic spin.[12]

Unlike Heisenberg's case, we do not have any theory about the handling of the highly divergent singularities inherent in nonlinear interactions. So we will introduce, as an additional and independent assumption, an *ad hoc* relativistic cutoff or form factor in actual calculations. Thus the theory may also be regarded as an approximate treatment of the intermediate-boson model with a large effective mass.

As will be seen in subsequent sections, the nonlinear model makes mathematics particularly easy, at least in the lowest approximation, enabling one to derive many interesting quantitative results.

III. THE SELF-CONSISTENT EQUATION FOR NUCLEON MASS

We will assume that all quantities we calculate here are somewhow convergent, without asking the reason behind it. This will be done actually by introducing a suitable phenomenological cutoff.

Without specifying the interaction, let Σ be the unrenormalized proper self-energy part of the fermion, expressed in terms of observed mass m, coupling constant g, and cutoff Λ. A real Dirac particle will satisfy the equation

$$i\gamma\cdot p+m_0+\Sigma(p,m,g,\Lambda)=0 \qquad (3.1)$$

for $i\gamma\cdot p+m=0$. Namely

$$m-m_0=\Sigma(p,m,g,\Lambda)\,|_{i\gamma\cdot p+m=0}. \qquad (3.2)$$

The g will also be related to the bare coupling g_0 by an equation of the type

$$g/g_0=\Gamma(m,g,\Lambda). \qquad (3.3)$$

Equations (3.1) and (3.2) may be solved by successive approximation starting from m_0 and g_0. It is possible, however, that there are also solutions which cannot thus be obtained. In fact, there can be a solution $m\neq 0$ even in the case where $m_0=0$, and moreover the symmetry seems to forbid a finite m.

This kind of situation can be most easily examined by means of the generalized Hartree-Fock procedure[8,13] which was developed before in connection with the theory of superconductivity. The basic idea is not new in field theory, and in fact in its simplest form the method is identical with the renormalization procedure of Dyson, considered only in a somewhat different context.

Suppose a Lagrangian is composed of the free and interaction part: $L=L_0+L_i$. Instead of diagonalizing L_0 and treating L_i as perturbation, we introduce the self-energy Lagrangian L_s, and split L thus

$$L=(L_0+L_s)+(L_i-L_s)$$
$$=L_0'+L_i'.$$

For L_s we assume quite general form (quadratic or bilinear in the fields) such that L_0' leads to linear field equations. This will enable one to define a vacuum and a complete set of "quasi-particle" states, each particle being an eigenmode of L_0'. Now we treat L_i' as perturbation, and determine L_s from the requirement that L_i' shall not yield additional self-energy effects. This procedure then leads to Eq. (3.2). The self-consistent nature of such a procedure is evident since the self-energy is calculated by perturbation theory with fields which are already subject to the self-energy effect.

In order to apply the method to our problem, let us assume that $L_s=-m\bar{\psi}\psi$, and introduce the propagator $S_F^{(m)}(x)$ for the corresponding Dirac particle with mass m. In the lowest order, and using the two alternative forms Eqs. (2.6) and (2.7), we get for Eq. (3.2)

$$\Sigma=2g_0[\mathrm{Tr}S_F^{(m)}(0)-\gamma_5\,\mathrm{Tr}S_F^{(m)}(0)\gamma_5 -\tfrac{1}{2}\gamma_\mu\,\mathrm{Tr}\gamma_\mu S_F^{(m)}(0)+\tfrac{1}{2}\gamma_\mu\gamma_5\,\mathrm{Tr}\gamma_\mu\gamma_5 S_F^{(m)}(0)] \qquad (3.4)$$

in coordinate space.

This is quadratically divergent, but with a cutoff can be made finite. In momentum space we have

$$\Sigma=-\frac{8g_0 i}{(2\pi)^4}\int\frac{m}{p^2+m^2-i\epsilon}d^4p\,F(p,\Lambda), \qquad (3.5)$$

where $F(p,\Lambda)$ is a cutoff factor. In this case the self-energy operator is a constant. Substituting Σ from Eq. (3.5), Eq. (3.2) gives $(m_0=0)$

$$m=-\frac{g_0 m i}{2\pi^4}\int\frac{d^4p}{p^2+m^2-i\epsilon}F(p,\Lambda). \qquad (3.6)$$

This has two solutions: either $m=0$, or

$$1=-\frac{g_0 i}{2\pi^4}\int\frac{d^4p}{p^2+m^2-i\epsilon}F(p,\Lambda). \qquad (3.7)$$

The first trivial one corresponds to the ordinary perturbative result. The second, nontrivial solution will determine m in terms of g_0 and Λ.

If we evaluate Eq. (3.7) with a straight noninvariant cutoff at $|\mathbf{p}|=\Lambda$, we get

$$\frac{\pi^2}{g_0\Lambda^2}=\left(\frac{m^2}{\Lambda^2}+1\right)^{\frac{1}{2}}-\frac{m^2}{\Lambda^2}\ln\left[\left(\frac{\Lambda^2}{m^2}+1\right)^{\frac{1}{2}}+\frac{\Lambda}{m}\right]. \qquad (3.8)$$

If we use Eq. (3.5) with an invariant cutoff at $p^2=\Lambda^2$ after the change of path: $p_0\rightarrow ip_0$, we get

$$\frac{2\pi^2}{g_0\Lambda^2}=1-\frac{m^2}{\Lambda^2}\ln\left(\frac{\Lambda^2}{m^2}+1\right). \qquad (3.9)$$

---

[12] This will be done in Part II.

[13] N. N. Bogoliubov, Uspekhi Fiz. Nauk 67, 549 (1959) [translation: Soviet Phys.-Uspekhi 67, 236 (1959)].

Since the right-hand side of Eq. (3.8) or (3.9) is positive and $\leqslant 1$ for real Λ/m, the nontrivial solution exists only if

$$0<2\pi^2/g_0\Lambda^2<1. \tag{3.10}$$

Equation (3.9) is plotted in Fig. 1 as a function of m^2/Λ^2. As $g_0\Lambda^2$ increases over the critical value $2\pi^2$, m starts rising from 0. The nonanalytic nature of the solution is evident as m cannot be expanded in powers of g_0.

In the following we will assume that Eq. (3.10) is satisfied, so that the nontrivial solution exists. As we shall see later, physically this means that the nucleon-antinucleon interaction must be attractive ($g_0>0$) and strong enough to cause a bound pair of zero total mass. In the BCS theory, the nontrivial solution corresponds to a superconductive state, whereas the trivial one corresponds to a normal state, which is not the true ground state of the superconductor. We may expect a similar situation to hold in the present case.

In this connection, it must be kept in mind that our solutions are only approximate ones. We are operating under the assumption that the corrections to them are not catastrophic, and can be appropriately calculated when necessary. If this does not turn out to be so for some solution, such a solution must be discarded. Later we shall indeed find this possibility for the trivial solution, but for the moment we will ignore such considerations.

Let us define then the vacuum corresponding to the two solutions. Let $\psi^{(0)}$ and $\psi^{(m)}$ be quantized fields satisfying the equations

$$\gamma_\mu\partial_\mu\psi^{(0)}(x)=0, \tag{3.11a}$$

$$(\gamma_\mu\partial_\mu+m)\psi^{(m)}(x)=0, \tag{3.11b}$$

$$\psi^{(0)}(x)=\psi^{(m)}(x) \quad \text{for} \quad x_0=0. \tag{3.11c}$$

According to the standard procedure, we decompose the ψ's into Fourier components:

$$\begin{aligned}\psi_\alpha{}^{(i)}(x)&=\frac{1}{V^{\frac12}}\sum_{\substack{\mathbf{p},s\\ p_0=(p^2+m^2)^{\frac12}}}\left[u_\alpha{}^{(i)}(\mathbf{p},s)a^{(i)}(\mathbf{p},s)e^{ip\cdot x}\right.\\&\qquad\left.+v_\alpha{}^{*(i)}(\mathbf{p},s)b^{(i)\dagger}(\mathbf{p},s)e^{-ip\cdot x}\right],\\ \psi_\alpha{}^{\dagger(i)}(x)&=\frac{1}{V^{\frac12}}\sum_{\substack{\mathbf{p},s\\ p_0=(p^2+m^2)^{\frac12}}}\left[u_\alpha{}^{(i)*}(\mathbf{p}\cdot s)a^{(i)\dagger}(\mathbf{p},s)\right.\\&\qquad\left.\times e^{-ip\cdot x}+v_\alpha{}^{(i)}(\mathbf{p},s)b^{(i)}(\mathbf{p},s)e^{ip\cdot x}\right],\\ &i=0 \text{ or } m,\end{aligned} \tag{3.12}$$

where $u_\alpha{}^{(i)}(\mathbf{p},s)$, $v_\alpha{}^{(i)}(\mathbf{p},s)$ are the normalized spinor eigenfunctions for particles and antiparticles, with momentum p and helicity $s=\pm1$, and

$$\begin{aligned}&\{a^{(i)}(\mathbf{p},s),a^{(i)\dagger}(\mathbf{p}',s')\}\\&\quad=\{b^{(i)}(\mathbf{p},s),b^{(i)\dagger}(\mathbf{p}',s')\}=\delta_{\mathbf{p}\mathbf{p}'}\delta_{ss'}, \text{ etc.}\end{aligned} \tag{3.13}$$

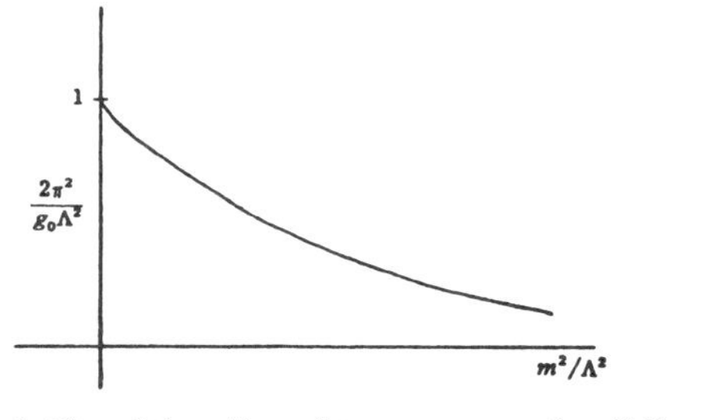

FIG. 1. Plot of the self-consistent mass equation (3.9).

The operator sets $(a^{(0)},b^{(0)})$ and $(a^{(m)},b^{(m)})$ are related by a canonical transformation because of Eq. (3.11c):

$$\begin{aligned}a^{(m)}(\mathbf{p},s)&=\sum_{\alpha,s'}\left[u_\alpha{}^{(m)*}(\mathbf{p},s)u_\alpha{}^{(0)}(\mathbf{p},s')a^{(0)}(\mathbf{p},s')\right.\\&\qquad\left.+u_\alpha{}^{(m)*}(\mathbf{p},s)v_\alpha{}^{(0)*}(-\mathbf{p},s')b^{(0)\dagger}(-\mathbf{p},s')\right],\\ b^{(m)}(\mathbf{p},s)&=\sum_{\alpha,s'}\left[v_\alpha{}^{(m)*}(\mathbf{p},s)v_\alpha{}^{(0)}(\mathbf{p},s')b^{(0)}(\mathbf{p},s')\right.\\&\qquad\left.+v_\alpha{}^{(m)*}(\mathbf{p},s)u_\alpha{}^{(0)*}(-\mathbf{p},s')a^{(0)\dagger}(-\mathbf{p},s')\right].\end{aligned} \tag{3.14}$$

Using Eq. (1.3), this is evaluated to give

$$\begin{aligned}a^{(m)}(\mathbf{p},s)&=\left[\tfrac12(1+\beta_p)\right]^{\frac12}a^{(0)}(\mathbf{p},s)\\&\qquad+\left[\tfrac12(1-\beta_p)\right]^{\frac12}b^{(0)\dagger}(-\mathbf{p},s),\\ b^{(m)}(\mathbf{p},s)&=\left[\tfrac12(1+\beta_p)\right]^{\frac12}b^{(0)}(\mathbf{p},s)\\&\qquad-\left[\tfrac12(1-\beta_p)\right]^{\frac12}a^{(0)\dagger}(-\mathbf{p},s),\\ \beta_p&=|\mathbf{p}|/(\mathbf{p}^2+m^2)^{\frac12}.\end{aligned} \tag{3.15}$$

The vacuum $\Omega^{(0)}$ or $\Omega^{(m)}$ with respect to the field $\psi^{(0)}$ or $\psi^{(m)}$ is now defined as

$$a^{(0)}(\mathbf{p},s)\Omega^{(0)}=b^{(0)}(\mathbf{p},s)\Omega^{(0)}=0, \tag{3.16}$$

$$a^{(m)}(\mathbf{p},s)\Omega^{(m)}=b^{(m)}(\mathbf{p},s)\Omega^{(m)}=0. \tag{3.16'}$$

Both $\psi^{(0)}$, $\bar\psi^{(0)}$ and $\psi^{(m)}$, $\bar\psi^{(m)}$ applied to $\Omega^{(0)}$ always create particles of mass zero, whereas the same applied to $\Omega^{(m)}$ create particles of mass m.

From Eqs. (3.15) and (3.16) we obtain

$$\begin{aligned}\Omega^{(m)}=\prod_{\mathbf{p},s}\{&\left[\tfrac12(1+\beta_p)\right]^{\frac12}\\&-\left[\tfrac12(1-\beta_p)\right]^{\frac12}a^{(0)\dagger}(\mathbf{p},s)b^{(0)\dagger}(-\mathbf{p},s)\}\Omega^{(0)}.\end{aligned} \tag{3.17}$$

Thus $\Omega^{(m)}$ is, in terms of zero-mass particles, a superposition of pair states. Each pair has zero momentum, spin and nucleon number, and carries ±2 units of chirality, since chirality equals minus the helicity s for massless particles.

Let us calculate the scalar product $(\Omega^{(0)},\Omega^{(m)})$ from Eq. (3.15):

$$\begin{aligned}(\Omega^{(0)},\Omega^{(m)})&=\prod_{\mathbf{p},s}\left[\tfrac12(1+\beta_p)\right]^{\frac12}\\&=\exp\{\sum_{\mathbf{p},s}\tfrac12\ln\left[\tfrac12(1+\beta_p)\right]\}.\end{aligned} \tag{3.18}$$

For large p, $\beta_p\sim1-m^2/2\mathbf{p}^2$, so that the exponent diverges as $V\pi m^2 \int d\mathbf{p}/(2\pi)^3$ (V = normalization volume). Hence

$$(\Omega^{(0)},\Omega^{(m)})=0. \tag{3.19}$$

It is easy to see that any two states $\Psi^{(0)}$ and $\Psi^{(m)}$, obtained by applying a finite number of creation operators on $\Omega^{(0)}$ and $\Omega^{(m)}$ respectively, are also orthogonal.

Thus the two "worlds" based on $\Omega^{(0)}$ and $\Omega^{(m)}$ are physically distinct and outside of each other. No interaction or measurement, in the usual sense, can bridge them in finite steps.

What is the energy difference of the two vacua? Since both are Lorentz invariant states, the difference can only be either zero or infinity. Using the expression

$$\begin{aligned} H^{(m)}&=\sum_{\mathbf{p},s}(\mathbf{p}^2+m^2)^{\frac{1}{2}}\{a^{(m)\dagger}(\mathbf{p},s)a^{(m)}(\mathbf{p},s)\\ &\qquad -b^{(m)}(\mathbf{p},s)b^{(m)\dagger}(\mathbf{p},s)\},\\ H^{(0)}&=\sum_{\mathbf{p},s}|\mathbf{p}|\{a^{(0)\dagger}(\mathbf{p},s)a^{(0)}(\mathbf{p},s)\\ &\qquad -b^{(0)}(\mathbf{p},s)b^{(0)\dagger}(\mathbf{p},s)\},\end{aligned} \tag{3.20}$$

we get for the respective energies

$$E^{(m)}-E^{(0)}=-2\sum\nolimits_{\mathbf{p}}[(\mathbf{p}^2+m^2)^{\frac{1}{2}}-|\mathbf{p}|], \tag{3.21}$$

which is negative and quadratically divergent. So $\Omega^{(m)}$ may be called the "true" ground state, as was expected.

There remains finally the question of γ_5 invariance. The original Hamiltonian allowed two conservations X and N, Eqs. (2.1) and (2.2). Both $\Omega^{(0)}$ and $\Omega^{(m)}$ belong to $N=0$, and their elementary excitations carry $N=\pm 1$. In the case of X, the same is true for the space $\Omega^{(0)}$, but $\Omega^{(m)}$ as well as its elementary excitations are not eigenstates of X, as is clear from the foregoing results. If the latter solution is to be a possibility, there must be an infinite degeneracy with respect to the quantum number X. A ground state will be in general a linear combination of degenerate states with different $X=0$, $\pm 2, \cdots$:

$$\Omega^{(m)}=\sum_{n=-\infty}^{\infty} C_{2n}\Omega_{2n}^{(m)}. \tag{3.22}$$

Equation (3.17) is in fact a particular case of this. The γ_5-gauge transformation Eq. (2.3) induces the change

$$\begin{aligned} a^{(0)}(\mathbf{p},\pm 1)&\rightarrow e^{\mp i\alpha}a^{(0)}(\mathbf{p},\pm 1),\\ b^{(0)}(\mathbf{p},\pm 1)&\rightarrow e^{\mp i\alpha}b^{(0)}(\mathbf{p},\pm 1),\\ a^{(0)\dagger}(\mathbf{p},\pm 1)&\rightarrow e^{\pm i\alpha}a^{(0)\dagger}(\mathbf{p},\pm 1),\\ b^{(0)\dagger}(\mathbf{p},\pm 1)&\rightarrow e^{\pm i\alpha}b^{(0)\dagger}(\mathbf{p},\pm 1),\end{aligned} \tag{3.23}$$

and the coefficients of Eq. (3.22) become

$$C_{2n}\rightarrow e^{-2ni\alpha}C_{2n}. \tag{3.24}$$

In particular

$$\begin{aligned} \Omega^{(m)}&\rightarrow\Omega_\alpha^{(m)}\\ &=\exp[-i\alpha X]\Omega^{(m)}\\ &=\prod_{\mathbf{p},\pm}\{[\tfrac{1}{2}(1+\beta_p)]^{\frac{1}{2}}-[\tfrac{1}{2}(1-\beta_p)]^{\frac{1}{2}}\\ &\qquad\times e^{\pm 2i\alpha}a^{(0)\dagger}(p,\pm)b^{(0)\dagger}(-p,\pm)\}\Omega^{(0)}.\end{aligned} \tag{3.25}$$

The Dirac equation (3.11b), at the same time, is transformed into

$$[\gamma_\mu\partial_\mu+m\cos 2\alpha+im\gamma_5\sin 2\alpha]\psi=0. \tag{3.26}$$

The moral of this is that the self-consistent self-energy Σ is determined only up to a γ_5 transformation. This can be easily verified from Eq. (3.4), in which the second term on the right-hand side is nonvanishing when a propagator corresponding to Eq. (3.26) is used. Although Eq. (3.26) seems to violate parity conservation, it is only superficially so since $\Omega_\alpha^{(m)}$ is now not an eigenstate of parity. We could alternatively say that the parity operator undergoes transformation together with the mass operator. Despite the odd form of the equation (3.26), there is no change in the physical predictions of the theory. We shall see more of this later.

Let us calculate, as before, the scalar product of $\Omega_\alpha^{(m)}$ and $\Omega_{\alpha'}^{(m)}$. From Eqs. (3.17) and (3.25) we get

$$\begin{aligned} (\Omega_\alpha^{(m)},\Omega_{\alpha'}^{(m)})&=\prod_{p,\pm}[\tfrac{1}{2}(1+\beta_p)-e^{\pm 2i(\alpha'-\alpha)}\tfrac{1}{2}(1-\beta_p)]\\ &=\prod_{p,\pm}[1+(e^{\pm 2i(\alpha'-\alpha)}-1)\tfrac{1}{2}(1-\beta_p)]\\ &=\exp\{\sum_{p,\pm}\ln[1+(e^{\pm 2i(\alpha'-\alpha)}-1)\tfrac{1}{2}(1-\beta_p)]\}.\end{aligned} \tag{3.27}$$

For large $|\mathbf{p}|$, the exponent goes like

$$\frac{V}{(2\pi)^3}\sum_{\pm}(e^{\pm 2i(\alpha'-\alpha)}-1)\int\frac{m^2}{4\mathbf{p}^2}d^3p.$$

The integral is again divergent. Hence

$$\begin{aligned}(\Omega_\alpha^{(m)},\Omega_{\alpha'}^{(m)})&=(\Omega^{(m)},\exp[-i(\alpha'-\alpha)X]\Omega^{(m)})\\ &=0,\quad \alpha'\neq\alpha(\mathrm{mod}\,2\pi),\end{aligned} \tag{3.28}$$

and, of course

$$(\Omega^{(0)},\Omega_\alpha^{(m)})=0. \tag{3.28'}$$

We can evaluate $(\Omega_\alpha^{(m)},\Omega_{\alpha'}^{(m)})$ alternatively from Eqs. (3.22) and (3.24). Then

$$\sum_{m=-\infty}^{\infty}|C_{2n}|^2e^{2ni(\alpha-\alpha')}=0,\quad \alpha\neq\alpha'(\mathrm{mod}\,2\pi), \tag{3.29}$$

implying that

$$|C_0|=|C_{\pm 2}|=|C_{\pm 4}|=\cdots=C. \tag{3.30}$$

Thus there is an infinity of equivalent worlds described by $\Omega_\alpha^{(m)}$, $0\leqslant\alpha<2\pi$. The states Ω_{2n} of Eq. (3.22) are then expressed in terms of $\Omega_\alpha^{(m)}$ as

$$C_{2n}\Omega_{2n}^{(m)}=\frac{1}{2\pi}\int_0^{2\pi}e^{2ni\alpha}\Omega_\alpha^{(m)}d\alpha, \tag{3.31}$$

which form another orthogonal set. Since the original total H commutes with X, it will have no matrix elements connecting different "worlds." Moreover, as was the case with $\Omega^{(m)}$ and $\Omega^{(0)}$, no finite measurement can induce similar transitions. This is a kind of superselection rule, which effectively avoids the apparent degeneracy to show up as physical effects.[14] The usual description of the world by means of $\Omega^{(m)}$ and ordinary Dirac particles must be regarded as only the most convenient one.

We still are left with some paradoxes. The X conservation implies the existence of a conserved X current:

$$j_{\mu 5}=i\bar{\psi}\gamma_\mu\gamma_5\psi, \tag{3.32}$$

$$\partial_\mu j_{\mu 5}=0, \tag{3.32'}$$

which can readily be verified from Eq. (2.6). On the other hand, for a massive Dirac particle the continuity equation is not satisfied:

$$\partial_\mu\bar{\psi}^{(m)}\gamma_\mu\gamma_5\psi^{(m)}=2m\bar{\psi}^{(m)}\gamma_5\psi^{(m)}. \tag{3.33}$$

If a massive Dirac particle has to be a real eigenstate of the system, how can this be reconciled? The answer would be that the X-current operator taken between real one-nucleon states should not be given simply by $i\gamma_\mu\gamma_5$ because of the "radiative corrections." We expect instead

$$\langle p'|j_{\mu 5}|p\rangle=\bar{u}(p')X_\mu(p',p)u(p), \tag{3.34}$$

where the renormalized quantity $X_{\mu 5}$ should be, from relativistic invariance grounds, of the form

$$X_\mu(p',p)=F_1(q^2)i\gamma_\mu\gamma_5+F_2(q^2)\gamma_5 q_\mu,$$
$$q=p'-p,\quad p^2=p'^2=-m^2. \tag{3.35}$$

The continuity equation (3.32′), together with Eq. (3.33), further reduces this to

$$F_1=F_2q^2/2m\equiv F,$$
$$X_\mu(p',p)=F(q^2)\left(i\gamma_\mu\gamma_5+\frac{2m\gamma_5q_\mu}{q^2}\right). \tag{3.36}$$

The real nucleon is not a point particle. Its X-current (3.36) is provided with the dramatic "anomalous" term.

To understand the physical meaning of the anomalous term, we have to make use of the dispersion relations. The form factors F_1 and F_2 will, in general, satisfy dispersion relations of the form

$$F_i(q^2)=F_i(0)-\frac{q^2}{\pi}\int\frac{\mathrm{Im}F_i(-\kappa^2)}{(q^2+\kappa^2-i\epsilon)\kappa^2}d\kappa^2, \tag{3.37}$$

assuming one subtraction. Each singularity at κ^2 corresponds to some physical intermediate state. Thus if $F(0)\neq 0$, Eq. (3.36) indicates that there is a pole at $q^2=0$ for F_2 (and no subtraction), which means in turn that there is an isolated intermediate state of zero mass.

[14] This was discussed by R. Haag, Kgl. Danske Videnskab. Selskab, Mat.-fys. Medd. 29, No. 12 (1955). See also L. van Hove, Physica 18, 145 (1952).

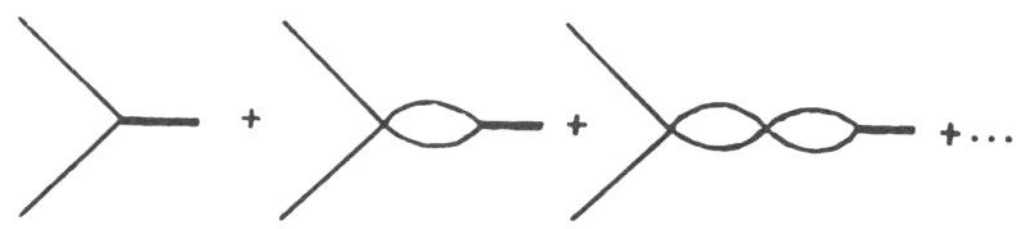

FIG. 2. Graphs corresponding to the Bethe-Salpeter equation in "ladder" approximation. The thick line is a bound state.

To see its nature, we take a time-like q in its own rest frame and go to the limit $q^2\to 0$. The anomalous term has then only the time component, and is proportional to the amplitude for creation of a nucleon pair in a $J=0^-$ state. Hence the zero mass state must have the same property as this pair. It belongs to nucleon number zero, so that we may call it a zero-mass pseudoscalar meson. *In order for a γ_5-invariant Hamiltonian such as Eq. (2.6) to allow massive nucleon states and a nonvanishing X current for $q=0$, it is therefore necessary to have at the same time pseudoscalar zero-mass mesons coupled with the nucleons.* Since we did not have such mesons in the theory, they must be regarded as secondary products, i.e., bound states of nucleon pairs. This conclusion would not hold if in Eq. (3.36) $F(q^2)=O(q^2)$ near $q^2=0$. A nucleon then would have always $X=0$. Such a possibility cannot be excluded. We will show, however, that the pseudoscalar zero-mass bound states do follow explicitly, once we assume the nontrivial solution of the self-energy equation.

IV. THE COLLECTIVE STATES

From the general discussion of Secs. 2 and 3, we may expect the existence of collective states of the fundamental field which would manifest themselves as stable or unstable particles. In particular we have argued that, as a consequence of the γ_5 invariance, a pseudoscalar zero-mass state must exist. We want now to discuss the problem in detail, trying to determine the mass spectrum of the collective excitations (at least its general features) and the strength of their coupling with the nucleons. These states must be considered as a direct effect of the same primary interaction which produces the mass of the nucleon, which itself is a collective effect. We will study the bound-state problem through the use of the Bethe-Salpeter equation, taking into account explicitly the self-consistency conditions. We first verify in the following the existence of the zero-mass pseudoscalar state.

The Bethe-Salpeter equation for a bound pair B deals with the amplitude

$$\Phi(x,y)=\langle 0|T(\psi(x)\bar{\psi}(y))|B\rangle. \tag{4.1}$$

As is well known, the equation is relatively easy to handle in the ladder approximation. In our case we have a four-spinor point interaction and the analog of the "ladder" approximation would be the iteration of the simplest closed loop (see Fig. 2) in which all lines represent dressed particles. We introduce the vertex function

Γ related to Φ by

$$\Phi(p)=S_F^{(m)}(p+\tfrac{1}{2}q)\Gamma(p+\tfrac{1}{2}q,\,p-\tfrac{1}{2}q)S_F^{(m)}(p-\tfrac{1}{2}q). \tag{4.2}$$

All we have to do then is to set up the integral equation generated by the chain of diagrams, looking for solutions having the symmetry properties of a pseudoscalar state. This means that our solutions must be proportional to γ_5. This requirement makes only the pseudoscalar and axial vector part of the interaction contribute to the integral equation. We have

$$\begin{aligned}\Gamma(p+\tfrac{1}{2}q,\,p-\tfrac{1}{2}q)&\\ =\frac{2ig_0}{(2\pi)^4}\gamma_5\int &\mathrm{Tr}[\gamma_5 S_F^{(m)}(p'+\tfrac{1}{2}q)\\ &\times\Gamma(p'+\tfrac{1}{2}q,\,p'-\tfrac{1}{2}q)S_F^{(m)}(p'-\tfrac{1}{2}q)]d^4p'\\ -\frac{ig_0}{(2\pi)^4}\gamma_5\gamma_\mu\int &\mathrm{Tr}[\gamma_5\gamma_\mu S^{(m)}(p'+\tfrac{1}{2}q)\\ &\times\Gamma(p'+\tfrac{1}{2}q,\,p'-\tfrac{1}{2}q)S_F^{(m)}(p'-\tfrac{1}{2}q)]d^4p'.\end{aligned} \tag{4.3}$$

For the moment let us ignore the pseudovector term on the right-hand side. It then follows that the equation has a constant solution $\Gamma=C\gamma_5$ if $q^2=0$. To see this, first observe that for the special case $q=0$, Eq. (4.3) reduces to

$$1=-\frac{8ig_0}{(2\pi)^4}\int\frac{d^4p}{p^2+m^2-i\epsilon}, \tag{4.4}$$

which is nothing but the self-consistency condition (3.7), provided that the same cutoff is applied. Since the pseudoscalar term of Eq. (4.3) gives a function of q^2 only, the same condition remains true as long as $q^2=0$.

When the pseudovector term is included, we have still the same eigenvalue $q^2=0$ with a solution of the form $\Gamma=C\gamma_5+iD\gamma_5\gamma\cdot q$, which is not difficult to verify (see Appendix).

We now add some remarks. First, the bound state amplitude for this solution spreads in space over a region of the order of the fermion Compton wavelength $1/m$ because of Eq. (4.2), making the zero-mass particle only partially localizable. We want also to stress the role played by the γ_5 invariance in the argument. We had in fact already inferred the existence of the pseudoscalar particle from relativistic and γ_5 invariance alone, and at first sight the same result seems to follow now essentially from the self-consistency equation. However, we must notice that only the scalar term of the Lagrangian appears in this equation while only the pseudoscalar part contributes in the Bethe-Salpeter equation. It is because of the γ_5-invariant Lagrangian that the Bethe-Salpeter equation can be reduced to the self-consistency condition.

Along the same line we could try to see whether other bound states exist in the "ladder" approximation. However, besides calculating the spectrum, it is also important to determine the interaction properties of these collective states with the fermions. For this purpose the study of the two-"nucleon" scattering amplitude appears much more suitable, as we shall realize after the following remark. Once we have recognized that in the ladder approximation the collective states would appear as real stable particles, we must expect to the same degree of approximation poles in the scattering matrix of two nucleons corresponding to the possibility of the virtual exchange of these particles. For definiteness we shall refer again as an example to the pseudoscalar zero-mass particle. Let us indicate by $J_p(q)$ the **analytical** expression corresponding to the graph whose iteration produces the bound state [Fig. 3(a)]. We construct next the scattering matrix generated by the exchange of all possible simple chains built with this element. This means that we consider the set of diagrams in Fig. 3(b). The series is easily evaluated and we obtain

$$2g_0 i\gamma_5\frac{1}{1-J_p(q)}i\gamma_5, \tag{4.5}$$

where the γ_5's refer to the pairs (1,1′) and (2,2′), respectively. The meaning of this result is clear: because of the self-consistent equation $J_p(0)=1$, Eq. (4.5) is equivalent to a phenomenological exchange term where the intermediate particle is our pseudoscalar massless boson (Fig. 4). The coupling constant G can now be evaluated by straightforward comparison. Before doing this calculation we need the explicit expression of $J_P(q)$. Using the ordinary rules for diagrams, we have

$$J_P(q)=-\frac{2ig_0}{(2\pi)^4}\times\int\frac{4(m^2+p^2)-q^2}{[(p+\frac{1}{2}q)^2+m^2][(p-\frac{1}{2}q)^2+m^2]}d^4p. \tag{4.6}$$

It is however more convenient to rewrite J_P in the form of a dispersive integral, and if we forget for a moment that it is a divergent expression, a simple manipulation gives

$$J_P(q)=\frac{g_0}{4\pi^2}\int_{4m^2}^{\Lambda^2}\frac{\kappa^2(1-4m^2/\kappa^2)^{\frac{1}{2}}}{q^2+\kappa^2}d\kappa^2. \tag{4.6'}$$

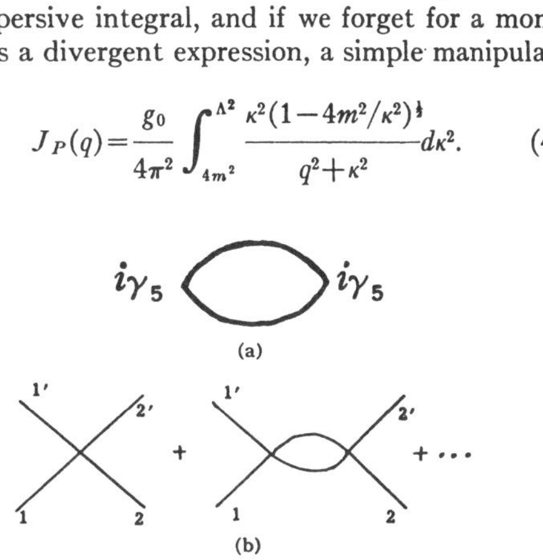

FIG. 3. The bubble graph for J_P and the scattering matrix generated by it.

In order for this expression to be meaningful, a new cutoff Λ must be introduced. There is no simple relation between this and the previous cutoffs. The dispersive form is more comfortable to handle and accordingly we shall reformulate the self-consistent condition $J_P(0)=1$, or

$$1=\frac{g_0}{4\pi^2}\int_{4m^2}^{\Lambda^2}(1-4m^2/\kappa^2)^{\frac{1}{2}}d\kappa^2. \qquad (4.7)$$

It may be of interest to remark at this point that Eq. (4.7) can be obtained also if we think of our theory as a theory with intermediate pseudoscalar boson in the limit of infinite boson mass. We are now in a position to evaluate the phenomenological coupling constant G. From Eqs. (4.6′) and (4.7) we have

$$J_P(q^2)=1-q^2\frac{g_0}{4\pi^2}\int_{4m^2}^{\Lambda^2}\frac{(1-4m^2/\kappa^2)^{\frac{1}{2}}}{q^2+\kappa^2}d\kappa^2, \qquad (4.8)$$

which leads immediately to the result

$$\frac{G_P{}^2}{4\pi}=2\pi\left[\int_{4m^2}^{\Lambda^2}\frac{(1-4m^2/\kappa^2)^{\frac{1}{2}}}{\kappa^2}d\kappa^2\right]^{-1}. \qquad (4.9)$$

This equation is interesting since it establishes a connection between the phenomenological constant G_P and the cutoff independently of the value of the fundamental coupling g_0. This fact exhibits the purely dynamical origin of the phenomenological coupling G_P. Actually g_0 is buried in the value of the mass m.

So far we have exploited only the γ_5 vertex. What happens then if the scalar part is iterated to form chains of bubbles similar to those we have already discussed? The procedure just explained can be followed again, and a quantity $J_S(q)$ can be defined similarly with the result

$$J_S(q)=\frac{g_0}{4\pi^2}\int_{4m^2}^{\Lambda^2}\frac{(\kappa^2-4m^2)(1-4m^2/\kappa^2)^{\frac{1}{2}}}{q^2+\kappa^2}d\kappa^2. \qquad (4.10)$$

It is immediately seen that because of Eq. (4.7)

$$J_S(-4m^2)=1, \qquad (4.11)$$

which causes a new pole to appear in the S matrix for $q^2=-4m^2$. This means that we have another collective state of mass $2m$, parity + and spin 0! We observe that it is necessary to assume the same cutoff as in the pseudoscalar case in order that this result may be obtained. The choice of the same cutoff in both cases seems to be suggested by the γ_5 invariance as will be seen later. We also notice the peculiar symmetry existing between the pseudoscalar and the scalar state: the first has zero mass and binding energy $2m$, while the opposite is true for the scalar particle. So in the bound-state picture the scalar particle would not be a true bound state and should be, rather, interpreted as a

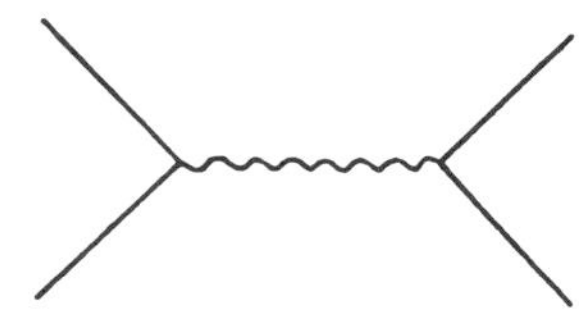

FIG. 4. The equivalent phenomenological one-meson exchange graph.

correlated exchange of pairs in the scattering process.[15] The "nucleon-nucleon" forces induced by the exchange of the scalar particle are, of course, of rather short range. The general physical implications of these results will be discussed more thoroughly later.

The phenomenological coupling constant G_S for the scalar meson is given by

$$\frac{G_S{}^2}{4\pi}=2\pi\left[\int_{4m^2}^{\Lambda^2}\frac{(1-4m^2/\kappa^2)^{\frac{1}{2}}}{(\kappa^2-4m^2)}d\kappa^2\right]^{-1}. \qquad (4.12)$$

Let us next turn to the vector state generated by iteration of the vector interaction. In this case we obtain for each "bubble" a tensor

$$J_{V\mu\nu}=(\delta_{\mu\nu}-q_\mu q_\nu/q^2)J_V,$$

$$J_V=-\frac{g_0}{4\pi^2}\frac{q^2}{3}\int_{4m^2}^{\Lambda^2}\frac{d\kappa^2}{q^2+\kappa^2}\times\left(1+\frac{2m^2}{\kappa^2}\right)(1-4m^2/\kappa^2)^{\frac{1}{2}}. \qquad (4.13)$$

Perhaps a remark is in order here regarding the evaluation of J_V. It suffers from an ambiguity of subtraction well known in connection with the photon self-energy problem. The above result is of the conventional gauge invariant form, which we take to be the proper choice.

Equation (4.13) leads to the scattering matrix

$$g_0\left[\gamma_\mu\frac{1}{1-J_V}\gamma_\mu-\gamma\cdot q\frac{J_V}{(1-J_V)q^2}\gamma\cdot q\right], \qquad (4.14)$$

where the second term is, of course, effectively zero. It can be easily seen that the denominator can produce a pole below $4m^2$ for sufficiently small Λ^2. In fact, from Eqs. (4.7) and (4.13), we find

$$(8/3)m^2<\mu_V{}^2. \qquad (4.15)$$

The coupling constant is given by

$$\frac{G_V{}^2}{4\pi}=3\pi\left[\int_{4m^2}^{\Lambda^2}d\kappa^2\frac{\kappa^2+2m^2}{(\kappa^2-\mu_V{}^2)^2}(1-4m^2/\kappa^2)^{\frac{1}{2}}\right]^{-1}. \qquad (4.16)$$

It must be noted that the mass of the vector meson now depends on the cutoff, unlike the previous two cases.

Finally we are left with the pseudovector state. We

[15] Of course this and other heavy mesons will in general become unstable in higher order approximation, which is beyond the scope of the present paper.

find for the bubble[16]

$$J_{A\mu\nu}=-J_{V\mu\nu}+J_A'\delta_{\mu\nu},$$
$$J_A'=\frac{g_0m^2}{2\pi^2}\int_{4m^2}^{\Lambda^2}\frac{d\kappa^2}{q^2+\kappa^2}(1-4m^2/\kappa^2)^{\frac{1}{2}}. \tag{4.17}$$

In view of the self-consistency condition (4.7), it can be seen that this does not produce a pole of the scattering matrix for $-q^2<4m^2$, corresponding to a pseudovector meson.

So far we have considered only iterations of the same kind of interactions. In the ladder approximation there is actually a coupling between pseudoscalar and pseudovector interactions as was explicitly considered in Eq. (4.3). However, the coupling between scalar and vector interactions vanish because of the Furry's theorem.

This coupling of pseudoscalar and pseudovector interactions does not change the pion pole of the scattering matrix, but it affects the coupling of the pion to the nucleon since a chain of the pseudoscalar can join the external nucleon with an axial vector interaction. In other words, the pion-nucleon coupling is in general a mixture of pseudoscalar and derivative pseudovector types (Appendix).

We would like to inject here a remark concerning the trivial solution of the self-energy equation, against which we had no decisive argument. So let us also try to apply our scattering formula to this solution. For the pseudoscalar state we now find $J_P(q=0)>1$, provided that the cutoff Λ is kept fixed and m is set equal to zero in Eq. (4.6′). (The pseudovector interference vanishes.) In other words, there will be a pole for some $q^2>0$ ($\mu^2<0$). This is again a supporting evidence that the trivial solution could be unstable, capable of decaying by emitting such mesons. The final answer, however, depends on the exact nature of the cutoff.

Finally we would like to discuss the nucleon-nucleon scattering in the same spirit and approximation as for the nucleon-antinucleon scattering. In order to make a correspondence with the previous cases, it is convenient to rewrite the Hamiltonian in the following way:

$$\begin{aligned}H_1&=-g_0[\bar\psi\psi\bar\psi^c\psi^c-\bar\psi\gamma_5\psi\bar\psi^c\gamma_5\psi^c]\\&=\tfrac{1}{2}g_0[\bar\psi\gamma_\mu\psi^c\bar\psi^c\gamma_\mu\psi-\bar\psi\gamma_\mu\gamma_5\psi^c\bar\psi^c\gamma_\mu\gamma_5\psi]\\&=-\tfrac{1}{2}g_0[\bar\psi\gamma_\mu C\bar\psi\psi C^{-1}\gamma_\mu\psi-\bar\psi\gamma_\mu\gamma_5C\bar\psi\psi C^{-1}\gamma_\mu\gamma_5\psi],\end{aligned} \tag{4.18}$$

where ψ^c, $\bar\psi^c$ are the charge-conjugate fields.

The last form of Eq. (4.18) is suitable for our purpose. We note first that the vector part of the interaction is identically zero because of the anticommutativity of ψ. Thus only the pseudovector part survives. A "bubble" made of this interaction then is seen to give rise to the same integral J_A, Eq. (4.17). Since the interfering pseudoscalar interaction is missing in the present case, we get the complete scattering matrix by iterating J_A:

$$\begin{aligned}&-\gamma_\mu\gamma_5C\left[\frac{\delta_{\mu\nu}-q_\mu q_\nu/q^2}{1-J_A}+\frac{q_\mu q_\nu/q^2}{1-J_A'}\right]C^{-1}\gamma\gamma_5\\&=\gamma_\mu\gamma_5C\frac{1}{1-J_A}C^{-1}\gamma_\mu\gamma_5\\&\qquad+\gamma\cdot q\gamma_5C\frac{J_V/q^2}{(1-J_A')(1-J_A)}C^{-1}\gamma\cdot q\gamma_5,\end{aligned} \tag{4.19}$$

$$J_A\equiv J_A'-J_V.$$

The first term, corresponding to a scattering in the $J=1^-$ state, does not have a pole. The second term can have one below $4m^2$ for $1=J_A'$. With Eqs. (4.7) and (4.17), this determines the mass μ_D:

$$2m^2<\mu_D^2. \tag{4.20}$$

In this second term of the scattering matrix, the wave function is proportional to $C\gamma\cdot q\gamma_5$, so that the bound state behaves like a scalar "deuteron" (a singlet S state). The residue of the pole determines the nucleon-"deuteron" coupling constant (derivative) G_D^2, which is positive as it should be.

TABLE I. Mass spectrum.

Nucleon number	Mass μ	Spin-parity	Spectroscopic notation
0	0	0^-	$^1S_0$
0	$2m$	0^+	$^3P_0$
0	$(8/3)m^2<\mu^2$	1^-	$^3P_1$
±2	$2m^2<\mu^2$	0^+	$^1S_0$

Table I summarizes the main results of this section. Although our approximation is a very crude one, we believe that it reflects the real situation at least qualitatively, because all the results are understandable in simple physical terms. Thus in the nonrelativistic sense, our Hamiltonian contains spin-independent attractive scalar and vector interactions plus a spin-dependent axial vector interaction between a particle and an antiparticle. Between particles, the vector part turns into a repulsion. Table I is just what we expect for the level ordering from this consideration.

V. PHENOMENOLOGICAL THEORY AND γ_5 INVARIANCE

In the previous section special subsets of diagrams were taken into account, and the existence of various boson states was established, together with their couplings with the nucleons. As was discussed there, we can reasonably expect that these results are essentially correct in spite of the very simple approximations. Because the bosons have in general small masses (compared to the unbound nucleon states), they will play important roles in the dynamics of strong interactions at least at energies comparable to these masses.

[16] We meet here again the problem of subtraction. Our choice follows naturally from comparison with the vector case, and is consistent with Eq. (3.33).

Thus if we are willing to accept the conclusions of our lowest order approximation, what we should do then is to study the dynamics of systems consisting of nucleons and the different kinds of bosons which all together represent the primary manifestation of the fundamental interaction. These particles will be now assumed to interact via their phenomenological couplings. So we may describe our purpose as an attempt to construct a theory in the conventional sense in which a separate field is introduced for each kind of particle. However, this is not a simple and unambiguous problem because our fundamental theory is completely γ_5 invariant and we must make sure that this invariance is preserved at any stage of our calculations in order that the results be meaningful. For a better understanding of the problem, let us consider our Lagrangian in the lowest self-consistent approximation. We have

$$L'=L_0'+L_I',$$

where

$$\begin{aligned} L_0'&=-(\bar\psi\gamma_\mu\partial_\mu\psi+m\bar\psi\psi),\\ L_I'&=g_0[(\bar\psi\psi)^2-(\bar\psi\gamma_5\psi)^2]+m\bar\psi\psi. \end{aligned} \tag{5.1}$$

L' is obviously γ_5 invariant. In order to preserve this invariance we must study the S matrix generated by L_I'. Some subsets of diagrams have been considered in the previous section and it will be shown now how those calculations comply with γ_5 invariance. This point must be understood clearly so that we shall discuss it in a rather systematic way. Let us recall first how we constructed the scattering matrix in the "ladder" approximation. The lowest-order contribution is certainly invariant as no internal massive line appears. But what will happen to the next-order terms [Fig. 3(b)]? To these diagrams corresponds the expression

$$J_S(q^2)-\gamma_5 J_P(q^2)\gamma_5+iJ_{SP}(q^2)\gamma_5+i\gamma_5 J_{PS}(q^2). \tag{5.2}$$

In the gauge in which our calculations were performed, the last two terms happened to be zero. We write down next the transformation properties of the quantities appearing above. By straightforward calculation we find

$$\begin{aligned} \gamma_5&\rightarrow\gamma_5\cos2\alpha+i\sin2\alpha,\\ 1&\rightarrow\cos2\alpha+i\gamma_5\sin2\alpha,\\ J_P&\rightarrow J_P\cos^2 2\alpha+J_S\sin^2 2\alpha,\\ J_S&\rightarrow J_S\cos^2 2\alpha+J_P\sin^2 2\alpha,\\ J_{SP}&\rightarrow(J_P-J_S)\sin2\alpha\cos2\alpha,\\ J_{PS}&\rightarrow(J_P-J_S)\sin2\alpha\cos2\alpha. \end{aligned} \tag{5.3}$$

By simple substitution the invariance follows easily. The argument can now be extended to all orders, provided at each order all the possible combinations of S and P are included. The invariance of the scattering in the "ladder" approximation is thus established. It may look surprising that the SP and PS contributions do not vanish identically. This can be understood by considering the fact that the γ_5 transformation changes the parity of the vacuum which will be in general a superposition of states of opposite parities. In this way products of fields of different parities (as the SP propagator) may have a nonvanishing average value in the vacuum state.

We may now attempt the construction of the phenomenological coupling by introducing two local fields Φ_P and Φ_S describing the pseudoscalar and the scalar particles, respectively. We start by observing that, in the same gauge in which the previous calculations were made, we can write the meson-nucleon interaction as

$$L_I=G_P i\bar\psi\gamma_5\psi\Phi_P+G_S\bar\psi\psi\Phi_S. \tag{5.4}$$

In order to find the general expression valid in any gauge, it is convenient to introduce the following two-dimensional notation

$$\varphi\equiv\begin{pmatrix} i\bar\psi\gamma_5\psi \\ \bar\psi\psi \end{pmatrix},\quad \Phi\equiv\begin{pmatrix}\Phi_P\\ \Phi_S\end{pmatrix},\quad G\equiv\begin{pmatrix} G_P & 0\\ 0 & G_S\end{pmatrix}. \tag{5.5}$$

The interaction Lagrangian Eq. (5.4) can be written in this notation in a compact form,

$$L_I=\varphi G\Phi. \tag{5.6}$$

The effect of the γ_5 transformation on φ is described with the aid of the matrix

$$U\equiv\begin{pmatrix}\cos2\alpha & -\sin2\alpha\\ \sin2\alpha & \cos2\alpha\end{pmatrix}, \tag{5.7}$$

which satisfies $UU^+=UU^{-1}=UU^T=1$. In other words, the γ_5 transformation induces a unitary transformation in the two-dimensional space, and Eq. (5.6) remains invariant if

$$G\rightarrow UGU^{-1},\quad \Phi\rightarrow U\Phi. \tag{5.8}$$

To complete the construction of the theory, the free Lagrangian for the fields Φ_P and Φ_S must be added. If we work again in the special gauge $\alpha=0$, we may write

$$L_0=-\tfrac{1}{2}\partial_\mu\Phi_P\partial_\mu\Phi_P-\tfrac{1}{2}\partial_\mu\Phi_S\partial_\mu\Phi_S-\tfrac{1}{2}\mu^2\Phi_S^2, \tag{5.9}$$

where $\mu^2=4m^2$. We use again the two-dimensional notation, and defining the mass operator

$$M^2\equiv\begin{pmatrix}0 & 0\\ 0 & \mu^2\end{pmatrix}, \tag{5.10}$$

we write Eq. (5.9) in the invariant form

$$L_0=-\tfrac{1}{2}\partial_\mu\Phi\partial_\mu\Phi-\tfrac{1}{2}\Phi M^2\Phi. \tag{5.11}$$

In this way we have given a formal prescription for the γ_5 transformation in the phenomenological treatment. We have to emphasize here that the Lagrangians (5.9) and (5.11) are *not* γ_5 invariant in the ordinary sense of the word. In our theory, where the mesons are only phenomenological substitutes which partially represent the dynamical contents of the theory, they may be, however, called γ_5 covariant. In other words, *the masses and the coupling constants are not fixed parameters, but rather dynamical quantities which are subject to transformations when the representation is changed.* It will be legitimate to ask whether this situation corresponds to the one obtained in the framework of the fundamental theory and discussed in the "ladder" approximation in the previous section. We shall examine the transformation rule for the mass operator M^2, since this illustrates the case in point. Let us calculate explicitly M^2 in an arbitrary gauge α. We have

$$M^2 \rightarrow UM^2U^{-1} = \mu^2 \begin{pmatrix} \sin^2 2\alpha & -\sin 2\alpha \cos 2\alpha \\ -\sin 2\alpha \cos 2\alpha & \cos^2 2\alpha \end{pmatrix}. \quad (5.12)$$

The meaning of this equation is that the pseudoscalar and the scalar particle will have generally different masses in different gauges. In particular we see that the pseudoscalar particle has in the gauge α a mass $\sin 2\alpha\mu$. If this is the case we must expect that after the transformation the pole in the corresponding propagator will move from $q^2=0$ to $q^2=-(\sin^2 2\alpha)\mu^2$. This actually may be verified directly in the "ladder" approximation which shows that the pion propagator changes according to

$$iG_P{}^2\Delta_{FP} = \frac{2g_0}{1-J_P} \rightarrow \frac{2g_0}{1-J_P\cos^2 2\alpha - J_S \sin^2 2\alpha}. \quad (5.13)$$

Using the results of the previous section, it is seen that the denominator of the right-hand side vanishes for $q^2=-(\sin^2 2\alpha)4m^2$. In this way we have seen how our γ_5-invariant theory can be approximated by a phenomenological description in terms of pseudoscalar and scalar mesons. Of course one may add the vector meson as well. Such a description does not look γ_5 invariant. It is only γ_5 covariant, and the masses and coupling constants must be understood to be matrices which, however, can be simultaneously diagonalized.

The reason for this situation is the degeneracy of the vacuum and the world built upon it. Only after combining all the equivalent but nonintersecting worlds labeled with different α do we recover complete γ_5 invariance. Nevertheless, even in a particular world we can find manifestations of the invariance, such as the zero-mass pseudoscalar meson and the conserved γ_5 current.

VI. THE CONSERVATION OF AXIAL VECTOR CURRENT

In this section we will discuss another paradoxical aspect of the theory regarding the γ_5 invariance. In Sec. 3 we argued that the X current should really be conserved, and that this is possible if a nucleon X current possesses a peculiar anomalous term. We now verify the statement explicitly in our approximation.

First we have to realize that the problem is again how to keep the γ_5-invariant nature of the theory at every stage of approximation. It is well known in quantum electrodynamics that, in order to observe the ordinary gauge invariance, a certain set of graphs have to be combined together in a given approximation. The necessity for this is based on a general proof which makes use of the so-called Ward identity. In our present case there also exists an analog of the Ward identity. In order to derive it, let us first consider the proper self-energy part of our fermion in the presence of an external axial vector field B_μ with the interaction $L_B=-j_{\mu5}B_\mu$. The self-energy operator is now a matrix $\Sigma^{(B)}(p',p)$ depending on initial and final momenta. Expanding Σ in powers of B, we have

$$\Sigma^{(B)}(p',p)=\Sigma(p)+\Lambda_{\mu5}(p',p)B_\mu(p'-p)+\cdots. \quad (6.1)$$

We readily realize that the coefficient of the second term gives the desired X-current vertex correction.

On the other hand, the entire Lagrangian remains invariant under a *local* γ_5 transformation if Eq. (2.3) is accompanied by

$$B_\mu \rightarrow B_\mu - \partial_\mu\alpha, \quad (6.2)$$

where α is now an arbitrary function. In other words,

$$e^{i\alpha\gamma_5}\Sigma^{(B-\partial\alpha)}e^{i\alpha\gamma_5}=\Sigma^{(B)} \quad (6.3)$$

in a symbolical way of writing.[17]

Expanding (6.3) after putting $B=0$, we get

$$i\alpha(p'-p)[\gamma_5\Sigma(p)+\Sigma(p')\gamma_5] = i\alpha(p'-p)(p'-p)_\mu\Lambda_{\mu5}(p',p),$$

or

$$\gamma_5\Sigma(p)+\Sigma(p')\gamma_5=(p'-p)_\mu\Lambda_{\mu5}(p',p). \quad (6.4)$$

The entire vertex $\Gamma_{\mu5}=i\gamma_\mu\gamma_5+\Lambda_{\mu5}$ then satisfies

$$\begin{aligned} \gamma_5L'(p)+L'(p')\gamma_5 &= -(p'-p)_\mu\Gamma_{\mu5}(p',p), \\ L'(p) &\equiv -i\gamma\cdot p - \Sigma(p), \end{aligned} \quad (6.5)$$

which is the desired generalized Ward identity.[18] The right-hand side of Eq. (6.5) is the divergence of the X current, while the left-hand side vanishes when p and p' are on the mass shell of the actual particle. The X-current conservation is thus established. Moreover, the way the anomalous term arises is now clear. For if we assume $\Sigma(p)=m$, Eq. (6.4) gives

$$2m\gamma_5=(p'-p)_\mu\Lambda_{\mu5}(p'-p), \quad (6.6)$$

so that we may write the longitudinal part of Λ as

$$\Lambda_{\mu5}{}^{(l)}(p',p)=2m\gamma_5q_\mu/q^2, \quad q=p'-p, \quad (6.7)$$

which is of the desired form.

Next we have to determine what types of graphs

[17] We assume here that $\alpha(x)$ is different from zero only over a finite space-time region, so that the gauge of the nontrivial vacuum, which we may fix at remote past, is not affected by the transformation. The limiting process of going over to constant α is then ill-defined as we can see from the fact that the anomalous term in $\Gamma_{\mu5}$ has no limit as $q\rightarrow 0$.

[18] See also J. Bernstein, M. Gell-Mann, and L. Michel, Nuovo cimento 16, 560 (1960).

should be considered for Γ_μ in our particular approximation of the self-energy. Examining the way in which the relation (6.3) is maintained in a perturbation expansion, we are led to the conclusion that our self-energy represented by Fig. 5(a) gives rise to the series of vertex graphs [Fig. 5(b)]. The summation of the graphs is easily carried out to give

$$\Lambda_{\mu 5}=i\gamma_5\frac{1}{1-J_P}J_{PA}, \tag{6.8}$$

where J_P was obtained before [Eq. (4.8)], and

$$J_{PA}=\frac{2ig_0}{(2\pi)^4}\int \mathrm{Tr}\gamma_5 S(p+q/2)\gamma_\mu\gamma_5 S(p-q/2)d^4p$$

$$=-\frac{g_0}{2\pi^2}imq_\mu\int_{4m^2}^{\Lambda^2}\frac{d\kappa^2}{q^2+\kappa^2}(1-4m^2/\kappa^2)^{\frac{1}{2}}. \tag{6.9}$$

Thus

$$\Gamma_{\mu 5}=i\gamma_\mu\gamma_5+\Lambda_{\mu 5}$$
$$=i\gamma_\mu\gamma_5+2m\gamma^5 q_\mu/q^2, \tag{6.10}$$

in agreement with the general formula. We see also that there is no form factor in this approximation.

This example will suffice to show the general procedure necessary for keeping γ_5 invariance. When we consider further corrections, the procedure becomes more involved, but we can always find a set of graphs which are sufficient to maintain the X-current conservation. We shall come across this problem in connection with the axial vector weak interactions.

VII. SUMMARY AND DISCUSSION

We briefly summarize the results so far obtained. Our model Hamiltonian, though very simple, has been found to produce results which strongly simulate the general characteristics of real nucleons and mesons. It is quite appealing that both the nucleon mass and the pseudoscalar "pion" are of the same dynamical origin, and the reason behind this can be easily understood in terms of (1) classical concepts such as attraction or repulsion between particles, and (2) the γ_5 symmetry.

According to our model, the pion is not the primary agent of strong interactions, but only a secondary effect. The primary interaction is unknown. At the present stage of the model the latter is only required to have appropriate dynamical and symmetry properties, although the nonlinear four-fermion interaction, which we actually adopted, has certain practical advantages.

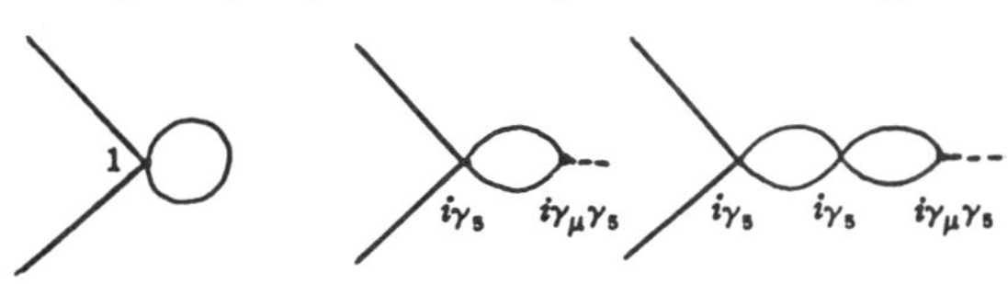

FIG. 5. Graphs for self-energy and matching radiative corrections to an axial vector vertex

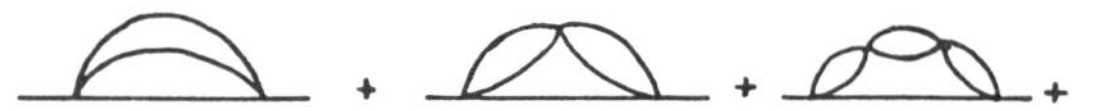

FIG. 6. A class of higher order self-energy graphs.

In our model the idealized "pion" occupies a special position in connection with the γ_5-gauge transformation. But there are also other massive bound states which may be called heavy mesons and deuterons. The conventional meson field theory must be regarded, from our point of view, as only a phenomenological description of events which are actually dynamic processes on a higher level of understanding, in the same sense that the phonon field is a phenomenological description of interatomic dynamics.

Our theory contains two parameters, the primary coupling constant and the cutoff, which can be translated into observed quantities: nucleon mass and the pion-nucleon coupling constant. It is interesting that the pion coupling depends only on the cutoff in our approximation. In order to make the pion coupling as big as the observed one (≈ 15) the cutoff has to be rather small, being of the same order as the nucleon mass.

We would like to make some remarks about the higher order approximations. If the higher order corrections are small, the usual perturbation calculation will be sufficient. If they are large compared to the lowest order estimation, the self-consistent procedure must be set up, including these effects from the beginning. This is complicated by the fact that the pions and other mesons have to be properly taken into account.

To get an idea about the importance of the corrections, let us take the next order self-energy graph (Fig. 6). This is only the first term of a class of corrections shown in Fig. 6, the sum of which we know already to give rise to an important collective effect, i.e., the mesons. It would be proper, therefore, to consider the entire class put together. The correction is then equivalent to the ordinary second order self-energy due to mesons, plus modifications arising at high momenta. Thus strict perturbation with respect to the bare coupling g_0 will not be an adequate procedure. Evaluating, for example, the pion contribution in a phenomenological way, we get

$$\frac{\delta m}{m}=\frac{G_P^2}{32\pi^2}\int_{m^2}^{\Lambda'^2}\frac{d\kappa^2}{\kappa^2}\left(1-\frac{m^2}{\kappa^2}\right), \tag{7.1}$$

where Λ' is an effective cutoff. Substituting G_P^2 from Eq. (4.9), this becomes

$$\frac{\delta m}{m}=\frac{1}{4}\int_{4m^2}^{4\Lambda'^2}\frac{d\kappa^2}{\kappa^2}\left(1-\frac{4m^2}{\kappa^2}\right)\Big/\int_{4m^2}^{\Lambda^2}\frac{d\kappa^2}{\kappa^2}\left(1-\frac{4m^2}{\kappa^2}\right)^{\frac{1}{2}}. \tag{7.2}$$

As Λ and Λ' should be of the same order of magnitude, the higher order corrections are in general not negligible. We may point out, on the other hand, that there is a tendency for partial cancellation between contributions from different mesons or nucleon pairs.

We already remarked before that the model treated here is not realistic enough to be compared with the actual nucleon problem. Our purpose was to show that a new possibility exists for field theory to be richer and more complex than has been hitherto envisaged, even though the mathematics is marred by the unresolved divergence problem.

In the subsequent paper we will attempt to generalize the model to allow for isospin and finite pion mass, and draw various consequences regarding strong as well as weak interactions.

APPENDIX

We treat here, for completeness, the problem created by the coupling of pseudoscalar and pseudovector terms encountered in the text. As we have seen, such an effect is not essential for the discussion of γ_5 invariance, but rather adds to complication, which however naturally appears in the ladder approximation.

First let us write down the integral equation for a vertex part Γ:

$$\begin{aligned}\Gamma(p+\tfrac{1}{2}q,\, p-\tfrac{1}{2}q) &= \gamma(p+\tfrac{1}{2}q,\, p-\tfrac{1}{2}q)+\frac{2ig_0}{(2\pi)^4}\gamma_5\int \mathrm{Tr}[\gamma_5 S(p'+\tfrac{1}{2}q) \\ &\quad\times\Gamma(p'+\tfrac{1}{2}q,\, p-\tfrac{1}{2}q)S_F(p-\tfrac{1}{2}q)]d^4p' \\ &\quad -\frac{ig_0}{(2\pi)^4}\gamma_5\gamma_\mu\int \mathrm{Tr}[\gamma_5\gamma_\mu S(p'+\tfrac{1}{2}q) \\ &\quad\times\Gamma(p'+\tfrac{1}{2}q,\, p-\tfrac{1}{2}q)S_F(p-\tfrac{1}{2}q)]d^4p'. \qquad (A.1)\end{aligned}$$

This embraces three special cases depending on the inhomogeneous term γ:

(a) $\gamma=0$ for the Bethe-Salpeter equation for the pseudoscalar meson;
(b) $\gamma=i\gamma_\mu\gamma_5$ for the pseudovector vertex function $\Gamma_{\mu5}$;
(c) $\gamma=2g_0(\gamma_5)_f(\gamma_5)_i-g_0(\gamma_\mu\gamma_5)_f(\gamma_\mu\gamma_5)_i$ for the nucleon-antinucleon scattering through these interactions.

Here i and f refer to initial and final states, and the integral kernel of Eq. (A.1) operates on the f part.

We will consider them successively.

(a) We make the ansatz $\Gamma=C\gamma_5+iD\gamma_5\gamma\cdot q$. The integrals in Eq. (A.1) then reduce to the standard forms considered in the text. Making use of Eqs. (4.9), (4.17), and (6.9), we get[16]

$$\begin{aligned}C&=C-(C+2mD)q^2I,\\ D&=(C+2mD)mI,\\ I(q^2)&=\frac{g_0}{4\pi^2}\int\frac{d\kappa^2}{q^2+\kappa^2}\left(1-\frac{4m^2}{\kappa^2}\right)^{\frac{1}{2}},\end{aligned}\qquad (A.2)$$

which lead to $q^2=0$, and $C:D=1-2m^2I(0):mI(0)$. From Eq. (4.8), we have $0<2m^2I(0)<\frac{1}{2}$.

(b) Put $$\begin{aligned}\Gamma_{\mu5}=&(i\gamma_\mu\gamma_5+2m\gamma_5q_\mu/q^2)F_1(q^2)\\ &+(i\gamma_\mu\gamma_5-i\gamma\cdot q\gamma_5q_\mu/q^2)F_2(q^2).\end{aligned}\qquad (A.3)$$

This is seen to satisfy the integral equation if

$$\begin{aligned}F_1&=1,\\ F_2&=J_A(q^2)/[1-J_A(q^2)],\\ J_A(q^2)&=2m^2I(q^2)-J_V(q^2),\end{aligned}\qquad (A.4)$$

where $J(q^2)$ was defined in Eq. (4.13).

On the mass shell, $\Gamma_{\mu5}$ reduces to

$$\begin{aligned}&(i\gamma_\mu\gamma_5+2m\gamma_5q_\mu/q^2)F(q^2),\\ &F(q^2)=1+F_2(q^2)=1/[1-J_A(q^2)].\end{aligned}\qquad (A.5)$$

For $q^2=0$, we have $J(q^2)=0$ so that $1<F(0)=1/[1-2m^2I(0)]<2$.

(c) From the structure of the inhomogeneous term, it is clear that the scattering matrix is given by

$$M=2g_0(\Gamma_5)_f(\gamma_5)_i+g_0(\Gamma_{\mu5})_f(i\gamma_\mu\gamma_5)_i,$$

where Γ_5 is the pseudoscalar vertex function.

Again, from Eq. (A.1), Γ_5 is determined as

$$\Gamma_5=\gamma_5[1-2m^2I(q^2)]/q^2I(q^2)-mi\gamma\cdot q\gamma_5/q^2, \qquad (A.6)$$

which has an entirely different behavior from the bare γ_5 for small q^2. The scattering matrix is then

$$\begin{aligned}M=&(\gamma_5)_f(\gamma_5)_i2g_0[1-2m^2I(q^2)]/q^2I(q^2)\\ &-[(i\gamma\cdot q\gamma_5)_f(\gamma_5)_i-(\gamma_5)_f(i\gamma\cdot q\gamma_5)_i]2mg_0/q^2\\ &-(i\gamma\cdot q\gamma_5)_f(i\gamma\cdot q\gamma_5)_ig_0J_A(q^2)/q^2[1-J_A(q^2)]\\ &+(i\gamma_\mu\gamma_5)_f(i\gamma_\mu\gamma_5)_ig_0/[1-J_A(q^2)].\end{aligned}\qquad (A.7)$$

The first three terms have a pole at $q^2=0$. The coupling constants of the pseudoscalar meson are then

pseudoscalar coupling:

$$G_p^2=2g_0[1-2m^2I(0)]/I(0),$$

pseudovector coupling:

$$\begin{aligned}G_{pv}^2&=g_0J_A(0)/[1-J_A(0)]\\ &=g_02m^2I(0)/[1-2m^2I(0)].\end{aligned}\qquad (A.8)$$

Their relative sign is such that the equivalent pseudoscalar coupling on the mass shell is

$$G_p'^2=4m^2g_0\left\{\left[\frac{1-2m^2I(0)}{2m^2I(0)}\right]^{\frac{1}{2}}+\left[\frac{2m^2I(0)}{1-2m^2I(0)}\right]^{\frac{1}{2}}\right\}^2. \qquad (A.9)$$

Reprinted from THE PHYSICAL REVIEW, Vol. 124, No. 1, 246–254, October 1, 1961

Dynamical Model of Elementary Particles Based on an Analogy with Superconductivity. II*

Y. NAMBU AND G. JONA-LASINIO†
Enrico Fermi Institute for Nuclear Studies and Department of Physics, University of Chicago, Chicago, Illinois

(Received May 10, 1961)

Continuing the program developed in a previous paper, a "superconductive" solution describing the proton-neutron doublet is obtained from a nonlinear spinor field Lagrangian. We find the pions of finite mass as nucleon-antinucleon bound states by introducing a small bare mass into the Lagrangian which otherwise possesses a certain type of the γ_5 invariance. In addition, heavier mesons and two-nucleon bound states are obtained in the same approximation. On the basis of numerical mass relations, it is suggested that the bare nucleon field is similar to the electron-neutrino field, and further speculations are made concerning the complete description of the baryons and leptons.

I. INTRODUCTION

IN Part I of this paper[1] we have proposed a model of strong interactions based on an analogy with the BCS-Bogoliubov theory of superconductivity. It is characterized by a nonlinear spinor field possessing γ_5 invariance, and simulates some important features of the meson-nucleon system. The basic principle underlying the model is the idea that field theory may admit, as a result of dynamical instability, extraordinary (nontrivial) solutions that have less symmetries than are built into the Lagrangian.[2] In fact we have obtained as an extraordinary solution a massive fermion and a massless pseudoscalar boson as idealized proton and pion, together with other heavy mesons.

If we now try to make our model more realistic, a number of problems spring up naturally. First of all, we would have to account for the isospin and strangeness quantum numbers. It seems rather obvious that these degrees of freedom have to be built into the theory from the beginning, although there may be some possibility of utilizing both the ordinary and extraordinary solutions to enlarge the Hilbert space, as will be discussed later.

These quantum numbers will not yet be enough to determine our theory satisfactorily, as we expect to have more additional symmetries which are at least approximately satisfied. Among other things, we have postulated the γ_5 invariance as a cornerstone of our previous model. What would be the proper generalization of the γ_5 invariance? Then there also arises the inevitable question of any possible symmetry among baryons of different strangenesses. Since such a symmetry is at any rate only approximate, the test of the theory will depend on its ability to account for the violation of the symmetry as well.

Finally, we face the problem of the baryon versus the lepton, the electromagnetism, and the weak processes. Here our theory creates a particular incentive for speculation concerning the baryon-lepton problem, since the ordinary and extraordinary solutions immediately remind us of these two families of particles.

We do not profess to have any clear-cut answers to these problems. In the present paper we shall again content ourselves with a rather modest task. We will first discuss a generalization of our model which incorporates the isospin for the nucleon and guarantees the existence of the pion. This can be done by demanding a $\gamma_5\times$isospin gauge group with a slight violation so as to give the pion its finite mass. We find that the bare mass necessary to achieve the latter end is at most several Mev. On this basis a suggestion is made that the bare nucleon field is essentially the same as the electron-neutrino field.

The complete picture of the baryon symmetries and the baryon-lepton problem is largely beyond the scope of the present paper, but some relevant discussions on this subject will also be presented, especially those concerned with the Sakata model and the general γ_5 symmetry

II. MODEL LAGRANGIAN FOR THE NUCLEON

First we would like to observe that the nonlinear spinor field adopted in I is not an essential element of our theory, as is the case with the Heisenberg theory[3] but is rather a model adopted to study our dynamical principles. At least in the present stage of the game, the controlling factors are the symmetry properties and qualitative dynamical characteristics of the basic fermion-fermion interaction, and whether the interaction is due to some fundamental boson, or fundamental nonlinearity (or something entirely new) is of secondary importance. Nevertheless, we have to choose

* This work was supported by the U. S. Atomic Energy Commission.

† Present address: Istituto di Fisica dell'Universita, Roma, and Istituto Nazionale di Fisica Nucleare, Sezione di Roma, Italy.

[1] Y. Nambu and G. Jona-Lasinio, Phys. Rev. **122**, 345 (1961); referred to hereafter as I. Y. Nambu, *Proceedings of the 1960 Annual International Conference on High-Energy Physics at Rochester* (Interscience Publishers, Inc., New York, 1960), p. 858.

[2] See also J. Goldstone, Nuovo cimento **19**, 154 (1961), N. N. Bogoliubov (to be published), V. G. Vaks and A. I. Larkin, *Proceedings of the 1960 Annual International Conference on High-Energy Physics at Rochester* (Interscience Publishers, Inc., New York, 1960), p. 871.

[3] H. P. Duerr, W. Heisenberg, H. Mitter, S. Schlieder, and K. Yamazaki, Z. Naturforsch. **14**, 441 (1959); W. Heisenberg, *Proceedings of the 1960 Annual International Conference on High-Energy Physics at Rochester* (Interscience Publishers, Inc., 1960), p. 851.

some model, and naturally there will arise certain predictions specific to the particular model. We take notice of the fact that the pion, the lightest of the meson family, is pseudoscalar and isovector, whereas its isoscalar counterpart of comparable mass does not seem to exist.[4] If the pion is to be intimately related to a symmetry property as in our previous model, this would imply that the model of nucleons should allow an (approximate) invariance under the $\gamma_5\times$isospin gauge group of Gürsey,[5] but not under the simple (Touschek) γ_5 gauge group, at least not so well as in the former case. For this reason, we would altogether consider the following gauge groups:

$$\psi\rightarrow e^{i\alpha}\psi,\qquad \bar{\psi}\rightarrow\bar{\psi}e^{-i\alpha}, \tag{2.1a}$$

$$\psi\rightarrow\exp(i\boldsymbol{\tau}\cdot\boldsymbol{\alpha}')\psi,\qquad \bar{\psi}\rightarrow\bar{\psi}\exp(-i\boldsymbol{\tau}\cdot\boldsymbol{\alpha}'), \tag{2.1b}$$

$$\psi\rightarrow\exp(i\gamma_5\boldsymbol{\tau}\cdot\boldsymbol{\alpha}'')\psi,\qquad \bar{\psi}\rightarrow\bar{\psi}\exp(i\gamma_5\boldsymbol{\tau}\cdot\boldsymbol{\alpha}''), \tag{2.1c}$$

where τ denotes the nucleon isospin matrices.

Obviously, the first two are generators of the nucleon number gauge and the isospin transformation, respectively. The second and third transformations combined form a four-dimensional rotation group on the four components composed by the proton and neutron of both handednesses.[5] Thus we may also replace Eqs. (2.1a) and (2.1b) by the following transformations

$$\psi_R\rightarrow\exp(i\boldsymbol{\tau}\cdot\boldsymbol{\alpha}_R)\psi_R,\qquad \psi_R^\dagger\rightarrow\psi_R^\dagger\exp(-i\boldsymbol{\tau}\cdot\boldsymbol{\alpha}_R),$$
$$\psi_L\rightarrow\exp(i\boldsymbol{\tau}1\boldsymbol{\alpha}_L)\psi_L,\qquad \psi_L^\dagger\rightarrow\psi_L^\dagger\exp(-i\boldsymbol{\tau}\cdot\boldsymbol{\alpha}_L), \tag{2.2}$$

where ψ_R and ψ_L are the right- and left-handed components.

As the simplest Lagrangian that meets our requirements, we adopt the form

$$L=-\bar{\psi}\gamma_\mu\partial_\mu\psi-\bar{\psi}M^0\psi$$
$$+g_0\Big[\bar{\psi}\psi\bar{\psi}\psi-\sum_{i=1}^{3}\bar{\psi}\gamma_5\tau_i\psi\bar{\psi}\gamma_5\tau_i\psi\Big]. \tag{2.3}$$

If the bare mass operator $M^0=0$, this Lagrangian possesses, in addition to Eq. (2.1), an invariance under the discrete "mass reversal" group:

$$\psi\rightarrow\gamma_5\psi,\qquad \bar{\psi}\rightarrow-\bar{\psi}\gamma_5. \tag{2.4}$$

The bare mass operator M^0 is a possible agent for the breakdown of the Gürsey group, and will be related to the finite pion mass.[6] For the moment, we will assume $M^0=0$. Before going to solve the self-consistent equation for the mass, we give the result of the Fierz transformation on Eq. (2.3): The interaction becomes

$$L_{\rm int}=\tfrac{1}{4}g_0[\bar{\psi}\psi\bar{\psi}\psi-\bar{\psi}\gamma_5\tau_i\psi\bar{\psi}\gamma_5\tau_i\psi]$$
$$+\tfrac{1}{4}g_0[\bar{\psi}\gamma_5\psi\bar{\psi}\gamma_5\psi-\bar{\psi}\tau_i\psi\bar{\psi}\tau_i\psi]$$
$$-\tfrac{1}{2}g_0[\bar{\psi}\gamma_\mu\psi\bar{\psi}\gamma_\mu\psi-\bar{\psi}\gamma_\mu\gamma_5\psi\bar{\psi}\gamma_\mu\gamma_5\psi]$$
$$+\tfrac{1}{8}g_0[\bar{\psi}\sigma_{\mu\nu}\psi\bar{\psi}\sigma_{\mu\nu}\psi-\bar{\psi}\sigma_{\mu\nu}\tau_i\psi\bar{\psi}\sigma_{\mu\nu}\tau_i\psi], \tag{2.5}$$

which is a rather complicated combination of all kinds of terms.

We now apply the linearization procedure of I to Eqs. (2.3) and (2.4), and obtain the self-energy

$$m=(1+\tfrac{1}{4})g_0\,{\rm Tr}S_F^{(m)}(0)$$
$$=-i\frac{10g_0}{(2\pi)^4}\int\frac{d^4pm}{p^2+m^2}F(p,\Lambda). \tag{2.6}$$

Note that the trace refers to both spin and isospin variables. This differs from Eq. (3.6) of I only by the change of the effective coupling $g_0\rightarrow 5g_0/2\equiv g_0'$. So we can simply take over the previous formulas, namely,

$$1=\frac{g_0'}{4\pi^2}\int_{4m^2}^{\Lambda^2}d\kappa^2\left(1-\frac{4m^2}{\kappa^2}\right)^{\frac{1}{2}}, \tag{2.7}$$

for the nontrivial solution if the dispersion integral (4.7) of I is used.

III. DETERMINATION OF MESON STATES

Since the interaction Lagrangian in Eqs. (1.1) and (1.3) contains a number of different couplings, we expect to get various kinds of "mesons" as bound nucleon-antinucleon pairs in our simple ladder approximation. As was explained in I, this is the proper approximation to match our self-energy equation at least for the pseudoscalar meson which is expected to have zero mass; moreover, even for other types of bound states we may reasonably trust its qualitative validity in predicting the existence and level ordering of possible bound states to the extent that our interaction is regarded basically as a short-range potential between spinor particles.

For general discussion, it is convenient to follow the procedure given in the Appendix of I. The basic equation to be considered is of the type

$$\Gamma(p+\tfrac{1}{2}q,\,p-\tfrac{1}{2}q)=\gamma+i\textstyle\sum_n g_nO_n$$
$$\times\int{\rm Tr}[O_nS_F(p'+\tfrac{1}{2}q)\Gamma(p'+\tfrac{1}{2}q,\,p'-\tfrac{1}{2}q)$$
$$\times S_F(p'-\tfrac{1}{2}q)]d^4p', \tag{3.1}$$

where the summation on the right-hand side is over the various tensor forms in the interaction Lagrangian. The "vertex function" $\Gamma(p+\frac{1}{2}q,\,p-\frac{1}{2}q)$ reduces to a bound state wave function when it becomes a homogeneous solution ($\gamma=0$) for a particular value of $q^2=-\mu^2$. We will briefly discuss those two-nucleon states for which there is a possibility of binding.

A. Pseudoscalar, Isovector Meson

Unlike the case in I, only the pseudoscalar interaction $\sim\bar{\psi}\gamma_5\tau_i\psi\bar{\psi}\gamma_5\tau_i\psi$ contributes to this state. Assuming

[4] It may not be impossible that the ordinary γ_5 invariance is violated more strongly than the Gürsey γ_5 invariance so that the mass of the π_0^0 meson may come sufficiently high. But to achieve this end by means of a bare mass does not seem to be feasible.

[5] F. Gürsey, Nuovo cimento **16**, 230 (1960).

[6] For its possible origin, see Sec. V.

$\Gamma_i{}^P=\gamma_5\tau_i\Gamma^P$, we obtain

$$\Gamma^P=\gamma^P+\Gamma^P[1-q^2I^P(q^2)],$$
$$I^P(q^2)=\frac{g_0'}{4\pi^2}\int_{4m^2}^{\Lambda^2}\frac{d\kappa^2}{q^2+\kappa^2}\left(1-\frac{4m^2}{\kappa^2}\right)^{\frac{1}{2}}, \tag{3.2}$$

where, of course, Eq. (2.7) was utilized. This has a homogeneous solution for $q^2=0$, corresponding to the zero-mass "pion." This pion-nucleon coupling is of pure pseudoscalar type, which can be calculated from the inhomogeneous equation with $\gamma^P=g_0'\gamma_5\tau_i$, as was done in the Appendix of I. We get, namely,[7]

$$G_P{}^2/4\pi=g_0'[4\pi I^P(0)]^{-1}$$
$$=\pi\left[\int_{4m^2}^{\Lambda^2}\frac{d\kappa^2}{\kappa^2}\left(1-\frac{4m^2}{\kappa^2}\right)^{\frac{1}{2}}\right]^{-1}. \tag{3.3}$$

B. Scalar, Isoscalar Meson

With the ansatz $\Gamma=\Gamma^S$ we have

$$\Gamma^S=\gamma^S+\Gamma^SI^S(q^2),$$
$$I^S(q^2)=\frac{g_0'}{4\pi^2}\int_{4m^2}^{\Lambda^2}\frac{\kappa^2-4m^2}{q^2+\kappa^2}d\kappa^2\left(1-\frac{4m^2}{\kappa^2}\right)^{\frac{1}{2}} \tag{3.4}$$
$$=1-(q^2+4m^2)I^P(q^2).$$

This leads to a zero-binding state: $q^2=-4m^2$ with the scalar nucleon coupling constant

$$G_S{}^2/4\pi=g_0'[4\pi I^P(-4m^2)]^{-1}. \tag{3.5}$$

C. Vector Mesons

There are two vector mesons, with isospin 1 and 0. The isovector meson arises from the tensor interaction $\sim\bar\psi\sigma_{\mu\nu}\tau_i\psi\bar\psi\sigma_{\mu\nu}\tau_i\psi$ with the wave function of the type

$$\Gamma_{\mu i}{}^V=\sigma_{\mu\nu}q_\nu\tau_i\Gamma^V.$$

The mass is determined from[8]

$$1=-\frac{g_0'}{60\pi^2}\int_{4m^2}^{\Lambda^2}\frac{d\kappa^2}{\kappa^2-\mu^2}\left(1-\frac{4m^2}{\kappa^2}\right)^{\frac{1}{2}}$$
$$\times\left[\kappa^2-4m^2-\mu^2\left(2+\frac{4m^2}{\kappa^2}\right)\right],$$

which has a solution (for sufficiently small Λ^2)

$$\mu^2\gtrsim 10m^2/3.$$

The coupling of this meson to the nucleon is necessarily of the derivative type.

[7] Note that this is half the value of I because a pion (e.g., π_0) consists of two substates $\bar pp$ and $\bar nn$, which changes the normalization of the pion wave function.

[8] The ambiguity about the subtraction of the most divergent part was discussed in I, section 4. The gross qualitative feature is not altered even if we do not make a subtraction.

To the isoscalar meson, both vector and tensor interactions contribute, the former being attractive and the latter repulsive. The wave function will have the form

$$\Gamma_\mu{}^V=\gamma_\mu\Gamma_1{}^V+\sigma_{\mu\nu}q_\nu\Gamma_2{}^V,$$

which yields a coupled equation for Γ_1 and Γ_2. This coupling, however, is rather small, so that we get a solution by neglecting Γ_2:

$$1=\frac{g_0'}{15\pi^2}\mu^2\int_{4m^2}^{\Lambda^2}\frac{d\kappa^2}{\kappa^2-\mu^2}\left(1-\frac{4m^2}{\kappa^2}\right)^{\frac{1}{2}}\left(1+\frac{2m^2}{\kappa^2}\right),$$
$$\mu^2\gtrsim 20m^2/7.$$

The nuclear coupling will be predominantly nonderivative.

D. The "Deuteron" States

As in I, we can discuss the nucleon-nucleon states in parallel with the meson states. The interaction may be written conveniently in the form

$$L_{\text{int}}=\tfrac{1}{4}g_0[\bar\psi\gamma_\mu\psi^c\bar\psi^c\gamma_\mu\psi-\bar\psi\sigma_{\mu\nu}\psi^c\bar\psi^c\sigma_{\mu\nu}\psi$$
$$+\bar\psi\gamma_\mu\gamma_5\tau_i\psi^c\bar\psi^c\gamma_\mu\gamma_5\tau_i\psi].$$

This is seen to lead to two bound states: a pseudovector, isoscalar ($J=1^+$, $T=0$) coming from the first two interaction terms, and a scalar, isovector ($J+0^+$, $T=1$), coming from the last term. For the $J=1^+$, $T=0$ state (deuteron) the main contribution comes from the attractive tensor interaction, and we get

$$\Gamma_\mu=\sigma_{\mu\nu}q_\nu\Gamma^{A'},$$
$$1=\frac{g_0'}{4\pi^2}\int_{4m^2}^{\Lambda^2}\frac{d\kappa^2}{\kappa^2-\mu^2}\left(1-\frac{4m^2}{\kappa^2}\right)^{\frac{1}{2}}\left[\mu^2+\frac{2}{15}(\kappa^2-4m^2)\right],$$
$$\mu^2\gtrsim 17m^2/5.$$

For the $J=0^+$, $T=1$ case we have

$$\Gamma=\gamma_5\gamma\cdot q\Gamma^{S'},$$
$$1=\tfrac{4}{5}m^2I^P(-\mu^2),$$
$$\mu^2\gtrsim 16m^2/5.$$

IV. VIOLATION OF γ_5 INVARIANCE

Let us now discuss the violation of the γ_5 invariance as indicated by the finite mass of the real pion. It would be senseless, of course, to talk about the invariance if the observed pion mass implied a large departure from our original Lagrangian, for example, due to a bare nucleon mass as large as the observed mass. So we need to estimate the amount of violation in the Lagrangian.

In general, the bare mass operator, which does not

violate nucleon number conservation,[9] can have the following form

$$M^0=m_1^0+m_2^0\boldsymbol{\tau}\cdot\mathbf{n}+m_3^0i\gamma_5+m_4^0i\gamma_5\boldsymbol{\tau}\cdot\mathbf{n}', \tag{4.1}$$

where $\mathbf{n}$ and $\mathbf{n}'$ are arbitrary unit vectors in the isospin space. The observed mass generated by Eq. (4.1) will also have a similar form. Because of the invariance of the rest of the Lagrangian under the transformations (2.1), we can choose it to be

$$M=m_1+m_2\tau_3+m_3i\gamma_5, \tag{4.2}$$

which gives two eigenmasses

$$\begin{aligned} m_p&\equiv[(m_1+m_2)^2+m_3^2]^{\frac{1}{2}},\\ m_n&\equiv[(m_1-m_2)^2+m_3^2]^{\frac{1}{2}}. \end{aligned} \tag{4.3}$$

The self-consistent self-energy equation to be solved is now

$$\begin{aligned} M=M^0+g_0\{&\mathrm{Tr}S_F^{(M)}(0)-\gamma_5\tau_i\,\mathrm{Tr}[\gamma_5\tau_iS_F^{(M)}(0)]\\ &+\tfrac{1}{5}\gamma_5\,\mathrm{Tr}[\gamma_5S_F^{(M)}(0)]-\tfrac{1}{5}\tau_i\,\mathrm{Tr}[\tau_iS_F^{(M)}(0)]\}. \end{aligned} \tag{4.4}$$

Equating the respective coefficients of both sides, we get

$$\begin{aligned} m_1&=m_1^0+\bar{I}m_1+\tilde{I}m_2,\\ m_2&=m_2^0-\tfrac{1}{5}\bar{I}m_2-\tfrac{1}{5}\tilde{I}m_1,\\ m_3&=m_3^0-\tfrac{1}{5}\bar{I}m_3,\\ 0&=m_4^0+\bar{I}m_3, \end{aligned} \tag{4.5}$$

where

$$\begin{aligned} \bar{I}&=\tfrac{1}{2}[I(m_p)+I(m_n)],\\ \tilde{I}&=\tfrac{1}{2}[I(m_p)-I(m_n)],\\ I(m)&=-\frac{8ig_0'}{(2\pi)^4}\int\frac{d^4p}{p^2+m^2}F(p,\Lambda). \end{aligned} \tag{4.5'}$$

We are interested in a small change of the non-trivial solution due to M^0. From Eq. (4.5) it is clear that $m_3=0$ unless

$$m_3^0=-(\tfrac{1}{5}+\bar{I}^{-1})m_4^0\neq0.$$

The term m_3 implies a violation of time and space reflections. Since we are not interested in such a violation, we will assume $m_3=m_3^0=m_4^0=0$ from now on. We further note that

$$\bar{I}=I(m_1)+O[(m_2/m_1)^2],\quad \tilde{I}=O[m_2/m_1].$$

In fact, up to the first order in m_2/m_1, we may put

$$m_1=m_1^0+I(m_1)m_1, \tag{4.6a}$$

$$m_2=m_2^0-\tfrac{1}{5}[I(m_1)+2I'(m_1)m_1^2]m_2, \tag{4.6b}$$

where

$$I'(m)=dI(m)/d(m^2)<0.$$

Equation (4.6a) determines m_1 in terms of m_1^0.

The self-consistency condition required, for $m_1^0=0$, is that $I(m)=1$. We may thus expand $I(m)$:

$$I(m_1)=1+I'(m)(m_1^2-m^2),$$

and obtain

$$\Delta m^2=m_1^2-m^2=-m_1^0/[mI'(m)]. \tag{4.7}$$

Since $I'(m)$ is of the order of $-I(m)/m^2$ (see below), this means

$$\Delta m=m_1-m\approx m_1^0. \tag{4.8}$$

From (4.6b), then

$$\begin{aligned} m_2&\approx m_2^0\{1-\tfrac{1}{5}[1+I'(m)m^2]\}^{-1}\\ &\approx m_2^0. \end{aligned} \tag{4.8'}$$

We note that originally there were two solutions $\pm|m|$, which now split into opposite directions according to Eq. (4.7) or (4.8). The meaning of this is as follows. Under the strict γ_5 invariance, there is a complete degeneracy with respect to the transformation (2.1c). The perturbation m_1^0 removes this degeneracy, so that the energy of the vacuum will depend on the orientation of the "γ_5 spin" of the negative energy fermions present in the "vacuum" with respect to this preferred direction. Obviously, the self-consistent procedure, which is similar to the variational method, gives the two extremum configurations corresponding to parallel ($m_0/m>0$) or antiparallel ($m_0/m<0$) γ_5-spin lineup. The parallel case has the larger "gap parameter" $|m|$ than the antiparallel case, so that the former will correspond to the stable ground state. The latter, on the other hand, should correspond to a metastable world.

It is perhaps interesting to see the general behavior of the self-consistency equation for arbitrary magnitude of m_1^0, assuming $m_2^0=0$ for simplicity. The relevant equation,

$$m[1-I(m)]=m^0,$$

is plotted schematically in Fig. 1.

Note that the trivial branch of the solution, which goes through the origin, has $m_0/m<0$. In other words, even in this case the self-consistent solution is qualitatively different from the simple perturbation result. As m^0 increases, it approaches the metastable nontrivial solution, and finally both go into the complex plane.

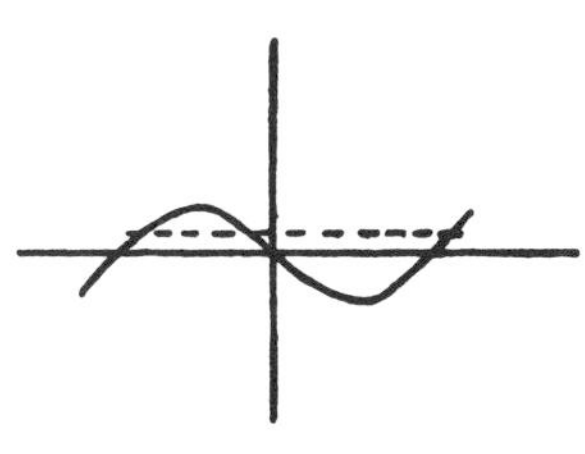

FIG. 1. The three self-consistent mass solutions m (ordinate) as a function of the bare mass m^0 (abscissa).

[9] The most general form of the self-energy Lagrangian (neglecting isospin dependence) is

$$\begin{aligned} &\bar{\psi}[i\gamma\cdot p\,\Sigma_1(p^2)+\Sigma_2(p^2)+i\gamma\cdot p\gamma_5\Sigma_3(p^2)+i\gamma_5\Sigma_4(p^2)]\psi\\ &\quad+\bar{\psi}^c[i\gamma\cdot p\,\Sigma_5(p^2)+\Sigma_6(p^2)+i\gamma\cdot p\gamma_5\Sigma_7(p^2)+i\gamma_5\Sigma_8(p^2)]\psi\\ &\qquad+\mathrm{H.c.} \end{aligned}$$

We do not attempt to study such a problem at this place.

We now come to the meson problem. The pion mass will be determined from

$$\Gamma_j = ig_0'\tau_i \int \mathrm{Tr}[\tau_i S_F(p'+\tfrac{1}{2}q)\Gamma_j S_F(p'-\tfrac{1}{2}q)]d^4p', \quad (4.9)$$

but

$$\underset{\text{isospin}}{\mathrm{Tr}}[\tau_i S_F^{(M)} \tau_j S_F^{(M)}]$$

$$= \underset{\text{isospin}}{\mathrm{Tr}}\left[\tau_i\left(S_F^{(m_p)}\frac{1+\tau_3}{2}+S_F^{(m_n)}\frac{1-\tau_3}{2}\right)\times\tau_j\left(S_F^{(m_p)}\frac{1+\tau_3}{2}+S_F^{(m_n)}\frac{1-\tau_3}{2}\right)\right]$$

$$=2\delta_{ij}\frac{S_F^{(m_p)}+S_F^{(m_n)}}{2}\frac{S_F^{(m_p)}+S_F^{(m_n)}}{2} +2(2\delta_{i3}\delta_{j3}-\delta_{ij})\frac{S_F^{(m_p)}-S_F^{(m_n)}}{2}\frac{S_F^{(m_p)}-S_F^{(m_n)}}{2}.$$

The second term yields convergent results, and is $O[(\Delta m/m)^2]$. To the order $\Delta m/m$, therefore, only the first term is important; moreover,

$$(S_F^{(m_p)}+S_F^{(m_n)})/2 \approx S_F^{(m_1)}.$$

In other words, there will be no first-order mass splitting of the pion. The mass is then determined from

$$1 = J_{p1}(-\mu^2) = \frac{g_0'}{4\pi^2}\int_{4m_1^2}^{\Lambda^2}\frac{\kappa^2 d\kappa^2}{\kappa^2-\mu^2}\left(1-\frac{4m_1^2}{\kappa^2}\right)^{\frac{1}{2}}. \quad (4.10)$$

For $m_1^0=0$, we had originally

$$1=J_p(0)=\frac{g_0'}{4\pi^2}\int_{4m^2}^{\Lambda^2} d\kappa^2\left(1-\frac{4m^2}{\kappa^2}\right)^{\frac{1}{2}},$$

which should now be replaced by

$$1=\frac{m_1^0}{m_1}+\frac{g_0'}{4\pi^2}\int_{4m_1^2}^{\Lambda^2} d\kappa^2\left(1-\frac{4m_1^2}{\kappa^2}\right)^{\frac{1}{2}}, \quad (4.11)$$

according to Eq. (4.6a).

From Eqs. (4.10) and (4.11) follows

$$\frac{m_1^0}{m_1}=\mu^2\frac{g_0'}{4\pi^2}\int_{4m_1^2}^{\Lambda^2}\frac{d\kappa^2}{\kappa^2-\mu^2}\left(1-\frac{4m_1^2}{\kappa^2}\right)^{\frac{1}{2}} \approx \mu^2\frac{g_0'}{4\pi^2}\int_{4m_1^2}^{\Lambda^2}\frac{d\kappa^2}{\kappa^2}\left(1-\frac{4m_1^2}{\kappa^2}\right)^{\frac{1}{2}} \leqslant \frac{\mu^2}{4m_1^2}\left(1-\frac{m_1^0}{m_1}\right). \quad (4.12)$$

For the observed value of $\mu^2/4m_1^2 \approx 1/200$ we then have, for the stable solution

$$m_1^0 \lesssim m_1/200 \approx 5 \text{ Mev}. \quad (4.13)$$

The amount of bare mass needed to produce the pion mass is thus surprisingly small.

On the other hand, the metastable solution $(m_1^0/m<0)$ produces an imaginary pion mass, indicating the unphysical nature of the solution.

The pion-nucleon coupling constant at the pion pole becomes [see Eq. (2.3)]

$$G_P^2/4\pi \approx g_0'[4\pi I_P(-\mu^2)]^{-1},$$

which is changed from the old one only by an order $\mu^2/m_1^2 \sim m_1^0/m_1$.

The other heavy meson states can be treated similarly. We see easily that the changes induced by m_1^0 are quite small: In general $\Delta\mu^2/m_1^2=O(m_1^0/m_1)$ and $\Delta G^2/G^2=O(m_1^0/m_1)$. Thus the effect of M^0 shows up dramatically only in the pion mass because it was originally zero.

Finally we remark that instead of a bare mass, we could assume slightly different coupling constants g_s and g_p $(<g_s)$ for the scalar and pseudoscalar interaction terms in the Lagrangian (2.3). The nature of the solution is somewhat different from the previous case because the Lagrangian still retains the mass reversal invariance $\psi \rightarrow \gamma_5\psi$, and the solution is twofold degenerate $(\pm m)$. The fractional change of the coupling necessary to produce the pion mass is again small: $|\Delta g/g| \approx \mu^2/4m_1^2$.

V. IMPLICATIONS OF THE MODEL

Let us now discuss the relevance of our present model to the physical realities of the nucleons and mesons.

1. We have seen that our Lagrangian (2.3) leads to the nucleon of isospin $\frac{1}{2}$ and the pion of isospin 1. The pion-nucleon coupling constant (pseudoscalar) depends on the cutoff parameter. For the observed large value (≈ 15) of $G_P^2/4\pi$, we see from Eqs. (1.5) and (2.3) that Λ must be of the same order of magnitude as the nucleon mass itself. This is not unreasonable, since the effective nucleon-nucleon interaction in higher approximations would proceed with the exchange of nucleon pairs.

A third parameter, the bare mass, enters our picture in order to make the meson mass finite. It would seem rather unsatisfactory and embarrassing that after all one has to break the postulated symmetry in an *ad hoc* manner. In order to clear up this point, the origin of the effective bare mass then becomes an interesting and important question. Since the required bare mass [Eq. (4.13)] seems to be quite small, a tempting possibility suggests itself that the bare nucleon field is the same as the electron-neutrino field. The electron mass itself could be either intrinsic or of electromagnetic origin.[10] Under this assumption, the bare mass operator would have the form $M^0=m_e^0(1+\tau_3)/2$, where the word "bare" is used relative to the interaction under consideration. According to the results of the previous section, it is only the isoscalar part of M^0 that produces the large shift of the pion mass, and the amount of violation of the isospin invariance will remain small.

2. Besides the pion, we have also derived vector mesons of both isoscalar ($T=0$) and isovector ($T=1$) types, and a scalar isoscalar meson which is actually unbound. No state corresponding to the isoscalar pion ($\pi_0{}^0$) is found. Of course, these results should depend sensitively on the choice of the interaction in the first place, and to a lesser extent also on the degree of approximation. At any rate, it seems to be a rather interesting and satisfactory feature of the model that these same vector mesons have been anticipated theoretically from various grounds,[11] even though there do not seem to be convincing experimental indications of their existence as yet.[12]

The mass values obtained here are rather high, and these mesons should actually decay into pions very quickly. The coupling constants are generally of the same order as the pion coupling constant, which means a very strong interaction for the vector and scalar mesons. These results, however, may be considerably altered in a better approximation. For one thing, the heavy mesons are coupled strongly to many-pion states which would make the former mere resonances of the latter. Moreover, the nucleon-nucleon and meson-meson interactions can go through long-range forces due to the exchange of these same mesons, which would in turn change the meson states themselves. These processes (the so-called left-hand cuts in the language of the dispersion theory) have not been taken into account in our ladder approximation.

This is a highly cooperative mechanism, and if one wants to handle it in a systematic way, one may be led to the same dispersion theoretical approach that is now widely pursued in pion physics. As a result of such effects, it is then conceivable that the masses of the vector mesons, for example, may come down.[13] Alternatively, it is also conceivable that we have more than one resonance having the same quantum numbers, of which we have obtained the higher ones. These high-energy poles may in turn determine the low-energy resonances.

In addition to the vector mesons, we expect a $T=0$, $J=0^+$ resonance, which has also been postulated by some people.[14] We should try to check these predictions against experimental evidence, such as the characteristic Q-value distributions and angular correlations in meson production processes.

Turning to the nucleon number 2 states, we expect two bound states ($T=0$, $J=1^+$ and $T=1$, $J=0^+$) with comparable masses to those for the vector mesons. This is a qualitatively satisfactory feature in view of the observed deuteron and the singlet virtual states, even though the actual binding is considerably weaker.[15]

3. As was already mentioned in I, our particular model was motivated by the approximate axial vector conservation observed in the nuclear β decay and the role of the pion in it.[5,16] The only difference from I is that (a) we now have the conservation of the isovector axial vector current $i\bar{\psi}\gamma_\mu\gamma_5\tau_i\psi$ instead of the simple axial vector current $i\bar{\psi}\gamma_\mu\gamma_5\psi$, and (b) a small violation of conservation is explicitly introduced. The general treatment of the problem will be completely analogous to the previous case.

Assuming that the β decay occurs through an additional term in the Lagrangian

$$L_\beta=g_\beta\bar{\psi}\gamma_\mu(1+\gamma_5)\tau_+\psi l_\mu{}_- +\text{H.c.}\quad [\tau_+=\tfrac{1}{2}(\tau_1+i\tau_2)],$$

where l_μ refers to the lepton current, the nuclear β-

[10] The electromagnetic interaction is invariant under the simple γ_5 transformation, but not under the Gürsey transformation since it fundamentally distinguishes between the charged and neutral components. Thus there is a built-in violation which can eventually produce the pion mass.

[11] W. R. Frazer and J. R. Fulco, Phys. Rev. **117**, 1609 (1960); Y. Nambu, *ibid.* **106**, 1366 (1957); G. Chew, Phys. Rev. Letters 4, 142 (1960); J. J. Sakurai, Ann. Phys. **11**, 1 (1960).

[12] J. A. Anderson, Vo X. Bang, P. G. Burke, D. D. Carmony, and N. Schmitz, Phys. Rev. Letters **6**, 365 (1961); A. Abashian, N. Booth, and K. M. Crowe, *ibid.* **5**, 258 (1960).

[13] A crude way to see the general tendency will be to argue as follows: The $T=1$ vector meson is coupled to the nucleon mainly through tensor coupling, so that it will cause a nucleon-antinucleon interaction of the type $-g^2(\boldsymbol{\sigma}_1\cdot\boldsymbol{\sigma}_2\boldsymbol{\tau}_1\cdot\boldsymbol{\tau}_2)e^{-\mu r}/r$. This tends to raise $T=1$, $J=0^+$ and $T=0$, $J=1^-$ meson states, and lower $T=0$, $J=0^+$ and $T=1$, $J=1^-$ states. Any change in the binding force, however, will be offset by the corresponding change in the nucleon mass, which automatically adjusts the pion mass to lie where it should be. The exchange of the $T=0$, $J=1$, and $J=0$ mesons, therefore, would not be so important in determining the relative shift of the meson levels.

[14] J. Schwinger, Ann. Phys. **2**, 407 (1957); M. Gell-Mann and M. Lévy, Nuovo cimento **16**, 705 (1960); S. Gupta, Phys. Rev. **111**, 1436 (1958), Phys. Rev. Letters **2**, 124 (1959); M. H. Johnson and E. Teller, Phys. Rev. **98**, 783 (1955); H. P. Duerr and E. Teller, *ibid.* **103**, 469 (1956). The σ meson mass obtained here is independent of the cutoff Λ, so that there may be some point in arguing that it is more reliable than for the vector mesons. If so, we may expect a nucleon-antinucleon resonance near zero kinetic energy (taking account of the mass shift due to M^0). The width may be quite broad.

[15] In fact, both $T=0$ and $T=1$ vector meson exchanges work in the direction to reduce the binding relative to the nucleon-antinucleon case.

[16] S. Bludman, Nuovo cimento **9**, 433 (1958); F. Gürsey, Ann. Phys. **12**, 91 (1961); Y. Nambu, reference 1; M. Gell-Mann and M. Levy, reference 14; J. Bernstein, N. Gell-Mann, and L. Michel, Nuovo cimento **16**, 560 (1960); J. Bernstein, S. Fubini, M. Gell-Mann, and W. Thirring, *ibid.* **17**, 757 (1960); Chou Kuang-Chao, J. Exptl. Theoret. Phys. (U.S.S.R.) **39**, 703 (1960) [Soviet Phys.—JETP **12**, 492 (1961)].

decay vertex becomes

$$\Gamma_\mu = g_\beta[i\gamma_\mu\tau_+ F_{V1}(q^2) - i\sigma_{\mu\nu}q_\nu\tau_+ F_{V2}(q^2) + \{i\gamma_\mu\gamma_5\tau_+ + [2m_1\gamma_5 q_\mu\tau_+/(q^2+\mu_\pi^2)]f(q^2)\}F_A(q^2)],$$

where q is the momentum change. In the ladder approximation, $F_{V1}(q^2)$ arises from the vector-type nucleon pairs, and $F_{V1}(0)=1$ (in accordance with the Ward identity, applicable to the isospin current, which shows that $F_{V1}(0)=1$ in general.[17])

In the axial vector part, $F_A(q^2)=1$ in our approximation. $f(q^2)$ arises because of the violation of the γ_5 invariance, but it deviates from 1 only to the order $m_1^0/m_1 \sim \mu^2/m_1^2$, as was already seen in the previous section. For practical purposes, therefore, the axial vector current has the desired form which would lead to the Goldberger-Treiman relation[18]

$$2m_1 g_A \approx \sqrt{2}G_\pi g_\pi,$$

where $g_A = g_\beta F_A(0)$ and G_π, g_π are, respectively, pion-nucleon and pion-lepton couplings.

In higher orders, however, $F_A(q^2)$ will be present, and in general $F_A(0)\neq 1$ even under the strict γ_5 invariance. People have conjectured in the past that $F_A(0)=g_A/g_V=1$ as $\mu_\pi\rightarrow 0$, but this does not seem to be easily guaranteed. The generalized Ward identity for the axial vector current[19] suffices to prove the Goldberger-Treiman relation, but is not enough to make $F_A(0)=1$. In order that the latter should come out rigorously, we would need a more subtle mechanism. Nevertheless, we can try a working hypothesis that $g_A/g_V=1$ under the strict invariance, and then estimate the deviation due to the violation. This scheme is carried out in the Appendix.

VI. FURTHER PROBLEM

We will consider here some of the general problems which have not been explored, but which seem to be important in a more comprehensive understanding of the elementary particles.

1. The hyperons. In order to incorporate the strange particles into our picture we would have to increase the dimensions of the fundamental field unless we do further unconventional things (see below). The simplest possibility from the point of view of quantum numbers would be to add a bare Λ-particle field as was originally proposed by Sakata.[20] We would then postulate, in addition, the generalized γ_5 symmetry, which would mean the invariance of the left-handed and right-handed components separately under the unitary transformation among the three fields or some subgroups of it. The mass splitting of the three baryons will be obtained from bare masses of similar magnitude, which destroys the otherwise rigorous symmetry.

This approach will produce easily the pions and K mesons and probably more, and their masses can again be related to the baryon bare masses. But we do not yet have a comparable dynamical method to predict Σ and Ξ particles. Consequently, we shall not be able to say whether or not the present model is dynamically satisfactory in this respect.

2. The leptons. In connection with the above model we are naturally led to the lepton problem. Gamba, Marshak, and Okubo[21] have pointed out an interesting parallelism between the $pn\Lambda$ and $\nu e\mu$ triplets. As was remarked in the beginning, our theory gives a special incentive for speculation about this relation because we have obtained two solutions: one ordinary and one extraordinary, differing in masses. Could they both be realized in nature simultaneously? According to our results in I, the answer is no because they belong to different Hilbert spaces. Moreover, the trivial solution gives rise to unphysical mesons at least under the assumption of fixed cutoff, with a large mass $(-\mu^2 \gtrsim \Lambda^2)$ but not necessarily a weak coupling $(G^2 \lesssim \Lambda^4/\mu^4)$. Nevertheless, it would seem too bad if Nature did not take advantage of the two solutions. A straightforward way to make the two solutions co-exist in the same world is obviously to postulate that the world is represented by the direct product of two Hilbert spaces[22]:

$$\mathcal{H} = \mathcal{H}^{(0)} \otimes \mathcal{H}^{(m)}, \tag{6.1}$$

built upon the vacuum state

$$\Omega = \Omega^{(0)} \otimes \Omega^{(m)}. \tag{6.1'}$$

It is true that this is effectively the same as doubling the fields, but here the choice of the two solutions (particles) is dictated by the dynamics of the original nonlinear theory. In order to describe this situation, we may adopt an *effective* Lagrangian

$$L = L^{(1)} + L^{(2)}, \tag{6.2}$$

where each of the $L^{(i)}$ has the same form, only differing in the charge assignments of the respective triplet fields. The Lagrangian obviously yields four subspaces

$$\mathcal{H}_1^{(0)} \otimes \mathcal{H}_2^{(m)}, \quad \mathcal{H}_1^{(m)} \otimes \mathcal{H}_2^{(0)}, \quad \mathcal{H}_1^{(0)} \otimes \mathcal{H}_2^{(0)}$$

and

$$\mathcal{H}_1^{(m)} \otimes \mathcal{H}_2^{(m)}.$$

According to our plan, we must say that we happen to live in the first subspace. [In the second space, the masses of $\nu e\mu$ and $pn\Lambda$ are interchanged, whereas in the third (fourth) case we have two kinds of leptons (baryons).]

[17] R. P. Feynman and M. Gell-Mann, Phys. Rev. **109**, 193 (1958).

[18] M. L. Goldberger and S. B. Treiman, Phys. Rev. **111**, 356 (1958).

[19] J. Bernstein *et al.*, reference 16.

[20] S. Sakata, Progr. Theoret. Phys. (Kyoto) **16**, 686 (1956).

[21] A. Gamba, R. E. Marshak, and S. Okubo, Proc. Natl. Acad. Sci. U. S. **45**, 881 (1959); Z. Maki, M. Nakagawa, Y. Ohnuki, and S. Sakata, Progr. Theoret. Phys. **23**, 1174 (1960).

[22] S. Okubo and R. E. Marshak, Nuovo cimento **19**, 1226 (1961), have independently proposed a similar idea. We thank the authors for valuable communications.

So far there is no interaction between leptons and baryons (except the electromagnetic, which is trivial). To introduce the weak interactions, we may, for example, add to Eq. (5.2) a third nonlinear term involving all the (left-handed) fields. This would complete our program of dealing with the strong and weak interactions.

But, of course, it is not yet a truly unified theory; the weak interaction is introduced only as an *ad hoc* additional process. Moreover, we do not know the mathematical consistency of such a procedure, because the additional interaction, if taken seriously, may qualitatively affect the baryon and lepton solutions we already have.

There is an alternative, but less drastic scheme; namely, to assume six different fields from the beginning, of which three (becoming eventually the baryon fields) have additional strong interactions in the Lagrangian. This may not be devoid of elegance if the interaction is mediated by a vector Bose field coupled to the baryon charge. The intermediate bosons, including the photons and possibly also the weak bosons, could then be interpreted as the agents that distinguish between different components of the bare fermions, which otherwise would enjoy a high degree of symmetry.

We would like to throw in another remark here that there may be also a possibility of utilizing the ordinary and extraordinary solutions in distinguishing between electron and muon, or baryons of different strangenesses.

3. The γ_5 invariance for general systems. In our theory the γ_5 invariance is a very essential element. It is a particular symmetry which exists in the Lagrangian, but is masked in reality because of the (approximate) degeneracy of the vacuum with respect to that symmetry. We have used the pion and the β decay in support of the assumption. In order to firmly establish its validity, however, we must try to find more evidences. For one thing, the induced pseudoscalar terms in nucleon β decay and μ capture should be examined more closely.

Furthermore, if such a symmetry is to have a general meaning, we must be able to consider partially conserved currents for processes such as

$$\begin{aligned} &H^3 \rightarrow He^3, \\ &He^6 \rightarrow Li^6, \\ &C^{14} \rightarrow N^{14}, \\ &\Sigma^- \rightarrow \Sigma^0, \\ &\Sigma^- \rightarrow \Lambda. \end{aligned} \qquad (6.3)$$

An elementary definition of the γ_5 transformation for the general system is obvious: When the wave function of a system is expressed in terms of the fundamental (bare) spinors obeying the rules Eq. (2.1), the transformation is unambiguously defined for each component, and thereby the total axial vector current is determined.

For superallowed transitions with spin $\frac{1}{2}$, the problem is particularly simple, since it is the same as for the neutron case. Thus for $H^3 \rightarrow He^3$ we have the same Goldberger-Treiman relation

$$(M_H+M_{He})g_A(\mathrm{H,He})/\sqrt{2}G_\pi(\mathrm{H,He}) \approx g_\pi, \qquad (6.4)$$

where g_A, G_π now characterize the β decay and the (unknown) pion coupling for the transition under consideration.

Similar relations hold for the Σ decays.[23] For the Σ-Σ case, we have

$$2m_\Sigma g_A(\Sigma\Sigma)/G_\pi(\Sigma\Sigma) \approx g_\pi. \qquad (6.5)$$

For the Σ-Λ case, the axial vector vertex becomes[24]

$$\begin{aligned} \Gamma_A \approx [i\gamma_\mu\gamma_5 + (m_\Sigma+m_\Lambda)\gamma_5 q_\mu/(q^2+\mu_\pi^2)]F_{1A}(q^2) \\ +i\gamma_5\sigma_{\mu\nu}q_\nu F_{2A}(q^2) \\ (m_\Sigma+m_\Lambda)F_{1A}(0)/G_\pi(\Sigma\Lambda) \approx g_\pi, \end{aligned} \qquad (6.6)$$

if the relative Σ-Λ parity is even. The vector current conservation is also violated because of the Σ-Λ mass difference, and it looks as though this would predict a corresponding scalar meson term. However, the analogy is rather superficial. Firstly, the violation disappears if $m_\Sigma = m_\Lambda$, in which case there would be no need for a scalar meson. The Σ-Λ mass difference itself might be due to the breakdown of the γ_5 symmetry. Secondly, it is an "unfavored" transition ($\Delta T = 1$), so that the vector part, corresponding to the off-diagonal element of the isospin current, should vanish in the ideal limit of strict isospin invariance and $q \rightarrow 0$. In other words, we expect

$$\Gamma_V \approx [q^2 i\gamma_\mu - (m_\Sigma - m_\Lambda)q_\mu]F_{1V}(q^2) + \sigma_{\mu\nu}q_\nu F_{2V}(q^2). \qquad (6.7)$$

In case the Σ-Λ parity is odd,[25] the vector and axial vector parts will interchange their roles. The vector part, which now looks like the axial vector current, would have the form

$$\begin{aligned} \Gamma_V \approx [q^2 i\gamma_\mu\gamma_5 + (m_\Sigma+m_\Lambda)\gamma_5 q_\mu]F_{1V}(q^2) \\ +i\gamma_5\sigma_{\mu\nu}q_\nu F_{2V}(q^2). \end{aligned} \qquad (6.8)$$

The axial vector part can similarly be put in the form

$$\begin{aligned} [i\gamma_\mu - q_\mu(m_\Sigma - m_\Lambda)/(q^2+\mu_\pi^2)f(q^2)]F_{1A}(q^2) \\ +\sigma_{\mu\nu}q_\nu F_{2A}(q^2). \end{aligned} \qquad (6.9)$$

But $f(q^2)$ need not be ≈ 1 if the Σ-Λ mass difference is also due to the violation of the γ_5 symmetry.

There are other processes for which the chirality conservation can be tested in a direct way. Although

[23] L. B. Okun', *Ann. Rev. Nuclear Sci.* **9**, 61 (1959); M. Gell-Mann, *Proceedings of the 1960 Annual International Conference on High-Energy Physics at Rochester* (Interscience Publishers, Inc., New York, 1960), p. 522.

[24] We have in this case three independent terms.

[25] See S. Barshay, Phys. Rev. Letters **1**, 97 (1958); Y. Nambu and J. J. Sakurai, *ibid.* **6**, 377 (1961).

extraordinary solutions are in general not eigenstates of chirality (even under strict γ_5 invariance), the conservation law should still apply to the expectation values of chirality. In fact, we can express the chirality conservation law $\langle X_i \rangle = \langle X_f \rangle$ for any reaction $i \rightarrow f$; for example

$$p+\pi \rightarrow p+\pi, \quad p+\pi+\pi', \text{ etc.,}$$
$$p+p \rightarrow p+p, \quad p+p+\pi, \text{ etc.,}$$

as a relation between the change of nucleon chirality and the magnitude of the pion production amplitude.

The ideas outlined in this section will be taken up in more detail elsewhere.

APPENDIX

We calculate here the renormalization of the axial vector (Gamow-Teller) coupling constant g_A for nuclear β decay under the following assumptions:

(1) Under strict γ_5 invariance (Gürsey type), there is no renormalization, namely $g_A = g_{A0}$ $(= g_{V0} = g_V)$, where g_{A0} is the bare coupling constant.

(2) The violation of the invariance gives rise to the finite pion mass as well as the deviation of the ratio $R = g_A/g_{A0} = g_A/g_V$ from unity, so that there is a functional relation between the two quantities.

Let us first consider the isovector axial vector vertex Γ_A in the usual perturbation theory. In our model, it consists of various graphs, some of which are shown in Fig. 2(a) and (b). The "ladder" graphs 2(a) have been considered in I as well as in the present paper, since they are intimately related to the γ_5 gauge transformation. In I (Appendix) we found that $R > 1$ when both pseudoscalar and pseudovector type interactions are present.[26] The graphs 2(b) have not been considered yet. These will come into our consideration as soon as we take corresponding higher-order approximations for the self-energy, which was briefly discussed in I. The chain of bubbles in these graphs will act like a meson when there is such a dynamical pole [Fig. 2(c)].

The (divergent) renormalization effect due to intermediate mesons is always negative,[27] irrespective of the type of the meson, so that the effect of these meson-like bubble graphs is also expected to be similar. When the chain does not produce a pole, however, the effect can be opposite.

Combining all these effects, we have no way to predict the resultant magnitude and sign of the renormalization correction. So we simply assume these contributions to cancel out under strict γ_5 invariance.

Next let us suppose that the invariance is slightly violated. This will cause changes in the propagators

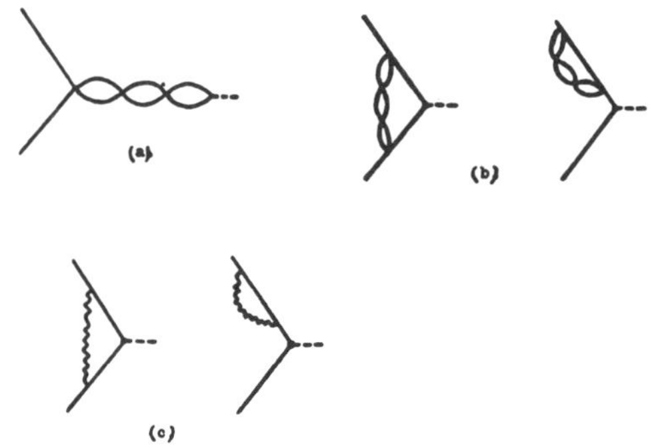

FIG. 2. Typical graphs considered in the evaluation of the axial vector vertex.

in all these graphs. Most of these changes are, however, quite small, being of the order of $m^0/m \approx \mu^2/4m^2$, as will be clear from the results of Sec. IV. The largest effect is naturally expected to come from the "pion" contribution in Fig. 2(b), as this is a change from zero mass (infinite range) to a finite one.

Let us accordingly take the effective pion graph from Fig. 2(c) with an arbitrary pion mass μ. Call its contribution to the vertex renormalization (for zero momentum transfer) $\Lambda(\mu)$. Then according to the above assumption

$$R = \Gamma_A(\mu_\pi)/g_A = \Gamma_A(\mu_\pi)/\Gamma_A(0) \approx 1 + \Lambda(\mu_\pi) - \Lambda(0). \tag{A1}$$

The difference $\Lambda(\mu) - \Lambda(0)$ is convergent, which turns out to be

$$\Lambda(\mu) - \Lambda(0) = \frac{G^2}{16\pi^2}\frac{\mu^2}{m^2}\left\{\left(3 - \frac{5\mu^2}{2m^2}\right)\ln\frac{m^2}{\mu^2} - 5 + \frac{16\mu}{\sqrt{3}m}\left[\tan^{-1}\left(\frac{4m^2}{3\mu^2} - \frac{2}{3}\right) + \tan^{-1}\left(\frac{2}{3}\right)\right]\right\}, \tag{A2}$$

where G is the phenomenological pion coupling constant.

As was expected, this goes like $(\mu^2/m^2)\ln(m^2/\mu^2)$ for small μ, which is more important than the contributions from the neglected processes behaving like μ^2/m^2.

With $(G^2/4\pi)(\mu^2/4m^2) = f^2/4\pi = 0.08$, Eq. (A2) gives

$$R - 1 = \begin{cases} 0.18 \\ 0.24. \end{cases} \tag{A3}$$

The first figure is the entire contribution from Eq. (A2), while the second is the contribution from the leading logarithmic term alone. Experimentally, R is estimated to be ≈ 1.25.[28]

[26] See also Z. Maki, Progr. Theoret. Phys. (Kyoto) 22, 62 (1959).
[27] We owe Dr. J. de Swart the mathematical check on this point.
[28] M. T. Burgy, V. E. Krohn, T. B. Novey, G. R. Ringo, and V. L. Telegdi, Phys. Rev. 120, 1829 (1961).

PHYSICAL REVIEW VOLUME 125, NUMBER 4 FEBRUARY 15, 1962

Chirality Conservation and Soft Pion Production*

Y. NAMBU AND D. LURIÉ
The Enrico Fermi Institute for Nuclear Studies and the Department of Physics, University of Chicago, Chicago, Illinois

(Received September 25, 1961)

A formally γ_5-invariant system consisting of a Dirac field and a massless pseudoscalar field allows chirality conservation in the sense that its expectation value is a constant of motion. This leads to the consequence that in any reaction a change in the fermion chirality (~helicity×velocity) is compensated for by the emission of a massless boson at zero energy, which can be expressed by a simple formula relating the radiative amplitude to the elastic amplitude. Assuming the pion-nucleon system to be γ_5-invariant when the pion mass can be neglected, the formula is applied to the processes $N+\pi \rightarrow N+\pi$ and $N+2\pi$. A reasonable agreement with experiment is obtained in a case dominated by the 3–3 resonance.

I. INTRODUCTION

THE observed equality of the vector coupling constant G_V in nuclear and muon β decays has led to the conserved current theory of Feynman and Gell-Mann,[1] according to which the vector part of the nuclear β-decay interaction is proportional to the total isotopic spin current. The nonrenormalization of G_V due to strong interactions is then guaranteed by the isospin conservation in strong interactions.

One expects further a close proportionality of the nucleon electromagnetic and β-decay matrix elements, including the Pauli magnetic term, and the pion β decay $\pi^\pm \rightarrow \pi^0$ would also proceed at the "universal" rate. Although these predictions are yet to be confirmed experimentally, there seems to be enough theoretical motivation for such speculations. For we would then be able to regard the weak interactions, like the electromagnetic interaction, as an agent which reveals the basic symmetries that might exist beneath the confusing effects of strong interactions.

As the nucleon β decay interaction also contains an axial vector part with a comparable strength ($G_A \sim 1.2 G_V$), several authors[2] have naturally tried to extend the principle by postulating axial vector current conservation or invariance under the so-called γ_5 transformation. In this case, however, it has not been found possible to guarantee the nonrenormalization of G_A by the axial vector conservation, and besides the conservation law seems to be only approximate under strong interactions. Nevertheless, it has led to one interesting result, namely the Goldberger-Treiman[3] relation between the Gamow-Teller constant G_A, the pion-nucleon constant, and the pion decay constant, which is known to be satisfied.

Recently one of us[4] has proposed a composite model of elementary particles based on an analogy with superconductivity. This model is built on the essential assumption that the Lagrangian describing the nucleons is invariant under the γ_5 transformation but that the physical vacuum state need not be so. As an interesting consequence, we observe that there exist nucleon-antinucleon bound states which behave as massless pseudoscalar mesons. The Goldberger-Treiman relation follows immediately from this if we identify them with the pions.

In the present paper we shall study another consequence of the axial vector conservation which can be used as an experimental test of the assumption. The main point is that for any reaction involving nucleons and mesons, the axial vector current conservation implies a close relation between the elastic amplitude and the "radiative" amplitude where an extra massless pion is emitted at zero energy.

Since the real pion has a finite mass, such a relation cannot actually be satisfied, but one may expect it to be approximately true at sufficiently high energies where the pion mass is negligible.

We shall first consider a simple model due to Nishijima and show how the above-mentioned relation is expected for a process involving the scattering of a nucleon by an external γ_5-invariant potential (Sec. II). We shall verify the relation explicitly in the lowest order perturbation (Sec. III), and then suggest a possible way of proving it in general (Sec. IV). Finally we shall discuss some experiments, in particular those involving pion-nucleon scattering, which would test the above-mentioned relation and hence the assumption of the (approximate) γ_5 invariance of strong interactions.

II. THE NISHIJIMA MODEL

In the actual β decay problem, the relevant symmetry to be associated with the axial vector part is the invariance under the $\gamma_5\times$isospin gauge transformation which acts on the nucleon (proton-neutron) field ψ as

* This work was supported by the U. S. Atomic Energy Commission.

[1] R. P. Feynman and M. Gell-Mann, Phys. Rev. 109, 193 (1958).

[2] S. Bludman, Nuovo cimento 9, 433 (1958); F. Gürsey, Nuovo cimento 16, 230 (1960); Ann. Phys. 12, 91 (1961); M. Gell-Mann and M. Lévy, Nuovo cimento 16, 705 (1960); J. Bernstein, M. Gell-Mann and L. Michel, Nuovo cimento 16, 560 (1960); J. Bernstein, S. Fubini, M. Gell-Mann, and W. Thirring, Nuovo cimento 17, 757 (1960); Chou Kang-Chao, J. Exptl. and Theor. Phys. (U.S.S.R.) 39, 703 (1960) [Soviet Phys. JETP 12, 492 (1961)]; Y. Nambu, Phys. Rev. Letters 4, 380 (1960).

[3] M. L. Goldberger and S. M. Treiman, Phys. Rev. 111, 354 (1958).

[4] Y. Nambu and G. Jona-Lasinio, Phys. Rev. 122, 345 (1961); 124, 246 (1961).

follows:

$$\psi(x)\to\exp(i\boldsymbol{\alpha}\cdot\boldsymbol{\tau}\gamma_5)\psi(x), \tag{2.1}$$

where $\boldsymbol{\alpha}$ is an arbitrary vector in isospace.

In this and following sections we shall study a simplified model proposed by Nishijima[5] because there would be no loss of the essential features.

The system consists of a single nucleon field ψ and a massless neutral pseudoscalar field ϕ ("pion") coupled through

$$L_{\rm int}=-m\bar{\psi}\exp[(ig/m)\gamma_5\phi]\psi, \tag{2.2}$$

which also incorporates the (bare) mass term for the nucleon. By expanding the exponential, we recognize that the first two terms are the mass and the ordinary meson-nucleon coupling terms. (It is also clear that with the transformation $\psi\to\exp[-(ig/2m)\gamma_5\phi]\psi$, the theory is equivalent to the simple derivative coupling model. We prefer the Nishijima representation because it brings in the nucleon γ_5 transformation explicitly.) In addition, we shall introduce an external vector (or axial vector) potential $V_\mu(A_\mu)$,

$$L_{\rm ext}=i\bar{\psi}\gamma_\mu\psi V_\mu \quad (\text{or } i\bar{\psi}\gamma_\mu\gamma_5\psi A_\mu). \tag{2.2'}$$

The entire Lagrangian is invariant under the γ_5 transformation,

$$\psi\to e^{i\alpha\gamma_5}\psi, \tag{2.3a}$$

and

$$\phi\to\phi-(2m/g)\alpha, \tag{2.3b}$$

for constant α. Of course it is essential for the invariance that the pion is massless. One easily verifies the axial vector current conservation which follows from the aforementioned invariance:

$$\partial_\mu j_\mu(x)=0,\qquad j_\mu=i\bar{\psi}\gamma_\mu\gamma_5\psi-(2m/g)\partial_\mu\phi. \tag{2.4}$$

An important feature of the theory is that the existence of the "pion" field is necessary to preserve the γ_5 invariance in the presence of finite nucleon mass. In this respect it is similar to the model proposed by Nambu and Jona-Lasinio.[4] There the pion field and finite bare nucleon mass are not assumed, but, in the end, we find that, if the nucleon has a finite observed mass, it must also be accompanied by a massless pion field which is to be interpreted as nucleon-antinucleon bound states, or a collective excitation of such pairs. In this case, the conserved current j_μ is simply

$$j_\mu=i\bar{\psi}\gamma_\mu\gamma_5\psi, \tag{2.5}$$

which, however, implicitly contains the pion contribution.

Let us now turn to the meaning of the conservation law associated with Eq. (2.4). We call this conserved quantity the chirality χ, defined by

$$\chi=-i\int j_4d^3x=\int\bar{\psi}\gamma_4\gamma_5\psi d^3x+(2im/g)\int\partial_4\phi d^3x$$
$$=\chi_N+\chi_\pi. \tag{2.6}$$

Equation (2.4) shows that χ is a constant of motion, i.e., a conserved quantity. However, we can easily see that a one-nucleon state or a one-meson state is not an eigenstate of χ. As was analyzed in the model of Nambu and Jona-Lasinio, this seemingly paradoxical situation is related to the fact that the γ_5 transformation generated by $\exp[i\chi\alpha]$ is not a proper operation in the Hilbert space of real particles, but it carries a Hilbert space into another which is orthogonal to it.

Since χ is thus not diagonal, we shall, in the following, work with the expectation value $\langle\chi\rangle$, which should be conserved in any reaction, i.e.,

$$\langle\alpha^{\rm in}|\chi|\alpha^{\rm in}\rangle=\langle\alpha^{\rm out}|\chi|\alpha^{\rm out}\rangle.$$

For a one-nucleon state of momentum p, in particular, we have

$$\langle\chi\rangle=Z\bar{u}_p\gamma_4\gamma_5u_p=-Zhv_p, \tag{2.7}$$

where u_p is the Dirac spinor normalized to $\bar{u}u=m/E$, h the helicity $\langle\boldsymbol{\sigma}\cdot\mathbf{p}/|\mathbf{p}|\rangle$, v_p the velocity $p/(p^2+m^2)^{\frac{1}{2}}$, and Z is a renormalization constant. For convenience, we can replace χ by $Z^{-1}\chi$ and call it chirality. Then Z will drop out in Eq. (2.7). For a one-pion state, we have, naturally, $\langle\chi\rangle=0$.

Let us next consider the scattering of a nucleon by the external potentials. For a static potential, the velocity will not change, so that one might conclude that the helicity remains unchanged: $h_i=h_f$. This, however, is clearly not the case as one can easily check by perturbation calculation. The contradiction is resolved by noting that any scattering can always be accompanied by emission of pions which are massless. The final asymptotic state is then a linear combination.

$$|f\rangle=C_0|N\rangle+C_1|N+\pi\rangle+C_2|N+2\pi\rangle+\cdots, \tag{2.8}$$

and the expectation value $\langle\chi\rangle_f$ must be taken with respect to this complete amplitude. From Eq. (2.6) we further recognize that $\langle\chi\rangle_f$ will have contributions both diagonal and nondiagonal in the decomposition (2.8), the latter being between states $|N+n\pi\rangle$ and $|N+(n\pm1)\pi\rangle$ differing by one zero-energy meson (spurion). We are led to the conclusion that taking the expectation value is essential in interpreting the conservation of chirality. In this sense, we can call it a weak conservation in contrast to the usual "strong" conservation such as charge conservation.

It is clear, furthermore, that one can distinguish various degrees of "weakness" depending on the number of final state degrees of freedom over which the expectation value $\langle\chi\rangle_f$ must be taken in order to satisfy

[5] K. Nishijima, Nuovo cimento 11, 698 (1959).

$\langle\chi\rangle_f=\langle\chi\rangle_i$. The weakest case would involve a summation over all spins, scattering angles, and many-pion production processes. A less weak case might involve only a spin summation and up to one-spurion emission processes. In the following section we shall verify the chirality conservation by perturbation calculation and show that the stronger type of weak conservation (which we might call detailed weak conservation) can sometimes be valid.

III. PERTURBATION CALCULATION

We shall consider the scattering of a nucleon by an external potential in the Born approximation and check the chirality conservation in the lowest order in the meson coupling constant g. For simplicity, the potential will be assumed to be of the vector type

$$L_V=i\bar{\psi}\gamma_\mu\psi V_\mu,$$

where V_μ is static and has an inversion symmetry: $V(\mathbf{x})=V(-\mathbf{x})$. (Of course one can work with an axial vector potential equally well.) We first note that the initial nucleon state can be specified by means of the covariant projection operator

$$\begin{aligned} &P(p,n)=\Lambda(p)S(n)=S(n)\Lambda(p),\\ &\Lambda(p)=(m-i\gamma\cdot p)/2E_p,\quad [\Lambda^2=(m/E_p)\Lambda],\\ &S(n)=(1+i\gamma\cdot n\gamma_5)/2. \end{aligned}\tag{3.1}$$

Here n is the covariant polarization vector which reduces to $(\mathbf{n}',0)$ in the nucleon rest system. In the laboratory frame one has $n=(\mathbf{n},n_0)$ where

$$\mathbf{n}=\mathbf{n}'+\mathbf{p}(\mathbf{p}\cdot\mathbf{n}')/m(E+m),\quad n_0=\mathbf{p}\cdot\mathbf{n}'/m.\tag{3.2}$$

Naturally

$$n\cdot p=0,\quad n^2=1.\tag{3.3}$$

The initial chirality of the nucleon is then given by

$$\begin{aligned}\chi_i=\bar{u}_i\gamma_4\gamma_5u_i&=\mathrm{Tr}[\gamma_4\gamma_5P(p,n)]\\ &=n_0m/E_p=\mathbf{n}'\cdot\mathbf{v}.\end{aligned}\tag{3.4}$$

It is clear from Eq. (2.6) that in our present approximation (which is the zeroth order in g) only the lowest order elastic and one-pion inelastic processes need be considered. The corresponding diagrams are shown in Fig. 1. These amplitudes are given by

$$\begin{aligned} &M_{\mathrm{el}}(p',p)=i\gamma\cdot V(q),\quad q=p'-p\\ &M_{\mathrm{rad}}(p',p;k)=i\gamma\cdot V(q+k)S(p-k)ig\gamma_5\\ &\qquad\qquad+ig\gamma_5S(p'-k)i\gamma\cdot V(q-k). \end{aligned}\tag{3.5}$$

The chirality of the scattered wave at a given angle may be evaluated with the aid of the formula

$$\begin{aligned}\langle O\rangle_f&=[\bar{u}_i\bar{M}\Lambda(p')O\Lambda(p')Mu_i]/[\bar{u}_i\bar{M}\Lambda(p')Mu_i]\\ &=\mathrm{Tr}[\bar{M}\Lambda(p')OMP(p,n)]/\mathrm{Tr}[\bar{M}\Lambda(p')MP(p,n)],\end{aligned}$$

where O is a Dirac matrix and $\bar{M}=\gamma_4M^\dagger\gamma_4$. Adapting

FIG. 1. Lowest order diagrams to be considered for chirality conservation in potential scattering. The point **X** is where the potential acts.

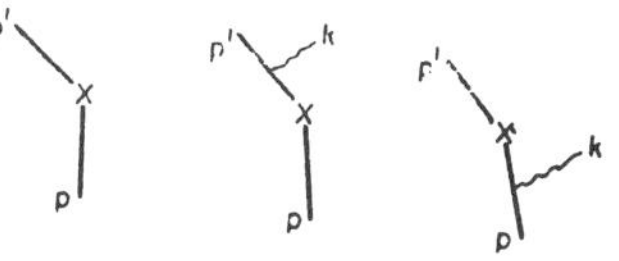

this to the present case, we get[6]

$$\begin{aligned}\langle\chi\rangle_f=\{&\mathrm{Tr}[\bar{M}_{\mathrm{el}}\Lambda(p')\chi_N'\Lambda(p')M_{\mathrm{el}}P(p,n)]\\ &+\mathrm{Tr}[\bar{M}_{\mathrm{el}}\Lambda(p')\chi_\pi'\Lambda(p')M_{\mathrm{rad}}P(p,n)]\\ &+\mathrm{Tr}[\bar{M}_{\mathrm{rad}}\Lambda(p')\chi_\pi'^\dagger\Lambda(p')M_{\mathrm{el}}P(p,n)]\}\\ &\qquad\times\{\mathrm{Tr}[\bar{M}_{\mathrm{el}}\Lambda(p')M_{\mathrm{el}}P(p,n)]\}^{-1},\end{aligned}$$

where

$$\begin{aligned}&\chi_N'=\gamma_4\gamma_5,\\ &\chi_\pi'=(2m/g)\gamma_4\left\langle 0\left|\int\dot{\phi}d^3x\,\phi(y)\right|0\right\rangle=-(im/g)\gamma_4.\end{aligned}\tag{3.6}$$

M_{rad} does not contribute to the denominator because of its vanishing weight. The elastic contribution can be easily evaluated. We find

$$\begin{aligned}&\mathrm{Tr}[\bar{M}_{\mathrm{el}}\Lambda(p')\chi_N'\Lambda(p')M_{\mathrm{el}}P(p,n)]\\ &\quad=(-im/2E^3)[V\cdot V(m^2+p\cdot p')-2p\cdot Vp'\cdot V]n_4\\ &\qquad+(im/E^3)[V\cdot n(m^2+p\cdot p')-p\cdot Vp'\cdot n]V_4,\\ &\mathrm{Tr}[\bar{M}_{\mathrm{el}}\Lambda(p')M_{\mathrm{el}}P(p,n)]\\ &\quad=(1/2E^2)[V\cdot V(m^2+p\cdot p')-2p\cdot Vp'\cdot V],\ V_\mu\equiv V_\mu(q),\end{aligned}$$

so that

$$\begin{aligned}\langle\chi_N\rangle_f=&(m/E)n_0\\ &-(2m/E)V_0\frac{V\cdot n(m^2+p\cdot p')-p\cdot Vp'\cdot n}{V\cdot V(m^2+p\cdot p')-2p\cdot Vp'\cdot V}.\end{aligned}\tag{3.7}$$

Here we have made use of the previous assumptions about V, which means that $E_f=E_i=E$, and the Fourier transform $V_\mu(\mathbf{q})$ is a real vector

$$\begin{aligned}&V_\mu(\mathbf{q})=V_\mu(-\mathbf{q})=\eta_\mu V_\mu^*(\mathbf{q}),\\ &\eta_\mu=1,\ \mu=1,2,3;\quad \eta_\mu=-1,\ \mu=4.\end{aligned}$$

For a general V_μ, the result is more complicated.

As for the inelastic contribution, we first note that although the matrix element of $\chi_\pi\propto\int\partial_4\phi d^3x$ vanishes like $k_0/(2k_0)^{\frac{1}{2}}$ as $k_0\to 0$, the inelastic amplitude contains a factor $1/(2k_0)^{\frac{1}{2}}$ which will cancel the former. Thus there is no trouble in this respect. A more serious difficulty is that the results of the calculation depend on the way the limit $k=0$ is defined. In fact we get different answers depending on whether (a) we approach the limit $k=0$ staying on the zero-mass shell ($k_0=|\mathbf{k}|$), or (b) first we allow a finite mass μ_0, and go to the limit $\mathbf{k}\to 0$, $k_0=\mu_0\to 0$ successively.

We shall show below that the second procedure gives the correct result. Note also that this is the more

---

[6] Note that χ_π acts as unity operator for the nucleon, which means $O=\gamma_4$. Hence the form of χ_π' below.

appropriate one for application to the real pion problem where μ_0 is indeed finite.

Accordingly we calculate the χ_π part thus:

$$\mathrm{Tr}[\bar{M}_{\mathrm{el}}\Lambda(p')\chi_\pi'\Lambda(p')M_{\mathrm{rad}}P(p,n)]$$

$$= \lim_{\mathbf{k}\to 0,\mu_0\to 0} m\left\{\int \mathrm{Tr}\left[\gamma\cdot V\frac{m-i\gamma\cdot p'}{2E}\gamma_5\frac{m-i\gamma\cdot(p'+k)}{(p'+k)^2+m^2}\gamma \cdot V\frac{m-i\gamma\cdot p}{2E}\frac{1+i\gamma\cdot n\gamma_5}{2}\right]k_0\delta(k^2+\mu_0^2)\theta(k_0)dk_0\right.$$

$$\left.+\int \mathrm{Tr}\left[\gamma\cdot V\frac{m-i\gamma\cdot p'}{2E}\gamma\cdot V\frac{m-i\gamma\cdot(p-k)}{(p-k)^2+m^2}\gamma_5\frac{m-i\gamma\cdot p}{2E}\times\frac{1+i\gamma\cdot n\gamma_5}{2}\right]k_0\delta(k^2+\mu_0^2)\theta(k_0)dk_0\right\},$$

which becomes

$$-(m/2E^3)[V\cdot n(m^2+p\cdot p')-p\cdot Vp'\cdot n]V_0.$$

The other interference term in Eq. (3.6) gives an equal contribution so that

$$\langle\chi_\pi\rangle_f=\frac{2m}{E}V_0\frac{V\cdot n(m^2+p\cdot p')-p\cdot Vp'\cdot n}{V\cdot V(m^2+p\cdot p')-2p\cdot Vp'\cdot V}. \quad (3.8)$$

Combining Eqs. (3.7) and (3.8) we get

$$\langle\chi\rangle_f=\langle\chi_N\rangle_f+\langle\chi_\pi\rangle_f=(m/E)n_0=\langle\chi\rangle_i. \quad (3.9)$$

It is interesting to see how the two contributions to $\langle\chi\rangle_f$ add up in the nonrelativistic approximation. The chirality of the nucleon is in this case

$$\langle\chi_N\rangle=-\langle\boldsymbol{\sigma}\rangle_i\cdot\mathbf{p}/m. \quad (3.10)$$

The scattering takes place through V_4 which does not flip the spin but changes the direction of motion so that, after the scattering, we have

$$\langle\chi_N\rangle_f=-\langle\boldsymbol{\sigma}\rangle_i\cdot\mathbf{p}'/m. \quad (3.11)$$

The difference is

$$\Delta\langle\chi_N\rangle=-\langle\boldsymbol{\sigma}\rangle_i\cdot\mathbf{q}/m. \quad (3.12)$$

On the other hand, the above-mentioned limiting procedure for $\langle\chi_\pi\rangle$ corresponds to the S-wave pion production process which goes through the negative energy states, and this is easily seen to cancel $\Delta\langle\chi_N\rangle$.

IV. GENERAL PROOF

The previous particular example shows that the weak conservation of chirality holds in detail, i.e., if summed only over the final nucleon spins, and over elastic and one-pion bremsstrahlung processes at a fixed scattering angle. This is a stronger result than the general statement $\langle\chi\rangle_i=\langle\chi\rangle_f$, and is due to the Born approximation and the special assumption made about V_μ. (In fact, for a completely general static potential V_μ, we do not obtain conservation unless we sum over all the scattered and unscattered waves.)

We expect that the chirality conservation in our sense will hold true to all orders in the pion coupling g, where many-pion processes will also come in. But the calculation is made difficult because of the self-energy effects. In the following, we will therefore try to derive a general relation which follows from the γ_5 invariance and certain other assumptions and is expressed in terms of directly observable (renormalized) quantities.

We assume that there exists a conserved quantity called chirality $\chi=-i\int\chi_4d^3x$ $(\partial_\mu\chi_\mu=0)$. This means that

$$\chi^{\mathrm{in}}=\chi^{\mathrm{out}}=S^{-1}\chi^{\mathrm{in}}S,$$

where S is the S matrix. We rewrite it as

$$S\chi^{\mathrm{in}}-\chi^{\mathrm{in}}S=0. \quad (4.1)$$

We will further assume, in accordance with the previous discussion, that χ consists of the nucleon part χ_N and the pion part $\chi_\pi=\lambda\int\dot{\phi}d^3x$. Asymptotically, χ_N will be an operator that does not change the number of pions, whereas χ_π, being linear in the pion field, will lead to the absorption or emission of a zero-energy pion. Accordingly Eq. (4.1) becomes

$$S\chi_N^{\mathrm{in}}-\chi_N^{\mathrm{in}}S=-S\chi_\pi^{\mathrm{in}}+\chi_\pi^{\mathrm{in}}S.$$

Let us apply this to the case of the potential scattering. Taking the matrix element between elastic states, we get

$$\begin{aligned} i\langle p'|M_{\mathrm{el}}\chi_N^{\mathrm{in}}-\chi_N^{\mathrm{in}}M_{\mathrm{el}}|p\rangle &=-\langle p'|S\chi_\pi^{\mathrm{in}}-\chi_\pi^{\mathrm{in}}S|p\rangle\\ &=\textstyle\sum_k[-\langle p'|S|pk\rangle\langle k|\chi_\pi|0\rangle\\ &\quad+\langle 0|\chi_\pi|k\rangle\langle p'k|S|p\rangle], \end{aligned} \quad (4.2)$$

because χ_π^{in} results only in the creation or absorption of a pion. Furthermore,

$$\begin{aligned}\langle p'|S|p,0\rangle=\langle p',0|S|p\rangle&=[i/(2k_0)^{\frac{1}{2}}]\langle p'|k^2\phi(k)|p\rangle|_{k=0}\\ &\equiv[i/(2k_0)^{\frac{1}{2}}]M_{\mathrm{rad}}(p',p;k)|_{k=0},\end{aligned}$$

and

$$\langle k|\chi_\pi|0\rangle=-\langle 0|\chi_\pi|k\rangle=\lambda i(k_0/2)^{\frac{1}{2}}\delta(\mathbf{k}). \quad (4.3)$$

Equation (4.2) thus may be written

$$\chi_NM_{\mathrm{el}}(p',p)-M_{\mathrm{el}}(p',p)\chi_N=i\lambda M_{\mathrm{rad}}(p',p;0). \quad (4.4)$$

If χ is so normalized that $\chi_N=\gamma_4\gamma_5$ for a free nucleon, then the continuity equation

$$\partial_\mu\chi_{N\mu}-\lambda\Box\phi=0,$$

which follows from χ conservation, means that $1/\lambda$ is more or less the conventional pion coupling constant

$$1/\lambda=f=g/2m. \quad (4.5)$$

It is not proven, however, that this agrees with the coupling constant defined in the dispersion theory. For the time being, we assume it to be the case.

Equation (4.4) represents a general relation between M_{el} and the one-pion emission amplitude M_{rad}. The origin of such a simple relation is easily traced back to our starting assumption about the asymptotic behavior of χ. From the above derivation, it is also clear that a similar relation will exist for any reaction amplitude M and the accompanying inelastic amplitude resulting in the emission of an extra zero-energy meson.

We shall next bring the relation (4.4) into a more explicit form. In the covariant form, M is usually defined without the projection operator Λ, so that the left-hand side of Eq. (4.4) should be written

$$\gamma_4\gamma_5\Lambda(p')M_{\mathrm{el}}(p',p)-M_{\mathrm{el}}(p',p)\Lambda(p)\gamma_4\gamma_5.$$

Observing that

$$\gamma_4\gamma_5\Lambda(p)-\Lambda(p)\gamma_4\gamma_5=-\gamma_5,$$

Eq. (4.4) becomes

$$iM_{\mathrm{rad}}(p',p;0)=(g/2m)[(\gamma_4\gamma_5m/E-\gamma_5)M_{\mathrm{el}}(p',p) +M_{\mathrm{el}}(p',p)(\gamma_5\gamma_4m/E-\gamma_5)]. \quad (4.6)$$

If we use the positive energy two-component spinors, the above equation simplifies to

$$iM_{\mathrm{rad}}(p',p;0) =-\frac{g}{2m}\left[\frac{\boldsymbol{\sigma}\cdot\mathbf{p}'}{E}M_{\mathrm{el}}(p',p)-M_{\mathrm{el}}(p',p)\frac{\boldsymbol{\sigma}\cdot\mathbf{p}}{E}\right]. \quad (4.7)$$

In this form the meaning of the relation is easy to understand. Suppose the initial nucleon is in an eigenstate of helicity: $(\sigma\cdot p/E)u_p=\pm vu_p$. The helicity change in the elastically scattered state p' is

$$\left(\bar{u}_pM_{\mathrm{el}}^\dagger\frac{\boldsymbol{\sigma}\cdot\mathbf{p}'}{E}M_{\mathrm{el}}u_p\right)\Big/(\bar{u}_pM_{\mathrm{el}}^\dagger M_{\mathrm{el}}u_p)-\left(\bar{u}_p\frac{\boldsymbol{\sigma}\cdot\mathbf{p}}{E}u_p\right)$$

$$=\left(\bar{u}_pM_{\mathrm{el}}^\dagger\left[\frac{\boldsymbol{\sigma}\cdot\mathbf{p}'}{E}M_{\mathrm{el}}-M_{\mathrm{el}}\frac{\boldsymbol{\sigma}\cdot\mathbf{p}}{E}\right]u_p\right)\Big/(\bar{u}_pM_{\mathrm{el}}^\dagger M_{\mathrm{el}}u_p)$$

$$=-i\lambda(\bar{u}_pM_{\mathrm{el}}^\dagger M_{\mathrm{rad}}u_p)/(\bar{u}_pM_{\mathrm{el}}^\dagger M_{\mathrm{el}}u_p).$$

The last expression just corresponds to the interference term $\langle\chi_\pi\rangle$ in the final state with a particular momentum p'. This shows that the weak chirality conservation holds *in detail*, i.e., after summing only over the final nucleon spins, and over elastic and associated inelastic amplitudes *provided that* the initial nucleon is in a helicity eigenstate. Otherwise a more general weak conservation will prevail in general.

V. $\gamma_5\times$ ISOSPIN INVARIANCE AND THE EXPERIMENTAL TEST

The results obtained in previous sections can be generalized to the $\gamma_5\times$ isospin gauge transformation [Eq. (2.1)]. Models that possess this invariance have been considered by Gürsey, Gell-Mann, *et al.*[2] and Nambu and Jona-Lasinio.[4] In this case there are three conserved quantities χ^i, $i=1, 2, 3$, corresponding to the three components of the isospin. The simplest example having such conservation is the conventional meson theory with derivative coupling. The results of Sec. IV can be directly taken over if we replace χ_N by $\chi_N{}^i=\chi_N\tau^i$, and χ_π by $\chi_\pi{}^i=\lambda\int\dot{\phi}^id^3x$.

In some of the models, the expression for χ is actually more complicated. In general it contains nonlinear terms in the pion field, and terms involving a neutral scalar meson field can also occur. We may assume, however, that these additional terms do not contribute to the asymptotic values $\langle\chi^{\mathrm{in}}\rangle$ and $\langle\chi^{\mathrm{out}}\rangle$ for the following reason. First, the contribution from non-linear terms $\sim\phi(x)^n$, $n>2$, which involves creation and annihilation of n mesons in the neighborhood of point x, will tend to zero when, before or after the scattering, the particles are well separated from each other, just as the interaction energy is supposed to vanish in this asymptotic region. As for the neutral scalar field, a meson of this kind (σ meson) may exist in nature, but it would be, in any case, quite massive and unstable. We can therefore exclude σ from the fields that contribute to χ^{in} and χ^{out}.

With this observation, Eqs. (4.6) and (4.7) are replaced in the present case by

$$iM_{\mathrm{rad}}{}^i=(g/2m)[(\gamma_4\gamma_5m/E-\gamma_5)\tau^iM_{\mathrm{el}} +M_{\mathrm{el}}\tau^i(\gamma_5\gamma_4m/E-\gamma_5)], \quad (5.1)$$

$$iM_{\mathrm{rad}}{}^i=-(g/2m)[(\boldsymbol{\sigma}\cdot\mathbf{p}'/E)\tau^iM_{\mathrm{el}}-M_{\mathrm{el}}\tau^i(\boldsymbol{\sigma}\cdot\mathbf{p}/E)]. \quad (5.2)$$

It is convenient to quantize the spins of the initial and final nucleons along their own direction of motion. Equation (5.2) then reduces simply to

$$iM_{\mathrm{rad}}{}^{h'h,i}=-(gv_p/2m)[h'\tau^iM_{\mathrm{el}}{}^{h'h}-M_{\mathrm{el}}{}^{h'h}\tau^ih], \quad (5.3)$$

where h and h' are the initial and final helicities, and $M^{h'h}$ are the corresponding amplitudes introduced by Jacob and Wick.[7]

We shall apply the above relation to the pion nucleon scattering, which seems to be the simplest case of physical interest. Naturally M_{el} has to be identified with the elastic scattering amplitude $M(p'q',pq)$ for the process $N_p+\pi_q\to N_{p'}+\pi_{q'}$, whereas $M_{\mathrm{rad}}{}^i(p'q',pq;k)$ corresponds to the production process $N_p+\pi_q\to N_{p'}+\pi_{q'}+\pi_k{}^i$. The fundamental assumption here is that the pion-nucleon system is γ_5 invariant to the extent that the pion mass can be neglected.

To exhibit the isotopic dependence of M_{el}, we write

$$iM_{\mathrm{el}\beta\alpha}=M^{(+)}\delta_{\beta\alpha}+M^{(-)}\tfrac{1}{2}[\tau_\beta,\tau_\alpha], \quad (5.4)$$

so that $M_{\mathrm{rad}}{}^{i=3}$ for the process $\pi^\pm+p\to\pi^\pm+p+\pi^0$ takes the form

$$iM_{\mathrm{rad}}{}^{h'h}(\pi^0)=-(gv_p/2m)(h'-h)M^{h'h}(\pi^\pm p\to\pi^\pm p) =-(gv_p/2m)(h'-h) \times[M^{(+)h'h}\mp M^{(-)h'h}]. \quad (5.5)$$

[7] M. Jacob and G. C. Wick, Ann. Phys. 7, 404 (1959).

The relation becomes somewhat more complicated when $\pi^\pm$ are produced. For $\pi^-+p\to\pi^-+n+\pi^+$, we have, for example

$$iM_{\rm rad}{}^{h'h}(\pi^+)=-\frac{\sqrt{2}gv_p}{2m}[h'M^{h'h}(\pi^-p\to\pi^-p)-hM^{h'h}(\pi^-n\to\pi^-n)]$$
$$=-\frac{\sqrt{2}gv_p}{2m}[h'(M^{(+)h'h}+M^{(-)h'h})-h(M^{(+)h'h}-M^{(-)h'h})]. \quad (5.6)$$

Let us now compare the elastic and inelastic cross sections. The simplest way to apply our formula will be to take the case where the meson k is produced with small energy ($|\mathbf{k}|\approx 0$). We easily find

$$\left(\frac{d^2\sigma_{\rm rad}}{d\omega_k d\Omega_{q'}}\right)\bigg/\left(\frac{d\sigma_{\rm el}}{d\Omega_{q'}}\right)=\frac{|\mathbf{k}|}{(2\pi)^2}|\bar{u}_{p'}M_{\rm rad}u_p|^2/|\bar{u}_{p'}M_{\rm el}u_p|^2. \quad (5.7)$$

For the process $\pi^\pm+p\to\pi^\pm+p+\pi^0$ [Eq. (5.5)] this reduces to

$$\left(\frac{d^2\sigma_{\rm rad}}{d\omega_k d\Omega_{q'}}\right)\bigg/\left(\frac{d\sigma_{\rm el}}{d\Omega_{q'}}\right)=\frac{1}{\pi}\frac{g^2}{4\pi}\frac{|\mathbf{k}|}{4m^2}v_p{}^2(h'-h)^2. \quad (5.8)$$

Averaging over the nucleon spin, we get

$$\langle(h'-h)^2\rangle_{\rm av}=2[1-A'(\Omega_{q'})], \quad (5.9)$$

where $A'\equiv\langle h'h\rangle_{\rm av}$ is one of the polarization parameters introduced by Wolfenstein.[8] Equations (5.7) and (5.8) are the relations that should hold at each scattering angle. If we integrate Eq. (5.8) over the angles, we get

$$\left(\frac{d^2\sigma_{\rm rad}}{d\omega_k}\right)\bigg/\sigma_{\rm el}=\frac{2}{\pi}\frac{g^2}{4\pi}\frac{|\mathbf{k}|}{4m^2}v_p{}^2(1-\bar{A}'), \quad (5.10)$$

where $\bar{A}'$ means an angular average.

In the actual case of finite meson mass, the low-energy limit $\mathbf{k}=0$ does not have a Lorentz invariant meaning, so that the analysis will depend on the choice of the coordinate system. For example, a meson at rest in the c.m. system will have a momentum $|\mathbf{k}_{\rm L}|=v_{\rm L}\mu/(1-v_{\rm L}{}^2)^{\frac{1}{2}}$ in the laboratory system. In order to keep such an ambiguity reasonably small, e.g., $|\mathbf{k}_{\rm L}|\lesssim\mu$, we must demand that $1/(1-v_{\rm L}{}^2)^{\frac{1}{2}}$ is not too large. In other words, the energy at which we carry out the experiment should not be large compared to the nucleon rest energy.

An alternative way to test our relation is to consider the energy distribution of the inelastically scattered meson near its maximum energy since the produced pion would come out with low energy. In this case, the ratio of cross sections is calculated to be

$$\left(\frac{d^2\sigma_{\rm rad}}{d\omega_{q'}d\Omega_{q'}}\right)\bigg/\left(\frac{d\sigma_{\rm el}}{d\Omega_{q'}}\right)=\frac{1}{(2\pi)^2}\left(\frac{E}{m}\right)^{\frac{1}{2}}[2\mu(\omega_{\rm m}-\omega_{q'})]^{\frac{1}{2}}\times|\bar{u}_{p'}M_{\rm rad}u_p|^2/|\bar{u}_{p'}M_{\rm el}u_p|^2, \quad (5.11)$$

where $\omega_{\rm m}$ is the maximum meson energy at the particular angle, and E is the total c.m. energy of the system. Again for the process $\pi^\pm+p\to\pi^\pm+p+\pi^0$, this yields

$$\left(\frac{d^2\sigma_{\rm rad}}{d\omega_{q'}d\Omega_{q'}}\right)\bigg/\left(\frac{d\sigma_{\rm el}}{d\Omega_{q'}}\right)=\frac{1}{\pi}\frac{g^2}{4\pi}\left(\frac{E}{m}\right)^{\frac{1}{2}}\frac{[2\mu(\omega_{\rm m}-\omega_{q'})]^{\frac{1}{2}}}{4m^2}v_p{}^2(h'-h)^2. \quad (5.12)$$

Equations (5.8) and (5.11) or the specific forms (5.8) and (5.12) give a relation between elastic and inelastic cross sections in terms of directly measurable quantities alone. When the polarization of the nucleon is not measured, they still can give an inequality since

$$(h'-h)^2\leq 4.$$

Unfortunately, these formulas are supposed to apply only at the extreme ends of the meson energy spectrum, for which experimental data are scarce and difficult to obtain.

On the other hand, if the elastic matrix elements are precisely known, we can directly calculate the inelastic amplitudes from Eq. (5.3) [or Eqs. (5.5) and (5.6)]. When, for example, the elastic scattering is dominated

TABLE I. Angular dependence and magnitude of the radiative cross section $\pi+N\to\pi+N+\pi'$, where the last pion is produced nearly at rest, for the case of $T=\frac{3}{2}$, $J=\frac{3}{2}^\pm$ elastic channel, $Z=\cos\theta_\pi$.

Process	$\dfrac{d^2\sigma_{\rm rad}}{d\Omega_{q'}d\omega_k}$	$\left(\dfrac{d\sigma_{\rm rad}}{d\omega_k}\right)\bigg/\left(\dfrac{g^2}{4\pi^2}\dfrac{v_p{}^2}{4m^2}k\sigma_{\rm el}\right)$
$\pi^\pm+p\to\pi^\pm+p+\pi^0$	$(1+3Z)^2(1-Z)$	2
$\pi^++p\to\pi^++n+\pi^+$ $\pi^-+p\to\pi^-+n+\pi^+$	$1+3Z+3Z^2-(27/5)Z^3$	20/9 20

[8] L. Wolfenstein, Phys. Rev. 96, 1654 (1954).

TABLE II. Angular dependence and magnitude of the radiative cross section for the case of $T=\frac{1}{2}$, $J=\frac{3}{2}^{\pm}$ and $\frac{5}{2}^{\pm}$ channels.

Process	$\dfrac{d^2\sigma_{\rm rad}}{d\Omega_q \cdot d\omega_k}$	$\left(\dfrac{d\sigma_{\rm rad}}{d\omega_k}\right) \Big/ \left(\dfrac{g^2}{4\pi^2}\dfrac{v_p^2}{4m^2}k\sigma_{\rm el}\right)$	
$\pi^- + p \rightarrow \pi^- + p + \pi^0$	$(1+3Z)^2(1-Z)$	2	$J=\frac{3}{2}$
	$(1-2Z-5Z^2)^2(1-Z)$		$J=\frac{5}{2}$
$\pi^- + p \rightarrow \pi^- + n + \pi^+$	$1+3Z^2$	2	$J=\frac{3}{2}$
	$1-2Z^2+5Z^4$		$J=\frac{5}{2}$

by a resonance, the associated production amplitude is easily obtained.

In Table I we give the results for the $\frac{3}{2}-\frac{3}{2}$ resonance. This is not an ideal case because the resonance energy is only twice the pion mass, which makes the neglecting of the pion mass somewhat dubious. But perhaps one could argue that we should compare the elastic scattering at energy T with the inelastic scattering at incident energy $T+\mu$. For the 3–3 resonance, the latter comes out to be ≈ 300 Mev.

In Fig. 2 we show the measured angular distribution of the inelastically scattered π^- from the reaction[9] $\pi^- + p \rightarrow \pi^- + n + \pi^+$ at 290 Mev laboratory energy which is to be compared with the theoretical curve. The agreement is quite reasonable. We note here that, although the experimental points are integral distributions over pion energy, the $\pi^- - n$ resonance in the final system would tend to produce low energy π^+ as is wanted in the comparison with theory.[9a]

As for the magnitude of the cross section, we can make only a crude comparison. We estimate the theoretical production cross section $\sigma(\pi^- p \rightarrow \pi^- n \pi^+)$ by extrapolating Eq. (5.7) over the entire energy range. This gives the ratio (with $g^2/4\pi \approx 15$)

$$r = \frac{\sigma(\pi^- p \rightarrow \pi^- n \pi^+)}{\sigma(\pi^- p \rightarrow \pi^- p)} \approx 1/100$$

at 290 Mev. Since the theoretical peak elastic cross section due to the 3–3 resonance is 20 mb, the above ratio means

$$\sigma(\pi^- p \rightarrow \pi^- n \pi^+) \approx 0.2 \text{ mb}.$$

The corresponding experimental value is 0.61 ± 0.13 mb at[9] 290 Mev and 0.71 ± 0.10 mb at 317 Mev.[10] These are quite compatible with the prediction considering the crudeness of the estimation. (We also note from Table I that the ratio r for the other production processes would be only 1/10 of the present case. In such an event there may be more contaminations from other partial waves.)

Similar considerations can be made for the higher resonances. We list in Table II the corresponding predictions for resonances with $T=\frac{1}{2}$, $J=\frac{3}{2}^{\pm}$ and $\frac{5}{2}^{\pm}$. Finally it is interesting to compare the preceding results with those of the statistical model. We may identify the Fermi interaction volume Ω with the ratio[11]

$$R = \frac{1}{2\mu}\frac{|M_{\rm rad}|^2}{|M_{\rm el}|^2} \approx \frac{g^2}{4\mu m^2}v_p^2 \approx \frac{v_p^2}{\mu^3},$$

as far as the soft meson emission is concerned. The corresponding interaction radius R is then

$$R \approx 0.6 v_p^{\frac{2}{3}} \mu^{-1},$$

which is energy dependent, and approaches $0.6\mu^{-1}$ at high energies.

VI. FURTHER REMARKS

The main theme of this paper is to show that, under the assumption of γ_5 invariance in strong interactions, the change of nucleon helicity in any reaction will result in the bremsstrahlung of soft pions. This relation is characterized specifically by Eq. (5.1) or (5.2), which expresses a kind of low energy limit theorem. A comparison with the pion-nucleon scattering data around the 3–3 resonance shows an agreement with the formula.

Other interesting applications or tests of our relations may be found in nucleon-nucleon and nucleon-antinucleon scattering and general high energy multiple production processes. Our formulas would give us a way to analyze these events by observing low-energy

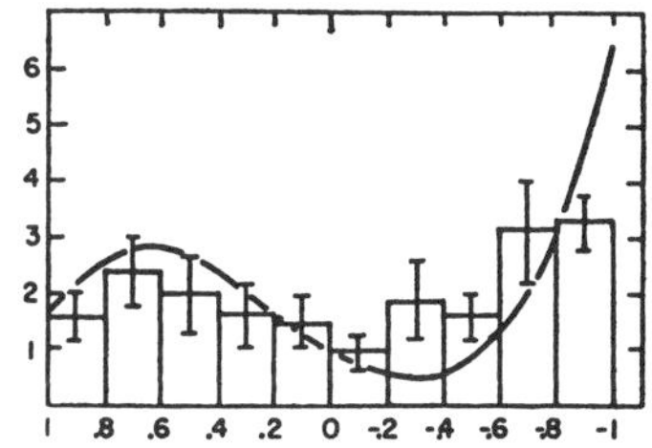

FIG. 2. Angular distribution of π^- from the reaction $\pi^- + p \rightarrow n + \pi^+ + \pi^-$ at 290 Mev.[9] The curve is calculated from the last line in Table I.

[9] Ya. A. Batusov, S. A. Bunyatov, V. M. Sidorov, and V. A. Yarba, *Proceedings of the 1960 International Conference on High-Energy Physics at Rochester* (Interscience Publishers, New York, 1960), p. 77; Ya. A. Batusov *et al.*, Doklady Nauk U.S.S.R., **133**, 52 (1960) [Soviet Phys.-Doklady **5**, 731 (1961)].

[9a] *Note added in proof.* However, the agreement may be fortuitous since there are various angular momentum channels that are neglected in this simplified approach. Our formula shows a characteristic dip in the forward direction. But the corresponding behavior of the experimental data may or may not be real.

[10] W. A. Perkins, III, J. C. Caris, R. W. Kenney, and V. Perez-Mendez, Phys. Rev. **118**, 1364 (1960).

[11] E. Fermi, Progr. Theoret. Phys. (Kyoto) **5**, 570 (1950).

mesons emitted. For example, in the case of nucleon-antinucleon annihilation at rest, the initial chirality is zero, and, since the final state contains no nucleons, we would have $\langle \chi_\pi \rangle_f = 0$, i.e., the amplitude for the emission of very low energy pions would be unusually small.

The notion of γ_5 invariance or chirality conservation can be extended to composite systems and strange particles.[4] There is also a possibility that the K meson plays a role similar to the pion in the conservation of strangeness-changing chirality current. It is likely, however, that even if such a symmetry existed in essence, the large mass of the K meson would tend to make it more approximate in nature than for the case involving pions, except perhaps at sufficiently high energies.

ACKNOWLEDGMENTS

One of the authors (Y.N.) thanks Dr. J. W. Calkin for his hospitality at the Brookhaven National Laboratory where part of the work was done.

Reprinted from THE PHYSICAL REVIEW, Vol. 139, No. 4B, B1006–B1010, 23 August 1965

Three-Triplet Model with Double $SU(3)$ Symmetry*

M. Y. HAN
Department of Physics, Syracuse University, Syracuse, New York

AND

Y. NAMBU
The Enrico Fermi Institute for Nuclear Studies, and the Department of Physics, The University of Chicago, Chicago, Illinois

(Received 12 April 1965)

With a view to avoiding some of the kinematical and dynamical difficulties involved in the single-triplet quark model, a model for the low-lying baryons and mesons based on three triplets with integral charges is proposed, somewhat similar to the two-triplet model introduced earlier by one of us (Y. N.). It is shown that in a $U(3)$ scheme of triplets with integral charges, one is naturally led to three triplets located symmetrically about the origin of I_3-Y diagram under the constraint that the Nishijima–Gell-Mann relation remains intact. A double $SU(3)$ symmetry scheme is proposed in which the large mass splittings between different representations are ascribed to one of the $SU(3)$, while the other $SU(3)$ is the usual one for the mass splittings within a representation of the first $SU(3)$.

I. INTRODUCTION

ALTHOUGH the $SU(6)$ symmetry strongly indicates that the baryon is essentially a three-body system built from some basic triplet field or fields, the quark model[1] is not entirely satisfactory from a realistic point of view, because (a) the electric charges are not integral, (b) three quarks in s states do not form the symmetric $SU(6)$ representation assigned to the baryons, and (c) a simple dynamical mechanism is lacking for realizing only zero-triality states as the low-lying levels.

These difficulties may be avoided if we introduce more than one basic triplet. Recently one of us (Y. N.) has attempted a two-triplet model[2] where the members of the triplets t_1 and t_2 had the charge assignment (1,0,0) and (0, -1, -1), as had been proposed earlier by Bacry *et al.*[3] The baryon would be represented by the combination $t_1t_1t_2$, whereas the mesons would correspond to some combination $\sim at_1\bar{t}_1'+bt_2\bar{t}_2'$. The triplets are assumed to have masses large compared to the baryon mass, which would mean that baryons and mesons have very large binding energies. A dynamical mechanism for this is provided by a neutral field coupled strongly to the "charm number"[4] C, which is 1 for t_1 and -2 for t_2, and therefore $C=0$ for baryons and mesons. In analogy with electrostatic energy, we can argue that the potential energy due to the charm field would be lowest when the system is "neutral," namely, $C=0$. Thus all other unwanted configurations with $C\neq 0$, which include among others triplet, sextet, etc. representations, would have high masses, and hence would not be easily observed.

There have been proposed two different ways in which to introduce basic triplet or triplets with integral charges. One approach essentially involves a modification of the Nishijima–Gell-Mann relation by way of introducing an additional quantum number, the triality quantum number,[5] and this has led to considerations of higher symmetry schemes based on rank-three Lie groups.[6] On the other hand, Okubo *et al.*[7] have recently shown that the minimal group required for this purpose is actually the group $U(3)$.[8] It is shown that a triplet scheme may be defined in $U(3)$ such that the triplet always possesses integral values of charge and hypercharge and satisfies the Nishijima–Gell-Mann relation without a modification. The $U(3)$ triplet considered by Okubo *et al.* is of Sakata type; i.e., it consists of an isodoublet and an isosinglet. Actually, the $U(3)$ scheme is much more appealing than those of the rank-three Lie groups on two accounts: firstly, the Nishijima–Gell-Mann relation is satisfied universally by triplets as by octets and decuplets, and secondly as far as the hitherto realized representations are concerned, $U(3)$ is equivalent to $SU(3)$.[9]

In what follows, we show that the $U(3)$ scheme, when fully utilized as described below, naturally and uniquely

* Work supported in part by the U. S. Atomic Energy Commission under the Contract No. AT(30-1)-3399 and No. AT(11-1)-264.

[1] M. Gell-Mann, Phys. Letters **8**, 214 (1964); G. Zweig, CERN (to be published).

[2] Y. Nambu, *Proceedings of the Second Coral Gables Conference on Symmetry Principles at High Energy* (W. H. Freeman and Company, San Francisco, 1965).

[3] H. Bacry, J. Nuyts, and L. van Hove, Phys. Letters **9**, 279 (1964).

[4] This name was originally used in connection with the $SU(4)$ symmetry. B. J. Bjørken and S. L. Glashow, Phys. Letters **11**, 255 (1964); A. Salam, Dubna Conference Report, 1964 (unpublished).

[5] G. E. Baird and L. C. Biedenharn, *Proceedings of the First Coral Gables Conference on Symmetry Principles at High Energy* (W. H. Freeman and Company, San Francisco, 1964); C. R. Hagen and A. J. Macfarlane, Phys. Rev. **135**, B432 (1964) and J. Math. Phys. **5**, 1335 (1964).

[6] For example, see I. S. Gerstein and M. L. Whippmann, Phys. Rev. **137**, B1522 (1965). Earlier references are given in this paper.

[7] S. Okubo, C. Ryan, and R. E. Marshak, Nuovo Cimento **34**, 759 (1964).

[8] The use of $U(3)$ in this connection has also been remarked by I. S. Gerstein and K. T. Mahanthappa, Phys. Rev. Letters **12**, 570, 656(E) (1964).

[9] S. Okubo, Phys. Letters **4**, 14 (1963).

leads to a set of three basic triplets with integral charges, namely an I-triplet (isodoublet and isosinglet), a U-triplet (U-spin doublet and U-spin singlet) and a V-triplet (V-spin doublet and V-spin singlet).[10] These triplets arise from three different ways of defining charge Q, hypercharge Y, and a displaced isospin I_3 in the $U(3)$ group as opposed to the $SU(3)$, in such a way that the charge and hypercharge have integral values, while keeping the Nishijima–Gell-Mann relation intact, and they differ from each other in their quantum-number assignments as well as in their transformation properties under the Weyl reflections.[11] This is described in Sec. II. In Sec. III, a double $SU(3)$ symmetry scheme is proposed based on the three-triplet model in which the large mass splittings between different representations are ascribed to one of the $SU(3)$, and the other $SU(3)$ is, as usual, responsible for the mass splittings within a representation. The low-lying baryon and meson states may be taken as singlets with respect to one of the $SU(3)$. The extended symmetry group with respect to the $SU(6)$ symmetry is briefly discussed.

II. THREE TRIPLETS

We shall denote the infinitesimal generators of $U(3)$ by $A_\nu{}^\mu$ which satisfies the following commutation relations:

$$[A_\beta{}^\alpha, A_\nu{}^\mu] = \delta_\beta{}^\mu A_\nu{}^\alpha - \delta_\nu{}^\alpha A_\beta{}^\mu, \tag{1}$$

where all indices take on the values 1, 2, and 3. The corresponding infinitesimal generators $B_\nu{}^\mu$ of $SU(3)$ are then given by

$$B_\nu{}^\mu = A_\nu{}^\mu - \tfrac{1}{3}\delta_\nu{}^\mu A_\lambda{}^\lambda \tag{2}$$

which satisfy the following equations:

$$[B_\beta{}^\alpha, B_\nu{}^\mu] = \delta_\beta{}^\mu B_\nu{}^\alpha - \delta_\nu{}^\alpha B_\beta{}^\mu \tag{3}$$

and

$$B_\lambda{}^\lambda = 0. \tag{4}$$

Furthermore, the unitary restriction gives

$$(A_\nu{}^\mu)^\dagger = A_\mu{}^\nu, \quad (B_\nu{}^\mu)^\dagger = B_\mu{}^\nu. \tag{5}$$

Let us now briefly summarize the relevant results of Okubo *et al.* In the $SU(3)$ scheme, the charge Q, the hypercharge Y and the third component of isospin I_3 are identified as follows[12]:

$$Q = -B_1{}^1, \tag{6a}$$

$$Y = B_3{}^3 = -B_1{}^1 - B_2{}^2 \quad [\text{by the relation (4)}], \tag{6b}$$

$$I_3 = \tfrac{1}{2}(B_2{}^2 - B_1{}^1). \tag{6c}$$

In the $U(3)$ scheme, the corresponding quantities $\tilde{Q}$, $\tilde{Y}$, and $\tilde{I}_3$ are defined as follows:

$$\tilde{Q} = -A_1{}^1 = Q - \tfrac{1}{3}\tau, \tag{7a}$$

$$\tilde{Y} = -A_1{}^1 - A_2{}^2 = Y - \tfrac{2}{3}\tau, \tag{7b}$$

$$\tilde{I}_3 = \tfrac{1}{2}(A_2{}^2 - A_1{}^1) = I_3, \tag{7c}$$

where

$$\tau = A_1{}^1 + A_2{}^2 + A_3{}^3. \tag{8}$$

With these definitions, the Nishijima–Gell-Mann relation is seen to be equally satisfied by the $U(3)$ and $SU(3)$ theories, i.e.,

$$Q = I_3 + \tfrac{1}{2}Y \tag{9}$$

and

$$\tilde{Q} = \tilde{I}_3 + \tfrac{1}{2}\tilde{Y}, \tag{10}$$

respectively. Since the generators $A_1{}^1$, $A_2{}^2$, and $A_3{}^3$ possess integral eigenvalues in any representation,[13] the identifications of $\tilde{Q}$ and $\tilde{Y}$ to be the charge and the hypercharge, respectively, in $U(3)$ theory shall always lead to integral values for the charge and the hypercharge. In particular, in the three-dimensional representation, the $U(3)$ triplet has the eigenvalues

$$\tilde{Q} = \begin{bmatrix} 1 & 0 & 0 \\ 0 & 0 & 0 \\ 0 & 0 & 0 \end{bmatrix}, \quad \tilde{I}_3 = \begin{bmatrix} \frac{1}{2} & 0 & 0 \\ 0 & -\frac{1}{2} & 0 \\ 0 & 0 & 0 \end{bmatrix}, \quad \tilde{Y} = \begin{bmatrix} 1 & 0 & 0 \\ 0 & 1 & 0 \\ 0 & 0 & 0 \end{bmatrix}. \tag{11}$$

This triplet corresponds to the Sakata triplet which we call an I triplet for short.

We can now generalize the above constructions of the $U(3)$ triplet in the following way. Comparing (6b) and (7b), we see that a particular choice has been made for $\tilde{Y}$. Had we defined $\tilde{Y}$ to be $A_3{}^3$, it would still have integral eigenvalues but the relation (10) would have been violated. This is because $B_\lambda{}^\lambda = 0$ in $SU(3)$ but $A_\lambda{}^\lambda \neq 0$ in general in $U(3)$ and thus some care is needed in defining corresponding quantities in $U(3)$. Making use of (4), the definition in (6) can be written more generally as

$$Q = -B_1{}^1 = B_2{}^2 + B_3{}^3, \tag{12a}$$

$$Y = B_3{}^3 = -B_1{}^1 - B_2{}^2, \tag{12b}$$

$$I_3 = \tfrac{1}{2}(B_2{}^2 - B_1{}^1) = \tfrac{1}{2}(2B_2{}^2 + B_3{}^3) = -\tfrac{1}{2}(2B_1{}^1 + B_3{}^3). \tag{12c}$$

As in (7), replacing $B_\nu{}^\mu$'s in (12) by corresponding $A_\nu{}^\mu$'s, we list all possible candidates for the corresponding quantities in $U(3)$ which are now however not equivalent to each other [they are equivalent, of course, when reduced to $SU(3)$], i.e.,

$$\tilde{Q}: \quad -A_1{}^1, \quad A_2{}^2 + A_3{}^3, \tag{13a}$$

$$\tilde{Y}: \quad A_3{}^3, \quad -A_1{}^1 - A_2{}^2, \tag{13b}$$

$$\tilde{I}_3: \quad \tfrac{1}{2}(A_2{}^2 - A_1{}^1), \quad \tfrac{1}{2}(2A_2{}^2 + A_3{}^3), \quad -\tfrac{1}{2}(2A_1{}^1 + A_3{}^3). \tag{13c}$$

---

[10] C. A. Levinson, H. J. Lipkin, and S. Meshkov, Nuovo Cimento **23**, 236 (1961); Phys. Letters **1**, 44 (1962) and Phys. Rev. Letters **10**, 361 (1963).

[11] A. J. Macfarlane, E. C. G. Sudarshan, and C. Dullemond, Nuovo Cimento **30**, 845 (1963).

[12] We use the sign convention of S. P. Rosen, J. Math. Phys. **5**, 289 (1964).

[13] For a derivation of this result, see Eq. (7) of Ref. 7.

To start with, the alternative choices in (13) provide twelve inequivalent ways in which to choose a set of three quantities $\tilde{Q}$, $\tilde{Y}$ and $\tilde{I}_3$ for the $U(3)$ scheme. In every choice $\tilde{Q}$ and $\tilde{Y}$ will have integral eigenvalues, but as can be easily checked the Nishijima–Gell-Mann relation will not be valid for all of them. In fact, there are only three cases for which it is valid and we are thus naturally led to three inequivalent triplets in the $U(3)$ scheme; they are defined by the following three choices:

$$t_I:\ \tilde{Q}=-A_1{}^1,\qquad \tilde{Y}=-A_1{}^1-A_2{}^2,\qquad \tilde{I}_3=\tfrac{1}{2}(A_2{}^2-A_1{}^1),\tag{14a}$$

$$t_U:\ \tilde{Q}=A_2{}^2+A_3{}^3,\qquad \tilde{Y}=A_3{}^3,\qquad \tilde{I}_3=\tfrac{1}{2}(2A_2{}^2+A_3{}^3),\tag{14b}$$

$$t_V:\ \tilde{Q}=-A_1{}^1,\qquad \tilde{Y}=A_3{}^3,\qquad \tilde{I}_3=-\tfrac{1}{2}(2A_1{}^1+A_3{}^3).\tag{14c}$$

Now the first one, t_I, for which

$$\tilde{Y}=-A_1{}^1-A_2{}^2,\tag{15}$$

$$\tilde{I}_3=\tfrac{1}{2}(A_2{}^2-A_1{}^1)=\tfrac{1}{2}(B_2{}^2-B_1{}^1)=I_3\tag{16}$$

corresponds to the I triplet mentioned above.

The structure of the remaining triplets t_U and t_V can be brought to much more transparent and symmetric forms in terms of the U-spin and V-spin subalgebras.[10] As in the case of relations (9) and (10) for $SU(3)$ and $U(3)$, we define the U and V spin of $U(3)$ in exactly the same forms as in $SU(3)$ except that all quantities are tilded quantities. From the $SU(3)$ definitions,[12] we then have

$$\tilde{Y}_U=-\tilde{Q}=-A_2{}^2-A_3{}^3,\tag{17}$$

$$\tilde{U}_3=\tilde{Y}-\tfrac{1}{2}\tilde{Q}=\tfrac{1}{2}(A_3{}^3-A_2{}^2)=\tfrac{1}{2}(B_3{}^3-B_2{}^2)=U_3\tag{18}$$

for (14b), and

$$\tilde{Y}_V=\tilde{Q}-\tilde{Y}=-A_3{}^3-A_1{}^1,\tag{19}$$

$$\tilde{V}_3=-\tfrac{1}{2}(\tilde{Y}+\tilde{Q})=\tfrac{1}{2}(A_1{}^1-A_3{}^3)=\tfrac{1}{2}(B_1{}^1-B_3{}^3)=V_3\tag{20}$$

for (14c). They correspond, therefore, to a U triplet and a V triplet, respectively, and hence the notations t_I, t_U, and t_V. With respect to the $SU(3)$ triplet (quark), these $U(3)$ triplets have their respective "hypercharges" (i.e., Y, Y_U, and Y_V) shifted by the amount of $\frac{2}{3}$ and as such they have quite different transformation properties under the Weyl reflections W_1, W_2, and W_3[11] which are reflections about the axis $I_3=0$, $U_3=0$, and $V_3=0$, respectively. Whereas the $SU(3)$ triplet is invariant under all three Weyl reflections, the $U(3)$ triplets are not. They transform according to

$$W_1:\quad t_I\rightarrow t_I,\quad t_U\leftrightarrow t_V;\tag{21a}$$

$$W_2:\quad t_U\rightarrow t_U,\quad t_I\leftrightarrow t_V;\tag{21b}$$

$$W_3:\quad t_V\rightarrow t_V,\quad t_I\leftrightarrow t_U.\tag{21c}$$

Figure 1 and Table I(a) list the quantum numbers $\tilde{I}_3$ and $\tilde{Y}$ for the single triplet (quark) model; a possible assignment implied by the two-triplet model[2] is shown in Fig. 2 and Table I(b); the corresponding quantum numbers for the three-triplet model are given in Fig. 3 and Table I(c).

TABLE I. Quantum-number assignments for (a) the quark model, (b) the two-triplet model, and (c) the three-triplet model.

(a)

	quark		
$\tilde{I}_3$	$\frac{1}{2}$	$-\frac{1}{2}$	0
$\tilde{Y}$	$\frac{1}{3}$	$\frac{1}{3}$	$-\frac{2}{3}$
$\tilde{Q}$	$\frac{2}{3}$	$-\frac{1}{3}$	$-\frac{1}{3}$

(b)

	t_1			t_2		
$\tilde{I}_3$	$\frac{1}{2}$	$-\frac{1}{2}$	0	$\frac{1}{2}$	$-\frac{1}{2}$	0
$\tilde{Y}$	1	1	0	−1	−1	−2
$\tilde{Q}$	1	0	0	0	−1	−1

(c)

	$t_1(t_I)$			$t_2(t_U)$			$t_3(t_V)$		
$\tilde{I}_3$	$\frac{1}{2}$	$-\frac{1}{2}$	0	0	−1	$-\frac{1}{2}$	1	0	$\frac{1}{2}$
$\tilde{Y}$	1	1	0	0	0	−1	0	0	−1
$\tilde{Q}$	1	0	0	0	−1	−1	1	0	0

III. DOUBLE $SU(3)$ SYMMETRY

Let us call the three triplets $t_1(=t_I)$, $t_2(=t_U)$, and $t_3(=t_V)$. Each triplet may be characterized in general by the average values, $\bar{I}_3$ and $\bar{Y}$, of $\tilde{I}_3$ and $\tilde{Y}$ for its three members. This specifies the location of the center of the triplet in the $\tilde{I}_3-\tilde{Y}$ diagram. Since $\bar{A}_1{}^1=\bar{A}_2{}^2=\bar{A}_3{}^3=\tilde{\tau}/3=\tau/3$, Eq. (14) gives for the three definitions of

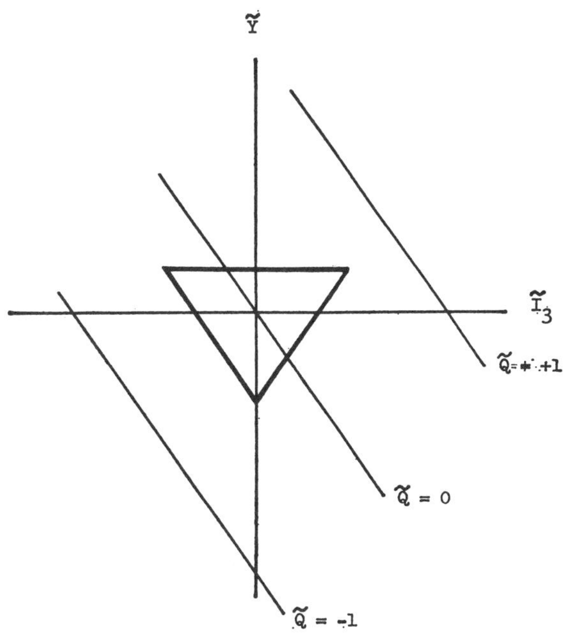

FIG. 1. The single-triplet (quark) model.

$\bar{I}_3$ and $\bar{Y}$,

$$\begin{aligned}\bar{I}_3&=0,\ \tfrac{1}{2}\tau,\ -\tfrac{1}{2}\tau,\\ \bar{Y}&=-\tfrac{2}{3}\tau,\ \tfrac{1}{3}\tau,\ \tfrac{1}{3}\tau,\end{aligned}\tag{22}$$

respectively, where $\tau=-1$ for all the triplets. We may define new quantities I_3, Y and $Q=I_3+\frac{1}{2}Y$ by the relations:

$$\begin{aligned}\tilde{I}_3&=\bar{I}_3+I_3,\\ \tilde{Y}&=\bar{Y}+Y,\\ \tilde{Q}&=\bar{I}_3+\tfrac{1}{2}\bar{Y}+I_3+\tfrac{1}{2}Y=\bar{Q}+Q.\end{aligned}\tag{23}$$

It is clear that I_3 and Y play the role of $SU(3)$ generators within each triplet. The charm number C defined in the two-triplet model[2] is then

$$\tfrac{1}{3}C=\bar{Q}=\bar{I}_3+\tfrac{1}{2}\bar{Y}.\tag{24}$$

Now it is interesting to note that according to Eq. (22) and Fig. 3, the centers of the three triplets form an antitriplet, equivalent to an antiquark, symmetrically located around the origin. Let us suppose that the nine members of the three triplets $t_{1\alpha}$, $t_{2\alpha}$, $t_{3\alpha}$, $\alpha=1, 2, 3$ be combined into a single multiplet $T=\{t_{i\alpha}\}$, $i=1, 2, 3$. We can then imagine two distinct sets of $SU(3)$ operations on T. One is the $SU(3)$ acting on the index α for each triplet, while the other $SU(3)$ acts on the index i, which mixes corresponding members of different triplets. T is then a representation (3,3*) of this group $G\equiv SU(3)'\times SU(3)''$.[14] The quantum numbers of $SU(3)'$ and

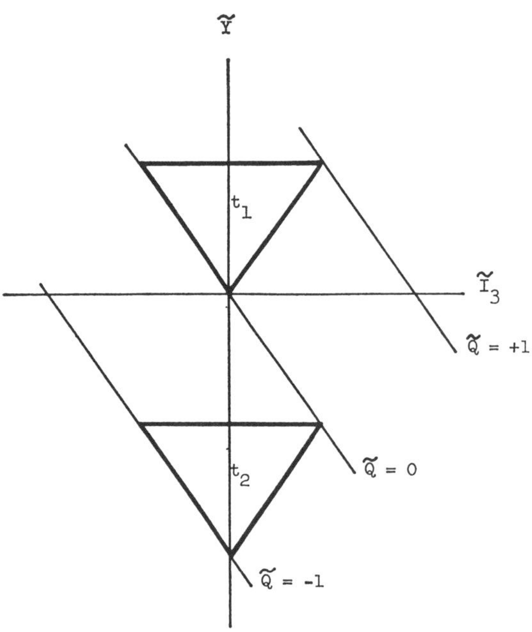

FIG. 2. The two-triplet model.

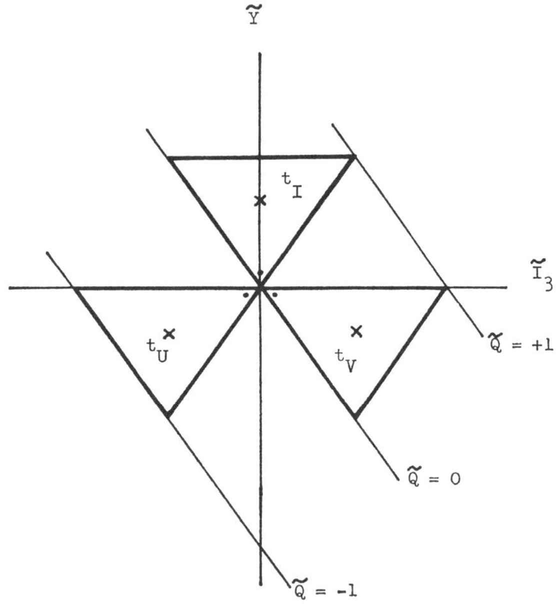

FIG. 3. The three-triplet model.

$SU(3)''$ are identified as $I_3'=I_3$, $Y'=Y$, $I_3''=\bar{I}_3$ and $Y''=\bar{Y}$ in Eq. (22), so that

$$\begin{aligned}&\tilde{I}_3=I_3'+I_3'',\quad \tilde{Y}=Y'+Y'',\\ &\tilde{Q}=I_3'+I_3''+\tfrac{1}{2}Y'+\tfrac{1}{2}Y'',\\ &\tfrac{1}{3}C=I_3''+\tfrac{1}{2}Y''.\end{aligned}\tag{25}$$

A general representation of G may be characterized by four numbers p', q', p'', q'' so that $D(p',q',p'',q'')\sim D(p',q')\times D(p'',q'')$, where $D(p,q)$ is a representation of $SU(3)$. However, in our scheme where the nonet T is the fundamental field, we do not get all the possible representations of G. This can be illustrated by means of the triality numbers[5] $t'=p'-q'\ \mathrm{mod}(3)$, $t''=p''-q''\ \mathrm{mod}(3)$. The nonet T has $t'=1$, $t''=-1$. All representations constructed out of T and T^* then satisfy $t'=-t''$.

Let us next consider the meson and baryon states $\sim TT^*$ and $\sim TTT$. The $SU(3)'\times SU(3)''$ contents of these 81- and 729-plets are

$$\begin{aligned}(3,3^*)\times(3^*,3)&=(8,1)+(1,1)+(1,8)+(8,8),\\ (3,3^*)\times(3,3^*)\times(3,3^*)&=(1,1)+2(8,1)+2(1,8)\\ &\quad+(1,10^*)+(10,1)+2(8,10^*)+2(10,8)\\ &\quad+4(8,8)+(10,10^*).\end{aligned}\tag{26}$$

It is an attractive possibility to postulate at this point that the energy levels are classified according to $SU(3)''$. The masses will then depend on the Casimir operators of $SU(3)''$. For example, a simple linear form will be

$$m=m_0+m_2C_2''+m_3C_3'',\tag{27}$$

[14] Such a nonet provides a natural basis for the symmetry of $SU(9)$. However, we will not consider it here.

where C_2'', C_3'' are the eigenvalues of quadratic and cubic Casimir operators of $SU(3)''$. In particular, we may assume that the main mass splitting comes from C_2''. Since this increases with the dimensionality of representation, the lowest mass levels will be $SU(3)''$ singlets. This selects the low-lying meson and baryon states to be (8,1), (1,1) and (8,1), (1,1), (10,1), respectively. In general, all low-lying states will have triality zero, $t'=t''=0$.

As for the baryon number assignment to the triplets, the simplest possibility would be to assign an equal baryon number, i.e., $B=\frac{1}{3}$, to them. In this case the triplets themselves would be essentially stable, and their nine members would behave like an octet plus a singlet of "heavy baryons" as may be seen from Fig. 3. Another simple possibility may be $B=\frac{1}{3}+Y''$, namely $B=(1,0,0)$ for (t_1,t_2,t_3). We expect a mass splitting depending on B or Y'', which may be the origin of the Okubo–Gell-Mann mass formula.

The advantage of the three-triplet model is that the $SU(6)$ symmetry can be easily realized with s-state triplets. The extended symmetry group becomes now $SU(6)'\times SU(3)''$. Since an $SU(3)''$ singlet is antisymmetric, the over-all Pauli principle requires the baryon states to be the symmetric $SU(6)$ 56-plet. Other $SU(6)$ representations such as the 70, will be obtained by bringing in either the orbital angular momentum or the "ρ spin" of the Dirac spinor triplets.

As in the two-triplet model mentioned in the Introduction, the mass formula of the type (27) may be derived dynamically. Instead of the charm number field, we introduce now eight gauge vector fields which behave as (1,8), namely as an octet in $SU(3)''$, but as singlets in $SU(3)'$. Since their coupling to the individual triplets is proportional to λ_i'' [the generators of $SU(3)''$], the interaction energy arising from the exchange of these vector fields will yield the first and second terms of Eq. (27). If these mesons obey again a similar type of mass formula, they will be expected to be massive compared to the ordinary mesons. However, it is not clear whether the resulting short-range character of the interaction can be readily reconciled with the postulated largeness of the interaction energy.

We may characterize the hierarchy of interactions and their symmetries implied by the above model as follows. First, the *superstrong* interactions responsible for forming baryons and mesons have the symmetry $SU(3)''$, and causes large mass splittings between different representations. The scale of mass involved would be comparable or large compared to the baryon mass, namely $\gtrsim 1$ BeV. The lowest states, i.e., $SU(3)''$ singlet states, would split according to $SU(3)'$, which would be the $SU(3)$ group observed among the known baryons and mesons, with their *strong* interactions. The scale of mass splitting would then be $\lesssim 1$ BeV.

When we go to the massive $SU(3)''$ nonsinglet states, there may very well be coupling between the two $SU(3)$ groups similar to the $L\cdot S$ coupling. The levels should be classified in terms of the three sets of Casimir operators formed out of λ_i', λ_i'', and $\lambda_i=\lambda_i'+\lambda_i''$, respectively. The splitting due to the coupling would naturally be intermediate between the above two splittings, namely ~ 1 BeV. Because of this coupling, the separate conservation of the two $SU(3)$ spins, I_3' and Y' on the one hand, and I_3'' and Y'' on the other, would be destroyed, and only the sums $I_3=I_3'+I_3''$ and $Y=Y'+Y''$ would be conserved under *strong* interactions. This in turn would mean that all the massive states are in general highly unstable, and decay strongly to the low-lying states. (In the two-triplet model, we considered only weak decays of $C\neq 0$ states. But strong decays are also a possibility as is contemplated here.)

We have discussed here a possible model of baryons and mesons based on three triplets. How can we distinguish this and other different models mentioned already? Certainly different models predict considerably different structure of massive states. These states are characterized by the triality for the quark model, by the charm number for the two-triplet model and by the $SU(3)''$ representation for the present three-triplet model. If we restrict ourselves to the low-lying states only, however, it seems difficult to distinguish them without making more detailed dynamical assumptions.

ACKNOWLEDGMENTS

One of us (M. Y. H.) wishes to thank Professor E. C. G. Sudarshan and Professor A. J. Macfarlane for their encouragement and useful discussions and Professor L. O'Raifeartaigh and J. Kuriyan for helpful comments.

A SYSTEMATICS OF HADRONS IN SUBNUCLEAR PHYSICS

YOICHIRO NAMBU

The Enrico Fermi Institute for Nuclear Studies and the Department of Physics, The University of Chicago, Chicago, Illinois

(Received May 3, 1965)

1.

With the recognition that the SU(3) symmetry is the dominant feature of the strong interactions, the main concern of the elementary particle theory has naturally become directed at the understanding of the internal symmetry of particles at a deeper level. An immediate question that arises in this regard is whether there are fundamental objects (such as triplets or quartets) of which all the known baryons and mesons are composed. These fundamental objects would be to the baryons and mesons what the nucleons are to the nuclei, and the electrons and nuclei are to the atoms. If that was really the case, it would certainly precipitate a new revolution in our conceptual image of the world. At the moment we can only hope that the question will be answered within the next ten to twenty years when the 100 GeV to 1000 GeV range accelerators will have been realized.

Even now, the amusing and rather embarassing success of the SU(6) theory [1] lends support to the existence of those fundamental objects. It is embarassing because this is basically a non-relativistic and static theory, and we do not know exactly how this can cover the realm of high energy relativistic phenomena.

Putting aside those theoretical difficulties mainly associated with relativity, let us make the working hypothesis that there are fundamental objects which are heavy ($\gg 1$ GeV), though not necessarily stable, and that inside each baryon or meson they are combined with a large binding energy, yet moving with non-relativistic velocities. Though this might look like a contradiction, at least it does not violate the uncertainty principle in non-relativistic quantum mechanics since the range of the binding forces ($10^{-14}-10^{-13}$ cm) is large compared

Reprinted from Preludes in Theoretical Physics, eds. A. De-Shalit, H. Feshbach, L. van Hove, © 1966 North-Holland, pp. 133–142.

to the Compton wave lengths of those constituents, and the strength of the forces can be arbitrarily adjusted. In other words, we have a model very similar to the atomic nuclei except for large binding energies. Theoretical justification of such a hypothesis must await future investigation.

In a previous article [2], we have put forward such a model with the following characteristic features.

1) There exist two fundamental fermion triplets t_1 and t_2 with charge assignments (1, 0, 0) and (0, -1, -1) for their three members. The baryons have the structure $\sim t_1 t_1 t_2$, and the mesons $\sim a t_1 \bar{t}_1 + b t_2 \bar{t}_2$.

2) To t_1 and t_2 are assigned "charm" charge $C = +1$ and -2 respectively. Thus the baryons and mesons (zero triality states) have $C = 0$. The primary binding forces acting on them are proportional to C. Let us imagine these forces to be mediated by a field (C-field). The resulting Coulomb-like energy though probably of finite range, then stabilizes the $C = 0$ ("uncharmed") systems against the $C \neq 0$ ("charmed") states, such as the triplets themselves.

3) The SU(6) symmetry can be brought in, with the Pauli principle taken into account, since the constituent particles are non-relativistic. In another paper, we also considered a three-triplet model, in which t_1, t_2 and t_3 have charge assignments (1, 0, 0), (1, 0, 0) and (0, -1, -1) respectively. This has the advantage that the baryon states (the 56-dimensional representation of SU(6)) may be realized with s-state triplets as $\sim t_1 t_2 t_3$.

The reasoning that has gone into the above stability problem is similar to the one used in nuclear physics in deriving the semi-empirical formula of Weizsäcker. The purpose of the present paper is to put this idea into a more precise form, even though the outcome should still be called at best semi-quantitative.

2.

Let us first consider states composed of an arbitrary number of t_1 and t_2, but without antiparticles $\bar{t}_1$ and $\bar{t}_2$. Their masses are M_1 and M_2, respectively, and the "charm" numbers 1 and -2, as was mentioned already. The pairwise interaction energy through the C-field will depend on the spatial configurations of the particles, but we will represent it, in the first approximation, by a constant V_c, as long as the size of the system is comparable with the range of the force. If the number of t_1's and t_2's are n_1 and n_2, respectively, the total energy of the system is

$$\begin{aligned} E(n_1, n_2) &= M_1 n_1 + M_2 n_2 + \\ &\quad + V_c \tfrac{1}{2} n_1(n_1-1) + 4V_c \tfrac{1}{2} n_2(n_2-1) - 2V_c n_1 n_2 \\ &= M_1 n_1 + M_2 n_2 + \tfrac{1}{2}V_c(n_1-2n_2)^2 - \tfrac{1}{2}V_c(n_1+4n_2) \\ &= (M_1 - \tfrac{1}{2}V_c)n_1 + (M_2 - 2V_c)n_2 + \tfrac{1}{2}V_c C^2, \\ C &= n_1 - 2n_2. \end{aligned} \tag{1}$$

As expected, the leading quadratic term depends only on the total charm C. If V_c is sufficiently large, this will favor $C = 0$ as the lowest states, which means $n_1 = 2n_2$. Restricting ourselves to $C = 0$ states now, the remaining terms are linear in n_1 and n_2, implying a saturation property. With $n_1 = 2n_2$, we have

$$E(2n_2, n_2) = (2M_1 + M_2 - 3V_c)n_2. \tag{2}$$

From the physical requirement that this increases with n_2 and that the baryon ($n_2 = 1$) be lighter than the triplets, we further need

$$M_1, M_2 > 2M_1 + M_2 - 3V_c > 0. \tag{3}$$

Thus the energy surface in the $n_1 - n_2$ plane has a valley running along the line $C = n_1 - 2n_2 = 0$, and its level rises linearly with increasing coordinates. However, it will be further necessary to make sure that the $C = 0$ states are actually lower than their neighbors even for small n's. Namely

$$\begin{aligned} E(2n_2 \pm 1, n_2) &> E(2n_2, n_2), \\ E(2n_2, n_2 \pm 1) &> E(2n_2, n_2). \end{aligned} \tag{4}$$

This gives two more conditions

$$V_c - M_1 > 0, \qquad 4V_c - M_2 > 0. \tag{5}$$

Combining Eqs. (3) and (5), we obtain

$$3V_c - 2M_1 > M_2 - M_1 > 3(V_c - M_1) > 0. \tag{6}$$

The second triplet, therefore, must be heavier than the first, but not too much heavier. This is because we have to maintain a balance between the energy due to rest masses and that due to interaction.

Eq. (1) may be expressed in terms of C and the baryon number B if we make an appropriate assignment: $B = x$ for t_1 and $B = y$ for t_2. Since the baryon $\sim t_1 t_1 t_2$ has $B = 1$, we require $2x+y = 1$. Possible choices given in ref. [2] are

$$(x, y) = (\tfrac{1}{3}, \tfrac{1}{3}) \quad \text{or} \quad (0, 1) \quad \text{or} \quad (1, -1). \tag{7}$$

The numbers n_1 and n_2 may be then expressed in terms of C and B as

$$\begin{aligned} n_1 &= 2B+yC \\ n_2 &= B-xC \end{aligned} \tag{8}$$

and thus

$$E(B, C) = \tfrac{1}{2}V_c C^2 + (2M_1+M_2-3V_c)B + [(M_1-\tfrac{1}{2}V_c)-(2M_1+M_2-3V_c)x]C. \tag{9}$$

At this point we should add a reservation that the linear terms in the above mass formula are not as meaningful as the leading quadratic terms since the effects depending on spatial configurations, such as those due to the finite range character of the C-field and the exchange energy, can be of the same order as the former.

3.

In order to consider the meson states, we will next bring in antiparticles as well in the picture. We make the basic assumption that a system consists of definite numbers of $n_1, \bar{n}_1, n_2, \bar{n}_2$ of $t_1, \bar{t}_1, t_2$ and $\bar{t}_2$. This means that *we regard pair creation and annihilation as forbidden processes*, which is consistent with our basic non-relativistic approach.

The formula corresponding to Eq. (1) becomes

$$E(n_1, \bar{n}_1, n_2, \bar{n}_2) = (M_1-\tfrac{1}{2}V_c)(n_1+\bar{n}_1)+(M_2-2V_c)(n_2+\bar{n}_2)+ +\tfrac{1}{2}V_c C^2, \tag{10}$$

$$C = n_1-\bar{n}_1-2(n_2-\bar{n}_2).$$

The requirement that $E > 0$ demands

$$M_1 - \tfrac{1}{2}V_c > 0, \qquad M_2 - 2V_c > 0 \tag{11}$$

in contrast to Eq. (5), which was derived for the special case $\bar{n}_1 = \bar{n}_2 = 0$. We find, together with Eqs. (3) and (5),

$$M_1 > V_c - M_1 > M_2 - 2V_c > 0 \tag{12}$$

which replaces Eq. (6).

We will now relate the constants M_1, M_2 and V_c to the baryon ($t_1 t_1 t_2$) and meson ($t_1 \bar{t}_1$ and $t_2 \bar{t}_2$) masses m, μ_1 and μ_2:

$$\begin{aligned} m &= 2M_1 + M_2 - 3V_c, \\ \mu_1 &= 2M_1 - V_c, \\ \mu_2 &= 2M_2 - 4V_c, \end{aligned} \tag{13}$$

from which we obtain an identity

$$2\mu_1 + \mu_2 = 2m. \tag{14}$$

Because of this, we cannot determine the three unknowns M_1, M_2, V_c uniquely. Instead, we can express Eq. (10) in terms of μ_1 and μ_2:

$$E(n_1, \bar{n}_1, n_2, \bar{n}_2) = \tfrac{1}{2}\mu_1(n_1 + \bar{n}_1) + \tfrac{1}{2}\mu_2(n_2 + \bar{n}_2) + \tfrac{1}{2}V_c C^2. \tag{15}$$

Turning to the relation (14), we put $m \sim 1.2$ GeV, $\mu_1 \sim 600$ MeV $= \tfrac{1}{2}m$ corresponding to the average baryon and meson masses, and *predict* a value

$$\mu_2 \sim m \sim 2\mu_1 \tag{16}$$

for the second meson. This is not an unreasonable value in view of the fact that a large number of unidentified meson resonances seem to exist in this energy range. Eq. (15) reduces then to the simple form

$$E(n_1, \bar{n}_1, n_2, \bar{n}_2) = \tfrac{1}{2}\mu_1[n_1 + \bar{n}_1 + 2(n_2 + \bar{n}_2)] + \tfrac{1}{2}V_c^2. \tag{17}$$

It is rather surprising that such a naive picture as ours can yield non-trivial and qualitatively reasonable results.

By way of a remark, we note from Eq. (13) that

$$\begin{aligned} M_1 &= \tfrac{1}{2}V_c + \tfrac{1}{2}\mu_1 \sim \tfrac{1}{2}V_c, \\ M_2 &= 2V_c + \tfrac{1}{2}\mu_2 \sim 2V_c \sim 4M_1 \end{aligned} \tag{18}$$

since $V_c \gg \mu_1, \mu_2$ by assumption. Interestingly enough, the above relation admits the interpretation that the mass of each triplet is made up of a self-energy due to the C-field plus a small "bare mass" $\frac{1}{2}\mu$.

4.

We will now turn to the three-triplet model [3] proposed as an alternative to the two-triplet model. The three triplets t_1, t_2 and t_3 altogether contain nine fermions $T_{i\alpha}$, $i, \alpha = 1, 2, 3$, where the index i distinguishes different triplets, and α the different members of a triplet. Two different SU(3) operations, called SU(3)′ and SU(3)″, are introduced, acting respectively on α and i, and in these spaces $T_{i\alpha}$ behave as a representation (3, 3*). The electric charge is assigned to each particle according to

$$Q = I_3' + \tfrac{1}{2}Y' + I_3'' + \tfrac{1}{2}Y'' \tag{19}$$

which takes integral values. In fact both $T_{1\alpha}$ and $T_{2\alpha}$ have the assignment (1, 0, 0), and $T_{3\alpha}$ have (0, −1, −1), exactly like t_1 and t_2 of the previous two-triplet model.

An important difference from the two-triplet case is that instead of the charm gauge group $U(1)$, we have the group SU(3)″. The charm gauge field C must then be replaced by an octet of gauge fields G_μ, $\mu = 1, \ldots, 8$, coupled to the infinitesimal SU(3)″ generators (currents) λ_μ'' of the triplets, with a strength g. For a system containing altogether N particles, the exchange of such fields between a pair then results in an interaction energy

$$V_G = +g^2 \sum_{n>m} \lambda_\mu''^{(n)} \cdot \lambda_\mu''^{(m)} = \tfrac{1}{2}g^2 \Big[\sum_{n=1}^{N} \lambda_\mu''^{(n)}\Big]\Big[\sum_{m=1}^{N} \lambda_\mu''^{(m)}\Big] - \tfrac{1}{2}g^2 \sum_{n=1}^{N} \lambda_\mu''^{(n)}\lambda_\mu''^{(n)}$$
$$= \tfrac{1}{2}g^2 [C_2 - NC_{20}], \tag{20}$$

where $\lambda_\mu^{(n)}$ refers to the n-th particle, C_2 is the quadratic Casimir operator of SU(3), and $C_{20} = 4/3$ is its value for a triplet representation $D(1, 0)$ or $D(0, 1)$. In general C_2 is given by

$$C_2(l_1, l_2) = \tfrac{1}{3}(l_1^2 + l_1 l_2 + l_2^2) + (l_1 + l_2) \tag{21}$$

for a representation $D(l_1, l_2)$.

Note that the only dependence on the total number N of constituents appears in the second term of Eq. (17).

We add to V_G the rest masses (M = common mass), and obtain the total energy

$$E = (M - \tfrac{1}{2}C_{20}g^2)N + \tfrac{1}{2}g^2C_2. \tag{22}$$

Bound states are characterized by $V_G < 0$, and the low lying states by the smallest value of C_2, namely $C_2 = 0$ for the singlet $D(0, 0)$. For the latter, E is simply proportional to the total number N of constituents, starting with the meson $(N = 2) \sim t_1\bar{t}_1 + t_2\bar{t}_2 + t_3\bar{t}_3$ and the baryon $(N = 3) \sim t_1t_2t_3$ (antisymmetric combination). Their masses are thus related by

$$\mu = 2(M - \tfrac{1}{2}C_{20}g^2) = \tfrac{2}{3}m, \tag{23}$$

and Eq. (22) becomes

$$E = \tfrac{1}{2}\mu N + \tfrac{1}{2}g^2C_2. \tag{24}$$

These are to be compared with Eqs. (14) and (15). Because of the high symmetry among the three triplets, we have found only one set of mesons with $N = 2$. In any case, the energy is simply proportional to the total number of constituents as long as $C_2 = 0$, as if it were made up of non-interacting basic units of mass $\tfrac{1}{2}\mu$.

5.

Having disposed of the gross mass spectrum of many-triplet compound systems, we now turn our attention to the "fine structure" of low lying states, which in our view comprise all the mesons and baryon resonances known so far. In all probability, however, our crude qualitative arguments are not really satisfactory for discussing these details. We will therefore restrict ourselves to general remarks only.

Because of our basic assumptions about the superstrong interactions and the static behaviour of particles, the dynamics we have been dealing with so far does not depend on the spin and the SU(3) spin variables, therefore the system possesses the symmetry of superstrong interactions, the SU(6) symmetry of combined spin and SU(3) spin, and the symmetry of orbital angular momentum. The overall Pauli principle imposes constraints among these symmetries, and thereby single out certain SU(6) and orbital states for the lowest configuration with respect to the superstrong interaction. The general classification of these states can be done as in the case of nuclear and atomic physics, but this will be beyond the scope of the present paper.

In the three-triplet model, however, the problem is relatively simple if we take only s-state triplets. The low lying three particle configuration is a SU(3)″ singlet, so the baryon must go into a complete symmetric SU(6) representation 56. No other states are possible without changing the spatial configuration, but this will cause some change in the superstrong interaction. For the mesons, we obviously obtain $36 = 35+1$ SU(6) states which are degenerate. These results are in accordance with those of the original SU(6) theory, as well as its "relativistic" version.

We must next discuss the two additional effects which do exist and tend to upset the symmetries. One arises from the internal motion of particles, and the other from the presence of virtual mesons. Contrary to the prevalent view, we regard the mesons as perturbing forces rather than the decisive factors in the physics of hadrons. Since the strong interactions are then merely first forbidden processes, so to speak, the meson and baryon resonances are really bound states decaying via violation of superstrong interaction symmetry. Nevertheless, these secondary effects can affect, and may even decide, the "fine structure" of low lying states. Perhaps we may compare the situation to the electronic levels of an atom where the main spectrum is determined by the static Coulomb force, and both the fine structure and the photon emission processes are higher order effects. In this sense, we do not necessarily find a contradiction between the present approach and the conventional strong interaction theory as far as the low lying states are concerned.

The reason we consider the strong interaction as generally symmetry breaking is that the virtual exchange of 36 virtual mesons do not possess an SU(6) symmetric form. An ideal SU(6) symmetric interaction would involve the 35 generators χ_μ as in Eq. (20):

$$V \sim \pm g^2 \sum_{n>m} \chi_\mu^{(n)} \chi_\mu^{(m)} = \pm \frac{g^2}{8} \sum_{n>m} [\tfrac{2}{3}\sigma_i^{(n)}\sigma_i^{(m)} + \lambda_\alpha^{(n)}\lambda_\alpha^{(m)} + \sigma_i^{(n)}\sigma_i^{(m)}\lambda_\alpha^{(n)}\lambda_\alpha^{(m)}]. \tag{25}$$

Viewed as a static force, this requires an exchange of 35 scalar and axial vector mesons.

if the relative signs of the various terms are to be correctly maintained for both particle-particle and particle-antiparticle interactions. [For processes involving meson-baryon scattering, however, Capps [4] and Belinfante and Cutkosky [5] have shown their compatibility with SU(6).]

Next consider the effect of the internal motion. This disturbs the basic symmetry in two senses. It mixes the Dirac spinor components, introducing corrections to the static superstrong forces. Further it simply adds the kinetic energy of orbital motion to the system. As far as the symmetry is concerned, these perturbations act like adding a neutral singlet meson with a suitable spin-parity. Its order of magnitude will depend on the internal velocity v of the particles, which should be of the order $1/MR$ where R is the size of the system. If we take this correction to be of the order $Mv^2 \sim 100$ MeV, and $R \sim 1/M$, we obtain the estimation

$$M \sim 10m, \qquad \frac{v}{c} \sim \frac{1}{10}$$

as we did before [2].

6.

Finally we would like to comment on some obvious difficulties and intriguing problems concerning our model of the subnuclear structure of hadrons.

a) What is the origin of superstrong interactions?

Are these another kind of vector fields or something entirely new? If they are ordinary fields, their range must be at least of the order of the baryon size, and moreover sufficiently smooth and well-behaved in order to keep the kinetic energy small. It is conceivable that no single or a relatively few well defined meson states are responsible for this. A direct confirmation of such interactions would be difficult.

b) The magnetic moments of baryons, for example, agree closely with the SU(6) symmetry, yet obviously the bulk of contributions come from the meson cloud. This means that regardless of whether the meson cloud obeys SU(6) symmetry or not, the baryon should not be considered as composed of three bare triplets without structure. How, then, can we justify our picture that each system, including the mesons, is composed of a definite number of triplets? The answer to this probably should be that the quantities like charm are at any instant well localized at a definite number of centers in space, and these centers are accompanied by large concentrations of energy, moving with slow velocities; whereas the quantities like ordinary charge are more uniformly spread out and carried by faster moving matter. In order to test such a picture experimentally, we would have to use some phenomena which depend on the energy distribution, the correlation functions of charges and energies at different points, the internal velocity of particles, etc.

c) The notion that decays and resonances are actually forbidden processes was first recognized as a surprising paradox in the process of adapting SU(6) to relativity. In our view, this is not only natural, but also simplifies the whole picture. We should be able to discuss the classes of first forbidden, second forbidden, etc. transitions, and they will be accessible to experimental test [6]. For this we should look especially for small, inconspicuous bumps in cross sections, many particle decay modes, and relatively rare events.

d) It has been widely speculated that an axial vector current conservation as relativistic chival symmetry has physical significance. If this is actually the case, it is probably beyond the capacity of our extreme static approach, since we have first to explain away the large masses of triplets, even though we can formally apply group theoretical arguments and the Goldberger-Treiman type relations to individual problems.

REFERENCES

1) F. Gürsey and L. A. Radicati, Phys. Rev. Letters, **13** (1964) 173;
A. Pais, Phys. Rev. Letters, **13** (1964) 175;
B. Sakita, Phys. Rev. **136** (1964) B1765.
2) Y. Nambu, Proc. of the Second Coral Gables Conference on Symmetry Principles at High Energy, University of Miami, January, 1965.
3) M. Y. Han and Y. Nambu, Syracuse University preprint 1206-SU-31.
4) R. Capps, Phys. Rev. Letters, **14** (1965) 31.
5) J. G. Belinfante and R. E. Cutkosky, Phys. Rev. Letters, **14** (1965) 33.
6) A more detailed study of this problem will be done elsewhere.

Volume 26B. number 10 PHYSICS LETTERS 15 April 1968

S-MATRIX IN SEMICLASSICAL APPROXIMATION *

Y. NAMBU
The Enrico Fermi Institute and Department of Physics,
The University of Chicago, Chicago, Illinois, USA

Received 18 March 1968

It is shown that the tree approximation currently utilized in connection with phenomenological chiral Lagrangians corresponds to a semiclassical approximation applied to the S-matrix in quantum field theory. Some invariance properties of the S-matrix are derived in this framework.

Recently the method of effective Lagrangians has been widely used in formulating the dynamical contents of chiral symmetry [1]. One writes down in this approach a non-linear Lagrangian invariant under the chiral SU(2) × SU(2) (or SU(3) × SU(3)) group, in which the pion field (or its SU(3) extention) transforming non-linearly under the group plays a crucial role [2]. It is characteristic of such a Lagrangian that the invariance does not apply to the free and the interaction part separately; these two will mix under the group transformation. Thus the asymptotic one-particle states defined by (appropriately renormalized) free fiels do not form a representation of the group, in sharp contrast to the ordinary case in which a symmetry implies a multiplet structure and associated quantum numbers for one-particle states. This situation may be interpreted in terms of a "spontaneous breakdown" of chiral symmetry; each time a chiral transformation is performed, one has to redefine the free and interaction parts, corresponding to a change of the vacuum state [3].

For the actual construction of the S-matrix, in chiral dynamics, one carries out a perturbation expansion with respect to a coupling parameter characterizing the degree of nonlinearity, collecting consistently all terms that contribute to a specific n-particle process in the lowest possible order. The corresponding Feynman diagrams turn out to be "tree diagrams", or simply connected diagrams having no closed loops [1]. Obviously it is an approximation which does not take into account unitarity (rescattering), self-energy and other higher order corrections to a given process. This may be reasonable in the low energy (soft pion) limit. On the other hand it is not clear whether the highly singular Lagrangian one uses here is really meaningful beyond such an approximation.

It is the purpose of this note to point out that the above procedure may be characterized as a semiclassical approximation to the S-matrix, which is an analog in quantum field theory of the WKB approximation to the Schrödinger equation. This statement is entirely general, being independent of the Lagrangian chosen. In the following we present the proof, and discuss its implications.

We start from the S-matrix in interaction representation, expressed as a time-ordered (T-) product ‡

$$S = T \exp[(i/\hbar c) \int L_{int}(\phi) \, d^4x] \, . \qquad (1)$$

Its contribution to individual processes can be obtained by converting it to a normal product, for which there is a compact formula [4] †

* Supported in part by the U.S. Atomic Energy Commission.

‡ When L_{int} contains derivatives, one should perform the differentiation after time-ordering.

† Eq. (10') could have been derived directly from Feynman's path integral formulation, or from the matrix representation

$$\langle \phi' | U | \phi'' \rangle \sim \exp[(i/2\hbar c) \int (\phi' - \phi'') K (\phi' - \phi'') \, d^4x] \, .$$

Applying this to eq. (2), we get

$$S(\phi) = U \exp[(i/\hbar c) \bar{L}_{int}(\phi)]$$

$$= \int \langle \phi | U | \tilde{\phi} \rangle \, D[\phi] \exp[(i/\hbar c) \bar{L}_{int}(\tilde{\phi})]$$

which is an exact result. Approximating the functional integral by a stationary "path" determined from the classical Euler equation for $\tilde{\phi}$, we arrive at the formulas (9') and (10'). However, their relation to the tree approximation would have been obscure.

$$T\exp[(i/\hbar c)\int L_{\rm int}(\phi)\,d^4x] = \\ = :U\exp[(i/\hbar c)\int L_{\rm int}(\phi)\,d^4x]: \equiv :S:,$$

$$U \equiv \exp[\tfrac{1}{2}\hbar c\iint \frac{\delta}{\delta\phi(x)}\Delta(x-x')\frac{\delta}{\delta\phi(x')}d^4x\,d^4x'] . \quad (2)$$

Here we have chosen for simplicity a neutral scalar field $\phi(x)$; $\Delta(x-x')$ is the corresponding free Feynman propagator $(1/\hbar c)\langle T(\phi(x),\phi(x'))\rangle_0$ ‡. The functional differentiation is carried out regarding all $\phi(x)$ with different x as if they were independent c-numbers; i.e., not subject to a free field equation. This formula can be readily generalized to all types of field. For a complex field ϕ_α, we replace U with

$$\exp[\hbar c\iint \frac{\delta}{\delta\phi_\alpha}\Delta_{\alpha\beta}(x-x')\frac{\delta}{\delta\phi^*_\beta(x')}d^4x\,d^4x'] .$$

When ϕ_α is a Fermi field, we make the further stipulation that $\phi_\alpha(x)$, $\delta/\delta\phi_\alpha(x)$, etc., at different points all anticommute. In order to built up the operation with U step by step, we may supply a fictitious parameter η in front of $\Delta(x-x')$, which will vary from 0 to 1. Differentiating with respect to η, we obtain

$$\frac{\partial}{\partial\eta}:S_\eta: = \\ = :\left(\tfrac{1}{2}\hbar c\iint\frac{\delta}{\delta\phi(x)}\Delta(x-x')\frac{\delta}{\delta\phi(x')}d^4x\,d^4x'\right)S_\eta: . \quad (3)$$

Now let us write

$$:S: = :\exp[-(i/\hbar c)\Theta(\phi)]: . \quad (4)$$

It is not difficult to see that $\Theta(\phi)$ corresponds to a collection of all connected Feynman diagrams; this will in fact become clear below. Substituting in eq. (3), we may drop the normal product symbol and manipulate S as a mere c-number functional. We get thus

$$\frac{\partial\Theta}{\partial\eta} + \tfrac{1}{2}i\iint\frac{\delta\Theta}{\delta\phi(x)}\Delta(x-x')\frac{\delta\Theta}{\delta\phi(x')}d^4x\,d^4x + \\ - \tfrac{1}{2}\hbar c\iint\frac{\delta^2\Theta}{\delta\phi(x)\,\delta\phi(x'x)}\Delta(x-x')\,d^4x\,d^4x' = 0 , \quad (5)$$

‡ The dimensional convention adopted here is such that the field equations are "classical" (i.e., use wave numbers rather than momenta), with the free Lagrangian given by

$-\tfrac{1}{2}(\partial_\mu\phi\partial_\mu\phi + \kappa^2\phi^2)$ for a scalar, and

$-(\bar\psi\gamma_\mu\partial_\mu\psi + \kappa\bar\psi\psi)$ for a spinor, where

$x_\mu = (x_1, x_2, x_3, x_4 = ict)$, $\kappa = mc/\hbar$.

with the initial condition

$$\Theta = \int L_{\rm int}(\phi)\,d^4x \equiv -\bar{L}_{\rm int} \quad \text{for} \quad \eta = 0 . \quad (6)$$

If Θ is expanded in a power series in η, the second term in eq. (5) means the contraction of two disconnected diagrams by means of a propagator to form a single connected diagram, whereas the third term corresponds to a contraction within a connected diagram, creating an internal loop. If we drop this third term, which is multiplied by $\hbar c$, eq. (5) takes the form of a classical "Hamilton-Jacobi equation" (writing Θ_c for Θ)

$$\frac{\partial\Theta_c}{\partial\eta} + \bar{H}(\pi(x)) = 0 ,$$

$$\bar{H} = \tfrac{1}{2}i\iint\pi(x)\Delta(x-x')\pi(x')\,d^4x\,d^4x' ,$$

$$\pi(x) \equiv \delta\Theta_c/\delta\phi(x) . \quad (7)$$

where $\phi(x)$ and $\pi(x)$ are canonical conjugates, and η plays the role of time. It is obvious that we generate only tree diagrams.

The "Hamiltonian" contains only the "kinetic energy" dependent on $\pi(x)$, so the solution can be immediately obtained by integrating the equations of motion

$$\frac{d\pi(x)}{d\eta} = 0 , \quad \frac{d\phi(x)}{d\eta} = i\int\Delta(x-x')\pi(x')\,d^4x ,$$

$$\phi(x) = \eta i\int\Delta(x-x')\pi(x')\,d^4x' + F(\pi) . \quad (8)$$

$F(\pi)$ is determined from the initial condition

$$\pi(x) = \delta\Theta/\delta\phi(x) = -\delta\bar{L}_{\rm int}/\delta\phi(x) \quad \text{for} \quad \eta = 0 .$$

Thus

$$\pi(x) = -\delta\bar{L}_{\rm int}(\tilde\phi)/\delta\tilde\phi ,$$

$$\tilde\phi(x) \equiv \phi(x) - \eta i\int\Delta(x-x')\pi(x')\,d^4x' , \quad (9)$$

which is an intrinsic equation relating ϕ and π. The Hamilton-Jacobi functional Θ_c is now determining from integration of the relation

$$\delta\Theta_c(\phi,\eta) = \int\pi(x)\delta\phi(x)\,d^4x - \bar{H}\delta\eta$$

We find readily

$$\Theta_c = -\bar{L}_{\rm int}(\tilde\phi) + \eta\bar{H}(\pi) . \quad (10)$$

This, together with eq. (9), gives the solution Θ_c when it is expressed in terms of $\phi(x)$, and η put equal to 1. Eqs. (9) and (10) can be brought to a more transparent form. From eq. (9) follows

$$\pi(x) = K(\tilde\phi(x) - \phi(x)) = -\delta\bar{L}_{\rm int}/\delta\tilde\phi(x) ,$$

or

$$\delta \bar{L}_{\rm tot}(\tilde{\phi})/\delta\tilde{\phi}(x) = K\phi(x) \;;$$

$$K \equiv i\Delta^{-1}\,, \quad \bar{L}_{\rm tot}(\tilde{\phi}) \equiv \bar{L}_{\rm int}(\tilde{\phi}) + \bar{L}_0(\tilde{\phi}) \qquad (9')$$
$$\equiv \bar{L}_{\rm int}(\tilde{\phi}) + \tfrac{1}{2}\int \tilde{\phi} K \tilde{\phi}\, d^4x,$$

and eq. (10) becomes (with $\eta = 1$)

$$\Theta_c(\phi, \tilde{\phi}) = -\bar{L}_{\rm int}(\tilde{\phi}) - \bar{L}_0(\tilde{\phi} - \phi) \qquad (10')$$
$$= -\bar{L}_{\rm tot}(\tilde{\phi}) + \int \phi K \tilde{\phi}\, d^4x - \tfrac{1}{2}\int \phi K \phi\, d^4x\,.$$

Here K is the Klein-Gordon operator $\Box - \kappa^2$, so $\bar{L}_{\rm tot}$ is the total Lagrange functional. Eq. (10') is nothing but a Lagrange functional for the field $\tilde{\phi}$, with ϕ regarded as an external source field, and eq. (9') is the resulting Euler equation. When the latter is solved for $\tilde{\phi}$ in terms of ϕ with the boundary condition characterized by the Feynman kernel Δ, and in the final result we let $K\phi = 0$, then we get a solution of the classical field equation which approaches asymptotically the free field ϕ. Θ_c becomes a functional of ϕ as the dummy (internal) field $\tilde{\phi}$ has been eliminated †.

This proves our contention about the nature of the approximation involved. Eqs. (9') and (10'), or (9) and (10), provide a concise formula for the construction of Θ_c. When there is more than one field, we can simply add a respective source term to each field. The whole thing is treated as a quasi-c number theory, in the sense that for Fermi fields anticommutatively is to be respected. We will now make use of the formalism to examine invariance properties of the S-matrix in our approximation.

1) Suppose

$$\Theta_c(\phi_\alpha, \tilde{\phi}) = -\bar{L}_{\rm int}(\tilde{\phi}_\alpha) - \sum_\alpha \bar{L}_{0\alpha}(\tilde{\phi}_\alpha - \phi_\alpha) \qquad (11)$$

is invariant under an infinitesimal change $d\phi_\alpha(x)$, $d\tilde{\phi}_\alpha(x)$, $\alpha = 1, 2, \ldots$, of the fields. This will be true of the free and interaction parts of $\bar{L}$ are separately invariant. After the $\tilde{\phi}$'s have been eliminated, $\Theta_c(\phi_\alpha)$ and the S-matrix remain invariant under the change $d\phi_\alpha$.

2) It is possible that $\Theta_c(\phi_\alpha, \tilde{\phi}_\alpha)$ is invariant under a change of the ϕ_α's alone due to a special feature of the free Langrangian. The S-matrix will again be invariant. A notable example is the familiar gauge invariance of the S-matrix under $dA_\mu(x) = \partial_\mu \lambda(x)$ (without making a phase transformation on charged fields). This is because $L_0(A_\mu)$ depends only on the antisymmetric tensor $F_{\mu\nu}$ *.

The case of a massless spin - 0 field does not belong to this class, although $L_0(\phi)$ is invariant under a constant displacement $d\phi(x) = c$. The reason is that the Klein-Gordon (or rather the d'Alembertian) operator appearing in the interference term in eq. (10') is spurious, and cannot be counted on to kill a constant, as actually $\tilde{\phi}$ has a singularity $\sim \Delta = iK^{-1}$ according to eq. (9).

3) Next let us assume that $\bar{L}_{\rm tot}(\tilde{\phi}_\alpha)$ as a whole is invariant, but not necessarily for its parts $L_{\rm int}$ and L_0 separately. We have then

$$\delta \bar{L}_{\rm tot}(\tilde{\phi}_\alpha) = \int d\tilde{\phi}_\alpha(x)\, \delta \bar{L}_{\rm tot}/\delta\tilde{\phi}_\alpha(x)\, d^4x = 0\,.$$

From the Euler equation for $\tilde{\phi}_\alpha$: $\delta \bar{L}_{\rm tot}/\delta\tilde{\phi}_\alpha(x) = K_\alpha \phi_\alpha(x)$, it follows that

$$\sum_\alpha \int d\tilde{\phi}_\alpha K_\alpha \phi_\alpha\, d^4x = 0\,.$$

As a result, the entire $\Theta_c(\phi_\alpha)$ is invariant. This, however, is not a statement about the symmetry of the S-matrix, as the asymptotic fields ϕ_α are not varied. It is rather a statement about the independence of the S-matrix (as well as its off-shell extrapolation $K_\alpha\phi_\alpha \neq 0$) on a redefinition of the dummy fields, which is a symmetry operation on $L_{\rm tot}$.

4) We can go even further, and subject the dummy field variables to an arbitrary substitution which is not a symmetry operation:

$$\tilde{\phi}_\alpha(x) = F_\alpha(\tilde{\phi}'_\beta(x)) = \tilde{\phi}'_\alpha(x) + f_\alpha(\tilde{\phi}'_\beta(x))\,, \qquad (12)$$

where f contains only nonlinear terms in the fields. We claim that the S-matrix (on the mass shell) is again invariant under such a redefinition of the $\tilde{\phi}$'s. To be more precise, we get the same result if we start from

$$\Theta_c(\phi_\alpha, \tilde{\phi}'_\alpha) = -\bar{L}_{\rm tot}(F_\alpha(\tilde{\phi}'_\beta)) - \sum_\alpha \bar{L}_{0\alpha}(\tilde{\phi}'_\alpha - \phi_\alpha) \quad (13)$$

instead of eq. (11), provided that we construct the solution as a power series in the fields. It will suffice to show this for infinitesimal f.

† See footnote † on page 626.

* We know from the Ward-Takahashi identity that gauge invariance in the present sense does not hold off the mass shell of charged particles in general, whereas the present argument would seem to apply also off the mass shell. The latter, however, is not true. In order for the Maxwell equations to be consistent, their currents have to be conserved. This would not be true if the external charged fields were arbitrary.

Volume 26B. number 10 PHYSICS LETTERS 15 April 1968

Comparing eqs. (11) and (13), we find

$$\Theta_c'(\phi,\tilde{\Phi}')-\Theta_c(\phi,\tilde{\Phi}) =$$

$$= \int \sum_\alpha f_\alpha(\tilde{\Phi}')K_\alpha \phi_\alpha = \Theta_c'(\phi)-\Theta_c(\phi)$$

when the respective Euler equations are satisfied (stationary values of $\tilde{\Phi}'$ and $\tilde{\Phi}$). But the f_α's have no mass shell singularities in a perturbative solution because they are non-linear. Thus the difference vanishes as $K_\alpha\phi_\alpha \to 0$. (The condition on the f's may be further relaxed.)

Summarizing, we found that the S-matrix in the "tree" approximation corresponds to a semi-classical (WKB) approximation applied to field theory, and that invariance properties expected of the exact S-matrix are also valid here. In particular, the item 4) above can be interpreted as the irrelevance of the choice of the "interpolating fields" $\tilde{\Phi}_\alpha$, a property implied by the Lehmann-Symanzik-Zimmermann formalism of quantum field theory [5]. This property has been explicitly noted in chiral dynamics [6] where the definition of the pion field is not unique, as there is no obvious principle by which to choose a particular nonlinear transformation.

1. S. Weinberg, Phys. Rev. Letters 18 (1967) 188;
J. Schwinger, Phys. Letters 24B (1967) 473;
J. A. Cronin, Phys. Rev. 161 (1967) 1483;
J. Wess and B. Zumino, Phys. Rev. 163 (1967) 1727;
L. S. Brown, Phys. Rev. 163 (1967) 1802;
P. Chang and F. Gürsey, Phys. Rev. 164 (1967) 1752;
T. Minamikawa and Y. Miyamoto, Prog. Theor. Phys. 38 (1967) 1195;
W. A. Bardeen and B. W. Lee, Canadian Summer Institute Lectures, to be published;
B. W. Lee, Proc. 1957 Conf. on Particles and fields (John Wiley and Sons, Inc., New York, 1957) p. 546;
Y. Ohnuki and Y. Yamaguchi, Inst. Nucl. Study (Tokyo) preprint (1967).
2. F. Gürsey, Nuovo Cimento 16 (1960) 230; Ann. Phys. (N.Y.) 12 (1961) 91;
M. Gell-Mann and M. Lévy, Nuovo Cimento 16 (1960) 705.
3. Y. Nambu and G. Jona-Lasinio, Phys. Rev. 122 (1961) 345.
4. S. Hori, Prog. Theor. Phys. (Kyoto) 7 (1952) 578.
5. H. Lehmann, K. Symanzik and W. Zimmermann, Nuovo Cimento 5 (1955) 205;
R. Haag, Phys. Rev. 112 (1958) 669;
K. Nishijima, Phys. Rev. 111 (1958) 995;
W. Zimmermann, Nuovo Cimento 10 (1958) 597.
6. J. Cronin, Ref. 1;
B. Zumino, private communication.
Actual differences that appear in π-π scattering, for example, are due to different choices of symmetry breaking terms in the Lagrangian, and not due to different choices of field variables in the present context.

* * * * *

Reprinted from Proceedings of the International Conference on Symmetries and Quark Models, ed. R. Chand, © 1970 Gordon and Breach, pp. 269–278.

*Quark model and the factorization of the Veneziano amplitude**

YOICHIRO NAMBU

The Enrico Fermi Institute
The University of Chicago, Chicago, Illinois

IF WE TAKE the quark model in the most realistic sense, a meson in the $q\bar{q}$ configuration, for example, would have a wave function which, beside the spin and the $SU(3)$ indices, depends on the relative coordinates of the quarks. We would therefore expect in general excitations in L as well as in the radial mode, and hence a family of Regge trajectories forming 36-plets of $U(6)$. So far we do not have a clearcut evidence for the radial excitations, but the apparent complexity and structure we see in higher mesonic levels suggest that there may well be excitations other than the orbital L type. This is also supported by the successes of the duality concept[1] and its elegant mathematical realization embodied in the Veneziano[2] amplitude. Although we do not yet know the exact meaning of duality, nor do we know whether it is to be regarded as a new fundamental principle, the Veneziano model seems to contain a lot of implications which have yet to be tested.

A close connection between duality and the quark model has been pointed out by a number of people recently.[3] It is based on a diagrammatic interpretation of hadronic reactions in terms of quark exchanges which satisfy simple topological rules, the most important of which is to first consider only "planar diagrams" with no crossed quark lines. These rules make qualitative but powerful predictions, and it should be highly interesting if we could

* This work supported in part by the U.S. Atomic Energy Commission.

make a more quantitative and precise fusion of the quark model and the analytic structure of the Veneziano representation. Work along this line seems to be going on already at various places. Our results which we would like to discuss here are in fact very similar to those of the Berkeley[4] and the MIT groups,[5] but it would be useful to put the problem in a clear, if not perfect, perspective.

The Veneziano amplitude for a two-body reaction is typically represented by the function (or more generally, a sum of functions)

$$V_{lmn}(s, t) = \Gamma(1 - \alpha_s)\,\Gamma(m - \alpha_t)/\Gamma(n - \alpha_s - \alpha_t) \qquad (1)$$

with $\alpha_s = s + a_s$, $\alpha_t = s + a_t$ and an appropriate set of integers l, m, n. V has the duality in the sense that it can be decomposed into an infinite series of resonances either in the s or in the t channel, which converges in its respective physical range, and exhibits the right Regge behavior but for the unphysical infinitely sharp resonance structure superimposed on it. This can probably be remedied by unitarizing V through iteration, perhaps at the risk of violating strict duality, but we will not worry about this question since we do not hold duality to be a sacred principle. Of course we expect that the Regge cuts and even the Pomeranchukon may be generated by starting from V as the input Born term.[6]

Our primary concern will be to find a self-consistent, factorizable set of Veneziano-type amplitudes for all hadronic processes. By this we mean that we want to find a set of hadronic states (represented e.g. by a master wave function with infinite components) and their basic coupling among themselves in such a way that if we build up "tree diagrams" one by one using the basic vertices we would obtain amplitudes for arbitrary many-particle processes and that the result will automatically exhibit duality. That the Veneziano amplitudes can be generalized to many-particle processes has been shown,[7] and we will use the general formula as the basis for our program. In order to build a realistic theory from the quark model it is of course important to discuss the way the $SU(3)$ and the spinor indices are to be coupled, but this is a straightforward problem once the basic topological rules are given.[4] On the other hand, the factorization of analytic functions of many variables requires introduction of a large number of internal coordinates, many more than would correspond to a naive bound state picture. A remarkable point, however, is that this can be done with a finite degree of degeneracy for each level, although the solution is not necessarily a unique one.

To focus our attention on the essential part, we will ignore the spins and consider an n-point Veneziano function for neutral scalar particles. Such a function can be written down according to the known rules, and it takes a particularly simple form in the multiperipheral configuration:

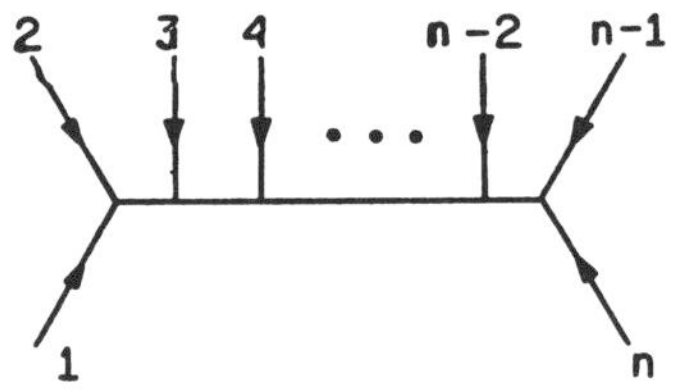

$$V = \int x^{(12)} y^{(123)} z^{(1234)} \cdots w^{(n-1,n)} \times$$
$$\times \left(\frac{1-x}{1-xy}\right)^{(23)} \left[\frac{(1-y)(1-xyz)}{(1-xy)(1-yz)}\right]^{(34)} \cdots \times$$
$$\times \left(\frac{1-xy}{1-xyz}\right)^{(234)} \left[\frac{(1-yz)(1-xyzu)}{(1-xyz)(1-yzu)}\right]^{(345)} \cdots \times$$
$$\times \cdots \times \frac{dx\, dy\, dz \cdots dw}{(1-xy)(1-yz)\cdots} \tag{2}$$

where

$$(12) \equiv -\alpha_2(s_{12}) - 1 = -s_{12} - a_2 - 1, \qquad s_{12} = (p_1 + p_2)^2, \quad \text{etc.}$$
$$(123) \equiv -\alpha_3(s_{123}) - 1 = -s_{123} - a_3 - 1, \quad s_{123} = (p_1 + p_2 + p_3)^2, \quad \text{etc.} \tag{3}$$

This function is invariant under the cyclic permutation $(12 \cdots n) \to (23 \cdots n1)$ and the inversion $(12 \cdots n) \to (n \cdots 21)$. The a's are the intercepts for the i-body Regge trajectories, which need not be equal to each other. Regrouping the individual factors and expanding the variables: $s_{12} = (p_1 + p_2)^2 = 2p_1 \cdot p_2 + 2m^2$, etc., we get

$$V = \int x^{(12)} y^{(123)} \cdots dx\, dy \cdots \times$$
$$\times (1-x)^{-2p_2 \cdot p_3 - c_2} (1-y)^{-2p_3 \cdot p_4 - c_2} \cdots \times$$
$$\times (1-xy)^{-2p_2 \cdot p_4 - c_3} (1-yz)^{-2p_3 \cdot p_5 - c_3} \cdots \times$$
$$\times (1-xyz)^{-2p_2 \cdot p_5 - c_4} (1-yzu)^{-2p_3 \cdot p_5 - c_3} \cdots \times$$
$$\times \cdots \times$$
$$\times (1 - xyz \cdots w)^{-2p_2 \cdot p_{n-1} - c_{n-2}} \tag{4}$$

where

$$c_2 = a_2 + 2m^2 + 1$$
$$c_3 = a_3 - 2a_2 - m^2 \tag{5}$$
$$c_n = a_n - 2a_{n-1} + a_{n-2}, \quad n \geq 4.$$

The last relations may be written in a uniform fashion for all c_n's by defining

$$a_0 \equiv 1, \quad a_1 \equiv -m^2. \tag{6}$$

Thus the external particles belong to the ground state $\alpha_1(m^2) = 0$ of "one-body" resonances.

When a chain of resonances occur simultaneously for $\alpha_2(s_{12}) = k$, $\alpha_3(s_{123}) = m$, $\alpha_4(s_{1234}) = l$, etc., the corresponding residue is given by the coefficient of $x^k y^m z^l \ldots$ in the product of factors $(1 - x)$, $(1 - xy)$, $(1 - xyz)$, ... in the integrand of Eq. (4), and it is this coefficient which we would like to factorize as

$$\sum \langle n| \cdots |l\rangle \langle l| \Gamma |m\rangle \langle m| \Gamma |k\rangle \langle k| \Gamma |1\rangle. \tag{7}$$

Here the summation is over an as yet unspecified set of states which make up each of the levels $k, m, l \ldots$ Note that we are regarding the process as a chain of transitions $1 \to k$, $k \to m$, $m \to l$, etc., induced by the external particles $2, 3, \ldots n - 1$. Thus we are not treating the lines 1 and n on an equal footing with the rest, but this is a natural consequence of the way Eq. (4) is constructed.

As the next step, consider the factor $(1 - x)^{-2p_2 \cdot p_3 - c_2}$. For a 4-point amplitude this would be the only factor to be dealt with. We rewrite it as

$$(1 - x)^{-2p_2 \cdot p_3 - c_2} = \exp\left[-(2p_2 \cdot p_3 + c_2) \ln(1 - x)\right]$$
$$= \exp\left[(2p_2 \cdot p_3 + c_2) \sum_{r=1}^{\infty} \frac{x^r}{r}\right]. \tag{8}$$

The coefficient of x^n is then to be equated with

$$\sum_{-k} \langle 4| \Gamma(p_3) |k\rangle \langle k| \Gamma(p_2) |1\rangle, \tag{9}$$

where p_2 and p_3 are the respective momentum transfers. Let us now introduce a set of Bose-like creation and annihilation operators $a_\alpha^{+(r)}$ $a_\alpha^{(r)}$, $(\alpha = 1, 2, \ldots, 5)$ $r = 1, 2, \ldots$; with the property

$$[a_\alpha^{(r)}, a_\beta^{+(s)}] = -\delta_{rs} g_{\alpha\beta} \quad (\text{metric } - - - + -)$$
$$[a_\alpha^{(r)}, a_\beta^{(s)}] = [a_\alpha^{+(r)}, a_\beta^{+(s)}] = 0. \tag{10}$$

For each r, $a_\alpha^{+(r)}$ and $a_\alpha^{(r)}$ correspond to a 5-dimensional harmonic oscillator in which the 5th dimension is purely fictitious, and the timelike (4th) excitations are "ghosts" with a negative metric. Observe then that the well known formula

$$e^A e^B = e^B e^A e^{[A,B]}$$

(valid for a c-number commutator) gives

$$\langle 0|\, e^{ik' \cdot a/\eta}\, e^{ik \cdot a^+ \xi}\, |0\rangle = e^{k' \cdot k\xi/\eta}\,, \tag{11}$$

where $k \cdot a^+ = k^\alpha a_\alpha^+$. It is obvious that if a complete set of intermediate states characterized by the occupation numbers $\{n_\alpha\}$ are inserted on the left, the states with $n = \Sigma\, n_\alpha$ contribute to the term $\sim(\xi/\eta)^n$ on the right.

Eq. (11) forms the basis of the factorization program. In fact Eq. (8) can be factorized as

$$\exp\,[(2p_2 \cdot p_3 + c_2) \sum x^r/r]$$
$$= \left\langle 0 \left| \exp\left[i\sum \sqrt{\frac{2}{r}}\, a^{(r)} \cdot k'\eta^{-r} \right] \exp\left[i\sum \sqrt{\frac{2}{r}}\, a^{+(r)} \cdot k\xi^r \right] \right| 0 \right\rangle, \tag{12}$$

$$k_\alpha = \left(p_2, i\sqrt{c_2/2}\right), \quad k'_\alpha = \left(p_3, i\sqrt{c_2/2}\right), \quad \xi/\eta = x.$$

The term $\sim x^N$ obviously comes from intermediate states such that

$$\sum_r \sum_\alpha r n_\alpha^{(r)} = -\sum_r r a^{+(r)} \cdot a^{(r)} = N. \tag{13}$$

We may therefore write down the vertex as

$$\langle N|\, \Gamma\, |0\rangle = \left\langle N \left| \exp\left[i\sum \sqrt{\frac{2}{r}}\, a^{+(r)} \cdot k \right| 0 \right\rangle,$$

$$k_\alpha = \left(p_N - p_0, i\sqrt{c_2/2}\right), \quad \langle 0|\, \Gamma\, |N\rangle = \langle N|\, \Gamma\, |0\rangle^+. \tag{14}$$

The level $s = s_N$ then consists of all possible excitations satisfying the condition (13). The 5th dimension was introduced for the purpose of handling the term c_2, assumed to be >0. If $c_2 < 0$ we have to change the metric of the 5th dimension. (The factor i in k_5 is not really necessary; its purpose is to bring Γ to the form (18) below.)

How will this scheme work out for the general n-point function? We can see without much difficulty that if all the c's are equal, we can factorize

Eq. (4) by the ansatz

$$\langle N'|\Gamma|N\rangle = \left\langle N' \left| \exp\left[i\sum\sqrt{\frac{2}{r}}\, a^{+(r)}\cdot k \right] \exp\left[i\dot{\sum}\sqrt{\frac{2}{r}}\, a\cdot k \right] \right| N \right\rangle ,$$

$$k = \left(p_{N'} - p_N, \quad i\sqrt{c_2/2}\right) . \tag{15}$$

This vertex operator is Hermitian except for the timelike excitations, and reduces to Eq. (14) when it operates on $|0\rangle$. In order to see how the ansatz works, we form a product of such Γ's according to Eq. (7), and take its vacuum expectation value. By shifting the annihilation operators step by step to the right we will note that the 5-momenta $k_1, k_2, \ldots$ associated with the external lines get contracted with one another, producing an exponential factor for each contraction. To extract a particular set of intermediate states $N_1, N_2, \ldots$, multiply $a^{(r)}$ and $a^{+(r)}$ in each vertex $\langle N_l|\,\Gamma\,|N_{l-1}\rangle$ by the power-counting variables $(\xi_{l-1})^{-r}$ and $(\xi_l)^r$ respectively, and pick up the coefficient of $\xi_1^{N_1}\,(\xi_2/\xi_1)^{N_2}\,(\xi_3/\xi_2)^{N_3}\ldots$ from the final result. We can verify raedily that the variables ξ_1, ξ_2/ξ_1, ξ_3/ξ_2, ... correspond to x, y, z, ... in Eq. (4).

In actuality, the constants c_n are not all equal. If all intercepts a_n are equal, i.e. all multiparticle resonances belong to the same trajectories, we have $c_m = 0$ for $m \geq 4$, but c_2 and c_3 are different. So the 5th dimension is necessary only for the nearest neighbor and the next nearest neighbor contractions, and we need to devise a scheme which treats these special cases. Another example is the case where we distinguish between even-n and odd-n resonances so that

$$a_0 = a_2 = \cdots = a_{\text{even}}, \quad a_1 = a_3 = \cdots = a_{\text{odd}}, \quad c_{\text{even}} = -c_{\text{odd}}$$

for all n. According to Eq. (5), however, this requires $a_{\text{even}} = 1$ and the external lines belong to α_{odd}. The alternating signs of c_n can be realized by a suitable definition of Γ. This example is relevant to the many-pion processes which have, however, $a_{\text{even}} = \frac{1}{2}$ rather than 1. We shall not go into the details of how one would handle these problems more precisely.

We have outlined above the basic procedure for factorizing n-point Veneziano amplitudes in a consistent way. For this we had to pay the price of introducing an infinite set of 5-dimensional harmonic oscillators, although this does not affect the leading trajectory. It is clear that each of them generates a tower of non-unitarity representations of $U(4, 1)$, leading to the unpleasant presence of ghosts. However, this does not necessarily mean that the amplitudes themselves are unphysical since there can be cancellations from various intermediate states.[5] For example, the elastic pion–pion scattering amplitude seems to be free of ghosts.[8] But the general answer to this question is not known yet.

The appearance of harmonic oscillators in our problem is intriguing since the simple bound-state picture of quarks with a harmonic oscillator potential would naturally give rise to linear trajectories and the $U(3, 1)$ level scheme. We can bring out this analogy more clearly in the following way. Let us introduce a Bose field $\phi_\alpha(\xi)$ and its canonical conjugate $\pi_\alpha(\xi)$, which are even and periodic with periodicity 2π in ξ. In analogy to the ordinary field theory, we decompose it into plane waves:

$$\phi_\alpha(\xi) = \sum_{r=1}^{\infty} \frac{1}{\sqrt{2r}} (a_\alpha^{(r)} + a_\alpha^{+(r)}) \cos r\xi$$

$$\pi_\alpha(\xi) = \sum_{r=1}^{\infty} i\sqrt{\frac{r}{2}} (a_\alpha^{(r)} - a_\alpha^{+(r)}) \cos r\xi \tag{16}$$

where we have excluded the constant mode $r = 0$. The a's and a^+'s are the operators we have defined above. In view of Eq. (13), the quantum number N which determines the resonance energy can be written

$$\begin{aligned} N &= -\sum_r r a^{+(r)} \cdot a^{(r)} \\ &= \frac{-1}{\pi} \int_0^{2\pi} :(\partial_\xi \phi(\xi) \cdot \partial_\xi \phi(\xi) + \pi(\xi) \cdot \pi(\xi)): d\xi. \end{aligned} \tag{17}$$

Furthermore, the vertex operator Γ can be written

$$\Gamma = \; :\exp[2ik \cdot \phi(0)]: \tag{18}$$

Eq. (17) suggests that the internal energy of a meson is analogous to that of a quantized string of finite length (or a cavity resonator for that matter) whose displacements are described by the field $\phi_\alpha(\xi)$. A master field Ψ representing the mesons will depend on the space-time coordinates x_μ as well as the internal variables $\phi_\alpha(\xi)$, and satisfy the free wave equation

$$(\Box - N - c)\, \Psi(x; \phi) = 0. \tag{19}$$

Its interaction is described by a Lagrangian $\sim \Psi^+(x)\, \Gamma \Psi(x)\, \varphi(x)$ where φ is the field for the external particles.

The form (18) has a striking resemblance to the form factor in a naive quark model, in which one of the quarks interacts with the external field carrying momentum k. If the $q\bar{q}$ system is described by a wave function $\Psi(x; r)$, where x and r are the center of mass and relative coordinates respectively, the interaction will carry a form factor exp $[ik \cdot r/2]$. An unsatisfactory feature of the result is that the external field φ and the master field Ψ have to be distinguished, whereas clearly the starting program was to include everything in Ψ. To achieve this end one would need to find a trilinear coupling of Ψ's which agrees with the coupling that arises when the n-point Veneziano amplitude is in a configuration in which three resonances meet. The above interpretation of the form factor suggests that one might be able to construct such a coupling geometrically from quark diagrams like

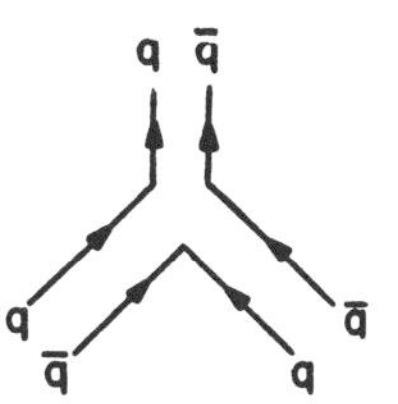

If this can be carried out, we are essentially led to a nonlocal picture of the Veneziano model, though the nonlocality may not manifest itself in a conspicuous way.

Another problem that confronts a field theory of duality is that of the multiple counting of diagrams. With a trilinear coupling of Ψ's one can construct diagrams of both s-channel and t-channel types. Because of duality they will of course give the same contributions, but the amplitude is doubled nevertheless. We can adjust for this by reducing the trilinear coupling parameter, but what about the higher order processes? Even if the factorization may work formally, it is not clear whether or not we should build a field theory in this fashion.[9] That will depend on how one interprets duality, and must eventually be solved through a confrontation with experiment.

Putting aside these problems, we can make some remarks about high energy phenomena based on our results. Since the internal excitations of hadrons resemble those of a cavity resonator, we are able to use statistical concepts like temperature, entropy and the Bose–Einstein distribution in a well defined way. For example, at extreme high energies ($\gg 1$ GeV) where the discreteness of eigenfrequencies becomes unimportant, we find

$$M^2 \sim E = TS + F \sim -F \sim (5\pi^2/6)\,(kT)^2, \qquad (20)$$

where M is the rest energy of the system, and the other symbols are the standard thermodynamic ones. Thus the density of the levels goes like exp $[S/k]$ $\sim$ exp $[2M\,(5\pi^2/6)^{1/2}]$.

References

1 R. Dolen, D. Horn and C. Schmid, *Phys. Rev.* **166**, 1768 (1968).
2 G. Veneziano, *Nuovo Cimento* **57A**, 190 (1968).
3 H. Harari, *Phys. Rev. Letters* **22**, 562 (1969); J. Rosner, *Phys. Rev. Letters* **22**, 689 (1969); J. Mandula, C. Rebbi, R. Slansky, J. Weyers and G. Zweig, *Phys. Rev. Letters* **22**, 1147 (1969); K. Matsumoto, *Lettere al Nuovo Cimento* **1**, 620 (1969); T. Matsuoka, K. Ninomiya and S. Sawada, Nagoya Univ. preprint.
4 S. Mandelstam, A Relativistic Quark Model Based on the Veneziano Representation, Berkeley preprint; K. Bardakci and M. B. Helpern, A Possible Born Term for the Hadron Bootstrap, Berkeley perprint; K. Bardakci and S. Mandelstam, Analytic Solution of the Linear-Trajectory Bootstrap, Berkeley preprint.
5 S. Fubini and G. Veneziano, Level Structure of Dual Resonance Models, MIT preprint.
6 P. G. O. Freund, *Phys. Rev. Letters* **22**, 565 (1969); K. Kikkawa, B. Sakita and M. A. Virasoro, University of Wisconsin preprint.
7 C. Goebel and B. Sakita, *Phys. Rev. Letters* **22**, 257 (1969); H. M. Chan, *Phys. Letters* **28B**, 425 (1969); H. M. Chan and T. S. Tsou, *Phys. Letters* **28B**, 485 (1969); J. F. L. Hopkins and E. Platt, *Phys. Letters* **28B**, 489 (1969); Z. Koba and H. B. Nielsen, *Nucl. Phys.* **B10**, 633 (1969).
8 Y. Nambu and P. Frampton, University of Chicago preprint EFI 69–56.
9 I thank P. Freund for a discussion on this point.

DISCUSSION

ROSNER Could you amplify that discussion in which you show by statistical argument the way the states should go considering the square root of N.

NAMBU You can use the standard formulae of statistical mechanics for a cavity resonator. The eigenfrequencies are quantized in units of $\omega_0 \simeq 1$ GeV, but a high temperatures $kT \gg \omega_0$ we may ignore the discreteness of energy, and integrate over the Bose–Einstein distribution. We find

$$E = TS + F \simeq -F \simeq (5\pi^2/6) \quad (kT)^2$$

where F is the Gibbs free energy, S is the entropy, and $E = M^2 \simeq N$, the square of the rest energy of the system since we are dealing with the Mandelstam variable s. This relation shows that $T \simeq S \simeq M \simeq \sqrt{N}$. So the level density goes like $\exp [S/k] \simeq \exp \left[c \sqrt{N}\right]$.

GUNZIK How do you get rid of the ghosts in your approach?

NAMBU I do not know. One thing I have thought about is the following. The vertex Γ contains the operators $a^+ \cdot k$ and $a \cdot k$ which cause excitations or dexcitations in the direction k of the oscillator space. If k is spacelike, $k \cdot k < 0$, you might say the excitation is normal; if it is timelike, $k \cdot k > 0$, you excite ghosts. This will give you a condition of the Regge intercept a since k depends on it. Applied to a_2, we get $1 + a_2 > 0$. But this is not what you get from the positivity of partial wave residues in a 4-point amplitude. Thus the above argument seems to be shaky.

GUNZIK Is k the momentum vector in the intermediate state?

NAMBU The first four components of k is the momentum transfer, the fifth is essentially the square root of the constant c, which depends on the Regge intercept.

BREIT Why do you have factorization?

NAMBU Well, without factorization, it would be hard to understand the Veneziano amplitudes in terms of the usual quantum mechanical description of resonances. I wanted to show that factorization is possible, but at the price of having ghost states.

Duality and Hadrodynamics†

Y.Nambu

Enrico Fermi Institute and

Department of Physics

University of Chicago

Notes prepared for the Copenhagen
High Energy Symposium, Aug. 1970
(Unpublished and undelivered)

(Supported in part by the U.S.Atomic Energy
Commission under the Contract No.AT(11-1)-264)

†Retypeset by M. Okai, Nov. 1993

1 Introduction

We would like to explore here possible dynamical properties of hadrons suggested by the duality principle of Dolen, Horn and Scmid, and in particular by its mathematical realization due to Veneziano. Our primary concern will be not in the very interesting details of the mathematical properties that the Veneziano model seems to possess in abundance, but rather in guessing at the internal structure and dynamics of hadrons which underlie the Veneziano model, recognizing that the latter is in all likelihood only an approximate and imperfect representation of the former.

On crucial step we take in interpreting the Veneziano model is factorization. Of course this is a trivial problem, at least in principle if not in practice, if one is dealing with a given four-point amplitude. But it becomes a very restrictive condition if one demands that the same set of resonances saturate all $n-$point amplitudes. Perhaps such a condition is unwarranted; in discussing as revolutionary a concept as duality one may have to give up all the conventional notions about resonances, e.g. that the intrinsic properties of a resonance is independent of how it is prepared, and a multiple resonance can be analyzed in terms of the individual resonances, etc. Nevertheless, the fact that a set of resonances have been found to saturate all the $n-$point Veneziano amplitudes of the standard variety is very significant. It suggests that our conventional notions about resonances still make some sense, and the know techniques of field theory can be applied to them.

The condition of factorizability immediately rules out an ad hoc addition of satellite terms to a scattering amplitude because it would in general destroy factorization unless more and more resonances are introduced. If satellite terms do arise, they must do so in a well defined and self-consistent way.

As it stands now, the Veneziano model is still beset with many difficulties. For all its mathematical elegance, its practical successes are few. It would therefore be appropriate to list here its basic predictions as well as difficulties without going into the details.

Basic predictions.

1) Linearly rising trajectories. This is in accordance with observation. Whether the trajectories really keep rising indefinitely or not is of course an open question, but it seems to be a valid simplification to assume that they do.

2) Regularly spaced daughter trajectories, implying a highly degenerate level structure. Again the actual degeneracy may be only approximate, but these

daughters must exist if the model makes sense at all. So far there is some evidence for ϵ (daughter of ρ), but none for ρ' and ϵ' (daughter of f).
3) The factorized Veneziano model implies an even higher degree of degeneracy. At each level many distinguishable states having the same spin exist. The number of states increases with energy as $\sim \exp(cE)$. This situation bears a striking similarity to Hagedorn's model of high energy reactions.

Unsolved practical problems.
1) Baryons. We do no yet have a quantitatively satisfactory picture of meson-baryon and baryon-baryon scattering based on the Veneziano model, in spite of the fact that duality was first discovered in meson-baryon scattering.
2) A general satisfactory unification of the quark model (or its $SU(3)$, $SU(6)$ and chiral $SU(3)$ aspects) with the Veneziano model does not exist yet.
3) No convincing theory of form factors exists.
4)We do not know what the Pomeron is within the framework of the Veneziano model.

We must emphasize, however, that there are numerous attempts and speculations regarding all these problems.

More fundamental difficulties.
1) Unitarity. The original Veneziano model is a zero width approximation. The amplitude wildly oscillates with energy, and only after averaging over an interval does it reproduce the smooth Regge behavior at high energy. If a basic Hamiltonian for the model is given, unitarization might be formally carried out by taking into account higher order processes, although the highly singular nature of the Hamiltonian casts doubt about its meaningfulness.
2) A more serious problem, however, is the existence of ghosts. Here we mean by ghosts unphysical particles having a) negative probability, or b) spacelike momenta (tachyons), or both. The levels of the factorized Veneziano model contain those of four-dimensional harmonic oscillators, where the timelike excitations have an intrinsic negative norm. This does not necessarily mean that the Breit-Wigner residue of a partial wave projection of a given amplitude is not positive. Contributions from many degenerate states can add up to a positive value, but there is no guarantee that this will always happen. Fubini and Veneziano have found a Ward-like identity which accomplishes this cancellation to a certain extent, and Virasoro has extended their result. They rely, however, on very special properties of the Hamiltonian, and it is not clear whether and how these can be preserved in general. Besides, Virasoro's scheme still leaves us with a tachyon ghost ($m^2 = -1$). The killing of

tachyons is easy in special cases like $\pi - \pi$ scattering (where the ρ trajectory has a potential tachyon), but a general prescription in a factorized model seems very complicated, if not impossible.

We have emphasized the ghost problem because this is not peculiar to the Veneziano model alone, but it is rather a common disease afflicting all attempts at a description of hadron states as infinite multiplets. The program of current algebra saturation, as well as the use of infinite-component wave equations, have floundered on the same difficulty. To a certain extent, the two kinds of ghosts seem to be complementary: The use of finite-dimensional Lorentz tensors (as in the factorized dual model) involves negative probabilities, whereas infinite-dimensional unitary representations in general lead to tachyons.

If these ghosts are so difficult to eliminate, why not accept them and look for them? Maybe ghosts of one kind or the other do exist, which would make either T.D.Lee or G.Feinberg happy (or both, plus Sudarshan and others). But the trouble is that the extent to which these ghosts appear in a particular problem seems to depend on one's cleverness and ability to avoid them. How many of the ghosts are "real"? This is the most serious question of principle that haunts us, especially us the theorists.

2 Factorized Veneziano model

As has been shown by various people, the $n-$point dual amplitude for scalar external particles can be factorized in terms of a set of harmonic oscillators corresponding to an elastic string of finite intrinsic length. In classical theory, the motion of a free mass point can be derived from the action integral

$$I = -m \int d\tau, \quad d\tau^2 = -dx_\mu dx^\mu \quad (\text{metric } (-+++)) \tag{1}$$

Alternatively one may take

$$I' = \frac{1}{2} m \int \left(\frac{dx_\mu}{d\tau} \frac{dx^\mu}{d\tau} - 1 \right) d\tau \tag{2}$$

τ being an independent parameter, and $x^\mu(\tau)$ the dynamical variables. (The constant in the integrand is added for convenience.) This form is more suitable for the transition to quantum theory. Because of the translational invariance under $\tau \to \tau + c$ and $x^\mu \to x^\mu + a^\mu$, both the Hamiltonian and the momenta

$$\begin{aligned} H &= (p_\mu p^\mu + m^2)/2m \\ p^\mu &= m dx^\mu/d\tau \end{aligned} \tag{3}$$

are conserved. By imposing the constraint

$$\frac{dx_\mu}{d\tau}\frac{dx^\mu}{d\tau} = -1 \tag{4}$$

we can normalize the parameter τ, which then becomes the proper time of the mass point. This condition (4) amounts to

$$H = 0. \tag{5}$$

In quantum mechanics, one postulates the commutation relations

$$[x^\mu, p^\nu] = i g^{\mu\nu} \tag{6}$$

and the Schrödinger equation

$$i\frac{\partial}{\partial\tau}\Psi = H\Psi \tag{7}$$

Eq.(5) is to be replaced by

$$H\Psi = 0 \tag{8}$$

which is nothing but the Klein-Gordon equation

$$(p^\mu p_\mu + m^2)\Psi = 0. \tag{9}$$

We see thus the usefulness of the proper-time formalism. If we just integrate Eq.(7), we get

$$\Psi(\tau) = \exp[-i(p^2 + m^2)\tau/2m]\Psi(0) \tag{10}$$

or

$$\begin{aligned} (\Psi(x';\tau), \Psi(x;0)) &= \langle x'|\exp[-i\tau(p^2+m^2)/2m]|x\rangle \\ &= (-im^2/4\pi^2)\exp[imx^2/2\tau - i\tau m/2] \\ &= (-im^2/4\pi^2)\exp[im\tau\{(x/\tau)^2 - 1\}/2] \end{aligned} \tag{11}$$

This is the transition amplitude $x \to x'$ after an elapsed "time" τ. Integrating over τ from 0 to ∞ (we might say it does not matter how much time it has elapsed between the events x and x'), we obtain the Feynman propagator

$$\frac{1}{2m}\int_0^\infty e^{-i\tau(p^2+m^2)/2m}d\tau = -i(p^2+m^2-i\epsilon)^{-1} \tag{12}$$

It is well known that the above results may be interpreted or derived from the path integration method.

After this digression, let us come to the elastic string. We can imagine it to be the limit of a chain of N mass points as $N \to \infty$. Each mass point will trace out a world line, so that in the limit we are dealing with a two-dimensional world sheet. This sheet may be parametrized by two intrinsic coordinates ξ ($0 \le \xi \le \pi$, let us say) and τ ($-\infty < \tau < \infty$), corresponding to spacelike and timelike coordinates. We assume the action integral

$$I = \frac{1}{4\pi}\iint \left(\frac{\partial x_\mu}{\partial \tau}\frac{\partial x^\mu}{\partial \tau} - \frac{\partial x_\mu}{\partial \xi}\frac{\partial x^\mu}{\partial \xi}\right) d\xi d\tau \tag{13}$$

from which follows the equation

$$(\partial^2/\partial\tau^2 - \partial^2/\partial\xi^2)x^\mu = 0 \tag{14}$$

with the boundary condition

$$\partial x^\mu/\partial\xi = 0 \qquad \text{at } \xi = 0,\ \pi. \tag{15}$$

Duality is essentially a result of the symmetry between ξ and τ though it is not yet a perfect symmetry because of the differences in domain and metric. We have chosen the hyperbolic form because only then can one formulate the Hamiltonian principle. Actually it turns out that in computing the scattering amplitudes a switch to an elliptic form through the change $\tau \to -i\eta$ brings out duality more explicitly.

Eq.(13) is invariant under the translations $\tau \to \tau + c$ and $x^\mu \to x^\mu + a^\mu$, which imply the conservation laws

$$\begin{aligned} H &= \int [\pi p_\mu p^\mu + \frac{1}{4\pi}\left(\frac{\partial x_\mu}{\partial\xi}\right)\left(\frac{\partial x^\mu}{\partial\xi}\right)]d\xi = \text{const.} \\ p^\mu &= (1/2\pi)(\partial x^\mu/\partial\tau), \\ P^\mu &= (1/2\pi)\int(\partial x^\mu/\partial\tau)d\xi = \int p^\mu d\xi = \text{const.} \end{aligned} \tag{16}$$

in direct analogy with Eq.(3). A normal mode decomposition of Eqs.(14) and (15) yields

$$\begin{aligned} x^\mu &= x_0^\mu + 2\sum_{n=1}^{\infty} x_n^\mu \cos n\xi \\ p_n^\mu &= \partial x_n^\mu/\partial\tau \ \ (n \neq 0), \ \ p_0^\mu = \frac{1}{2}\partial x_0^\mu/\partial\tau \\ H &= p_{0\mu}p_0^\mu + \frac{1}{2}\sum_{n=1}^{\infty}(p_{\mu n}p_n^\mu + n^2 x_{\mu n}x_0^\mu \\ P^\mu &= p_0^\mu \end{aligned} \tag{17}$$

We may interpret x_0^μ and p_0^μ as the center-of-mass coordinates and momenta.

When the system is quantized, we get the familiar expression

$$\begin{aligned} &H = P_\mu P^\mu + H_0, \ \ H_0 = \sum_{n=1}^{\infty} n a_{\mu n}^\dagger a_n^\mu (+\text{c number}) \\ &x_n^\mu = (a_n^\mu + a_n^{\mu\dagger})/\sqrt{2n} \\ &p_n^\mu = -i(a_n^\mu - a_n^{\mu\dagger})\sqrt{n/2} \ \ (n \neq 0) \\ &[a_n^\mu, a_n^{\nu\dagger}] = g^{\mu\nu}\delta_{nm} \end{aligned} \tag{18}$$

By imposing the subsidiary condition

$$(H - \alpha)\Psi = 0 \tag{19}$$

we can single out an infinite tower of states with the mass spectrum

$$M^2 = H_0 + \alpha = \sum_{\mu,n} nN_{\mu n} + \alpha \tag{20}$$

In order to construct dual scattering amplitudes we introduce an external scalar field φ and postulate

$$H = P_\mu P^\mu + H_0 + g : \varphi(x(\xi = 0)) : \tag{21}$$

We will not discuss how this leads to $n-$point dual amplitudes in the multi-peripheral configuration since it is well known.

We now add several remarks.

1) The condition $0 \leq \xi \leq \pi$ fixes a fundamental scale of length. ξ is here

measured in $(GeV/c)^{-1}$ to fit the trajectory slope of ~ 1. Whether the system is to be interpreted as a rubber string or a rubber band depends on the boundary condition to be imposed. The rubber band, as twice as many modes as the string, but the difference does not show up in a case like Eq.(21) because those modes which have nodes at $\xi = 0$ cannot be excited. Clearly there will be differences if one tries to extend the model, and this will be an important point in constructing a general theory of hadrons.
2) For a point particle Eq.(1) has a purely geometric meaning (as the length of a world line), but Eq.(2) does not since it depends on the scale (gauge) of the unphysical parameter τ. In the case of the string, on the other hand, Eq.(13) is invariant under the scaling $\tau, \xi \to \lambda\tau, \lambda\xi$. This is one aspect of the conformal invariance of two-dimensional Laplace equations, a property which has widely been utilized to study the Veneziano model.

Nonetheless, Eq.(13) is not a purely geometrical quantity. For curiosity, then, let us try to construct a geometric action integral as one does in general relativity. Obviously a natural candidate for it is the surface area of the two-dimensional world sheet; another would involve its Riemann curvature. The sheet is imbedded in the Minkowskian 4-space, so one can parametrize its points as $y^\mu(\xi^0, \xi^1)$, $(\xi^0 \sim \tau, \xi^1 \sim \xi)$. The surface element is a $\sigma-$tensor

$$\begin{aligned} d\sigma^{\mu\nu} &= G^{\mu\nu} d^2\xi, \\ G^{\mu\nu} &= \partial(y^\mu, y^\nu)/\partial(\xi^0, \xi^1) \end{aligned} \tag{22}$$

whereas its line element is

$$\begin{aligned} ds^2 &= g_{\alpha\beta} d\xi^\alpha d\xi^\beta \qquad (\alpha, \beta = 0, 1) \\ g_{\alpha\beta} &= (\partial y_\mu/\partial\xi^\alpha)(\partial y^\mu/\partial\xi^\beta) \end{aligned} \tag{23}$$

A possible action integral would be

$$I = \int |d\sigma_{\mu\nu} d\sigma^{\mu\nu}|^{1/2} = \iint |2\det g|^{1/2} d^2\xi \tag{24}$$

to be compared with the old one (13) which can be written $(y \to x)$

$$I = -\frac{1}{4\pi} \iint g_{\alpha\beta} \overset{\circ}{g}{}^{\alpha\beta} d^2\xi, \quad \overset{\circ}{g}{}^{\alpha\beta} = \begin{pmatrix} -1 & 0 \\ 0 & 1 \end{pmatrix} \tag{25}$$

It is obvious that Eq.(24) leads to nonlinear equations. More complicated equations involving curvature would be not only nonlinear, but also have

higher derivatives.
3) It is sometimes useful to consider the "energy-momentum tensor" in the (ξ, τ) space:

$$T_{\alpha\beta} = \frac{1}{2\pi}(g_{\alpha\beta} - \frac{1}{2} \overset{\circ}{g}_{\alpha\beta}\, g_{\gamma\delta}\, \overset{\circ}{g}^{\gamma\delta}) \tag{26}$$

In particular, let us take its space integral over a test function $f(\xi)$,

$$\bar{T}_{\alpha\beta}[f] = \int_0^\pi T_{\alpha\beta}(\xi) f(\xi) d\xi. \tag{27}$$

By virtue of the canonical commutation relations, they generate a commutator algebra

$$\begin{aligned} &\left[(\bar{T}_{00} \pm \bar{T}_{01})[f], \quad (\bar{T}_{00} \pm \bar{T}_{01})[g] \right] = -2i(\bar{T}_{00} \pm \bar{T}_{01})[h], \\ &\left[(\bar{T}_{00} \pm \bar{T}_{01})[f], \quad (\bar{T}_{00} \mp \bar{T}_{01})[g] \right] = 0, \\ &h = f'g - fg' \end{aligned} \tag{28}$$

as an integral form of the Schwinger conditions. These relations, when applied to the set $f_n = 1 - e^{-2in\xi}$, amount to

$$\begin{aligned} &\left[L_n^\pm, \quad L_m^\pm \right] = 2(n-m)L_{n+m}^\pm \\ &\left[L_n^\pm, \quad L_m^\mp \right] = 0 \\ &L_n^\pm = (\bar{T}_{00} \pm \bar{T}_{01})[e^{2in\xi}], \quad L^\pm[f_n] = L_0^\pm - L_{-n}^\pm \end{aligned} \tag{29}$$

These operators $L^\pm[f_n]$ have been found useful in generating the various gauge operations.
4) As we have mentioned already, the most serous defect of the above formulation is the indefinite metric that appears in defining the covariant commutation relations (18). The mass operator (20), however, acquires as a result the nice property of being positive. The transition from a classical to quantum picture of 4-dimensional harmonic ocsillators is a drastic one. A wave function in coordinate space would behave like $\exp[-c^2(\underline{x}^2 - x_0^2)]$, which explodes in the timelike direction. We could actually insist that it should behave instead like $\exp[-c(\underline{x}^2 + x_0^2)]$. This would amount to interchanging the creation and annihilation operators and using positive metric for the time component. But then the mass operator would not be positive, and moreover

each level would become infinitely degenerate. In group theoretical terms, the former corresponds to non-unitary, and the latter to unitary representations of $U(3,1)$. The former have negative probability ghosts while the latter have negative mass squared ghosts. The Veneziano model seems to prefer the former. Such a choice is necessary to ensure a Regge behavior à la Van Hove, but runs into trouble with form factors. This is a general agony of making the choice, not restricted to Mr.Veneziano alone.
5) We have ignored the problems of the extra scalar excitations which are needed to incorporate the proper trajectory intercepts in the dual channel. This is another rather unphysical aspect of factorization. These extra modes may be taken either as a set of harmonic oscillators in a fifth dimension or as a modification of the propagator. Whether these states have positive metric or not depends on the intercept and the external masses. Actually a sixth dimension would be necessary to take care of two-particle trajectories correctly. We have no illuminating interpretations to offer on this subject.

3 Quarks and the dual model

What we propose here is a program of building a general picture of the structure of hadrons on the basis of the factorized Veneziano model. It has been noted by Harari and Rosner that the duality may be interpreted schematically in terms of quark diagrams, which have a strong predictive power, albeit of qualitative nature. These diagrams are indeed very suggestive. First of all they agree with the foregoing picture that the hadrons form two-dimensional sheets in space-time. Furthermore, they imply that quarks and antiquarks form the boundary lines of the sheets. A meson system, for example, is then a $q - \bar{q}$ molecule bound by an elastic string. We could also imagine a rubber band in which q and $\bar{q}$ are attached to diametrically opposite points. To go on further, we have to make a choice.

On the basis of the duality diagram picture, we will adopt the linear molecule rather than the benzene ring. An advantage of the linear picture is that a linear chain can be broken in two linear chains, thereby accounting for the production mechanism

$$A \longrightarrow B + C \tag{30}$$

In fact the duality diagrams can be interpreted exactly in this way.

To make things a bit more sophisticated, we will present a modified version of the quark model. This is the three-triplet model proposed by Dr. Han and me some time ago. It had the advantage of a) having integral charges, b) naturally accounting for the zero triality of known hadrons, as well as c) for the $SU(6)$ classification of the baryons. In this scheme there are nine fundamental fermions grouped into three $SU(3)$ triplets. We may use the notation T_i^n; $i, n = 1, 2, 3$, where n distinguishes between different triplets. T_i^n behaves like a triplet representation in the ordinary $SU(3)$ space (lower index), and like an antitriplet representation in the new $SU(3)$ space (upper index). These $SU(3)$ spaces are denoted as $SU(3)'$ and $SU(3)''$ respectively. Each space has its own isospin and hypercharge, and the electric charge is the sum of two charge operators $\lambda'_Q + \lambda''_Q$. We call $3\lambda''_Q$ the charm number C. We will also make things a bit more exciting by naming the three different triplets D, N, and A. Their quantum number assignments are given in the table.

	C		Q	
D_i	1	1,	0,	0
N_i	-2	0,	-1,	-1
A_i	1	1,	0,	0

In the lowest approximation, $SU(3)'$ and $SU(3)''$ are separately good symmetries. In particular, all low lying hadron states are assumed to belong to $SU(3)''$ singlets, which turn out to correspond to only zero triality states in $SU(3)'$. (The same is accomplished also by assuming zero charm for hadrons.) Baryons and mesons have the usual pattern TTT and $T\bar{T}$. More precisely,

$$\begin{aligned} &B \sim DNA \\ &M \sim D\bar{D} + N\bar{N} + A\bar{A} \end{aligned} \tag{31}$$

where B is completely antisymmetric in the $SU(3)''$ space in order to be a singlet, and this takes care of the Pauli principle.

The preference of zero triality is thus reduced in this model to the $SU(3)''$ symmetry, which one may attribute to a dynamical property of superstrong interactions having a larger scale of masses ($\gtrsim 1GeV$) than for the strong interactions ($\sim 1GeV$).

The combination of duality and the three-triplet model will then produce the following picture. The hadrons are "molecules" bound by superstrong interactions, with the bond structure

$$\begin{aligned} B &\sim T - T - T \\ M &\sim T - \bar{T} \end{aligned} \tag{32}$$

The bonds must have a saturation property for zero triality. In the original scheme demanding perfect $SU(3)''$ symmetry, the baryon would have to have either a ring structure

$B \sim$ 

(33)

or a resonating linear structure

$$B \sim D - N - A \quad + \quad \text{permutations.} \tag{34}$$

If $SU(3)''$ symmetry is abandoned and only the neutral charm condition is imposed, we may simply assume

$$B \sim D - N - A \tag{35}$$

In the latter case the charm number is equivalent to valency. But then we lose the distinction between D and A, drifting back to a two-triplet model. Our tentative preference is in Eq.(34) though the other two possibilities should not be ignored. The meson scheme would follow Eq.(31).

What can we do with this model? We have now hadrons endowed with $SU(3)$ and Dirac spin. These are more or less localized at certain points along the string whose function is to carry bulk of the energy and momentum of the system. The triplets (or simply quarks) themselves are massless, or have only small masses. Several remarks are in order.

1) The interaction process (30) is viewed as a creation of pair $T\bar{T}$ at a point where the break occurs. After the cut, each portion subsequently grows into a full grown string, like an earthworm! This would not be possible if the

string is made up of a fixed number of mass points. We must conclude that the number is not only large but also indefinite. Let us examine this situation a little further. Take a string stretched between two fixed points in space with a distance L apart. If the number of mass points is N, the potential energy is

$$V \sim N(L/N)^2 = L^2/N \tag{36}$$

which depends on N. However, if the number of discrete steps in the "time" direction τ also increases with N to sustain duality, the action integral $I \sim NV \sim L^2$ will be independent of N. This is the scale invariance we have discussed. In units of the fictitious time τ, a system lives longer the larger the number N. The actual state of a hadron would be a linear superposition of configurations having different values of N, but their contributions are all proportional to each other.

If this view is accepted, we can define the Hamiltonian (or vertex) responsible for the process (30) as an overlap integral of the three wave functions corresponding to the states A, B, and C. There is a selection rule

$$N_A = N_B + N_C \quad (N_i > 0) \tag{37}$$

and its two cyclic permutations, either one of which must be satisfied. A more explicit expression satisfying (37) would look like

$$\sum_{0<N'<N} C_{N'N} \int \cdots \int \Psi_B^*(x^{(1)}, \cdots, x^{(N')}) \Psi_C^*(x^{(N'+1)}, \cdots, x^{(N)}) \\ \Psi_A(x^{(1)}, \cdots, x^{(N)}) F(x^{(1)}, \cdots, x^{(N)}) \prod d^4x^{(i)} \tag{38}$$

where F is some scalar function. This integral can be appropriately rewritten in terms of the wave functions $\Psi[x(\xi)]$ in the limit N, $N' \to \infty$.
In the actual Veneziano model, the interaction is such that the breaking of a string occurs only at one of the ends, which may be interpreted to mean the limit

$$N \to \infty; \quad N'/N \to 0 \quad \text{or} \quad N'/N \to 1$$

In general, however, there is no reason to impose such a condition. We would still get a dual theory of sorts; an amplitude corresponding to one

configuration will have singularities in the crossed channels too, though there may not be a symmetry between dual channels in individual amplitudes.
2) The triplet or the quark fields introduce extra spins to the system in conformity with the $SU(6)$ type theories, so π and ρ mesons belong to the s wave states of the string. But this also brings in the old headaches of relativistic $SU(6)$ theories as well. How can one eliminate half of the Dirac components in order to avoid parity doubling and ghost states? The difficulty is compounded by the fact that we would like to maintain duality too. A possible scheme based on the Carlitz-Kisslinger type cut mechanism has been developed by Freund et al. We will not discuss it here. Instead, we would like to propose a general formalism which tries to accomplish this in a dynamical way. The basic idea is as follows. Instead of regarding the triplets and the strings as separate entities, let us take a unified picture and replace the string with a chain of $T\bar{T}$ pairs, so that it would look like a polarized medium with opposite charges created at its ends. There will be interactions between neighbors which depend on their Dirac, $SU(3)'$, and $SU(3)''$ spins. To be dual, these interactions must occur in the "time" direction too, thus forming a two-dimensional polarizable medium. These interactions would contribute to the action integral in addition to the kinetic term represented by the string Hamiltonian. Roughly speaking, that would produce a spin and $SU(3)$ dependence of various trajectories. It is conceivable that parity doubling and other problems can be reduced in the same way to ones of dynamical stability. An example of $T\bar{T}$ interaction might be

$$I = g \sum_{(n,n')} (\gamma_\mu^{(n)} \gamma^{\mu(n')} + \text{const}) \tag{39}$$

where $\gamma_\mu^{(n)}$ refers to a triplet sitting on a site n of a two-dimensional lattice, and n' refers to one of its neighbors. We must choose the constant g in such a way that a chain $T\bar{T}T\bar{T}\cdots\bar{T}$ will have the lowest energy (which may be adjusted to zero). To be dual, the same patter must be repeated in the time direction too. We thus end up with a pattern

$$\begin{matrix} T\bar{T}T\bar{T}..........\bar{T} \\ \bar{T}T\bar{T}T..........T \\ T\bar{T}T\bar{T}..........\bar{T} \\ \bar{T}T\bar{T}T..........T \end{matrix}$$

....

....

In other words, it is a two-dimensional antiferromagnet or ionic crystal! External particles should couple to it like a magnetic field couples to the spin:

$$I' = \sum_n \gamma_\mu^{(n)} \phi^{\mu(n)} \tag{40}$$

The scattering amplitude would then be obtained, following the Feynman principle, from an expression like

$$\text{Tr} \quad \exp[I + I'] \tag{41}$$

the trace being taken with respect to the γ matrices. Eq.(41) is nothing but a partition function! It is not scale invariant, but rather an extensive quantity proportional to the number of constituents (unless $I = 0$). Thus the probability of creating exotic states having many disordered atoms (Large $|I|$) would be severely cut down.

The simplest mathematical model of the above type is the well known Ising model with its glorious Onsager solution. Perhaps we can adopt the Ising model here as a prototype. The transition from a Lagrangian to a Hamiltonian formalism is accomplished by means of the so-called transfer matrix.

3) The world sheet formalism presented here may accommodate Dirac's monopoles, because the monopoles can be described, according to Dirac, in terms of the same kind of world sheets swept out by strings attached to them. If this is the case, the triplets can have magnetic charges. Since all hadrons are magnetically neutral, these charges must add up to zero, which reminds us of the fact that the charm number is also zero for them. Thus we are tempted to identify the charm with the magnetic charge, which would cause strong binding between opposite charges. The two spaces $SU(3)'$ and $SU(3)''$ may be called electric and magnetic $SU(3)$ respectively, corresponding to the 3×3 ways of assigning electric and magnetic quantum numbers.

Such a formalism has been independently proposed by Schwinger from a different motivation. He calls these nine objects dyons. The direct parallelism between dyons and the triplets has been pointed out by Biedenhahn and Han.

Of course there are all sorts of problems associated with monopoles. The most serious and perhaps most intriguing is the large P and T violation one must expect off hand. This could give a natural explanation for CP violation, but the problem is how to suppress it to a degree $\lesssim 10^{-10}$ which is required by neutron electric dipole moment.

At any rate, the close mathematical connection between the dual model and the monopoles lies in the fact that the Maxwell field due to a pair of monopoles is given by

$$F_{\mu\nu}(x) = g\epsilon_{\mu\nu\lambda\rho} \iint \delta^4(x-y)G^{\lambda\rho}(y)d^2\xi \tag{42}$$

where $G_{\mu\nu}d^2\xi = d\sigma_{\mu\nu}$ is the surface element, Eq.(22), of a world sheet spread between the world lines of the pair. Eq.(42) is independent of the choice of the sheet; one gets the correct equations if Eq.(42) is substituted in the Maxwell Lagrangian and the y's are varied, including the end points.

4 Statistical approximation

We will briefly discuss here a high energy approximation to dual amplitudes which was originally based on an intuitive argument, but can also be justified more rigorously. The point is that in a high energy process a very large number of states are available according to the factorized dual theory. In fact the number of states $\rho(s)$ increases like $\exp[c\sqrt{s}]$, which one can easily derive from the Stephan-Boltzmann law for a one-dimensional black body radiation. The only difference is that $s =$ (center-of mass energy)^2 takes the place of the ordinary energy. The constant c is numerically

$$c = 2\pi\sqrt{n/6} \tag{43}$$

where n is the dimensionality (4, or 5, or more) of the oscillator vectors. As has been pointed out by Fubini et al, this happens to give the same number ($c \approx 1/160MeV^{-1}$ for $n = 6$) as the Hagedorn constant.

The absorptive part of a scattering amplitude consists of a sum over intermediate states with fixed s, i.e. over a microcanonical ensemble:

$$A = \sum_{s_n = s} \langle p'|j(-q')|n\rangle\langle n|j(q)|p\rangle \tag{44}$$

If s is large, one is tempted to replace it with a sum over a canonical ensemble. More precisely

$$\begin{aligned} A \;\sim\; & c(\beta)\sum_n \langle p'|j(-q')|n\rangle e^{-\beta s_n}\langle n|j(q)|p\rangle \\ & \equiv c(\beta)\sum_n F_n e^{-\beta s_n} \end{aligned} \tag{45}$$

Here the s_n's are the eigenvalues of the mass operator (20), not the actual $s = (p+q)^2$. This amounts to relaxing the subsidiary condition (19). $c(\beta)$ is a normalization factor. If the sum (45) had a sharp peak around $s_n = s$, it would be a good approximation to Eq.(44) as in the usual statistical mechanics. The parameter β should then be the inverse temperature ($\sim 1/\sqrt{s}$) of the string. Actually, things do not work out that way because F_n does not grow like $\rho(s_n)$ but much more slowly like s^α, which is the Regge behavior. If $\alpha > 0$, still there will be a peak around $s_n = s$ where

$$\alpha/s \approx \beta \tag{46}$$

So we can use this as the definition of β.

In the operator formalism, Eq.(45) can be explicitly evaluated from

$$A = c(\beta)\langle 0|\Gamma' e^{-\beta H_0}\Gamma|0\rangle \tag{47}$$

where Γ and Γ' are appropriate interaction vertices. We find

$$A \sim (1 - e^{-\beta})^{-1-\alpha(t)} \sim \beta^{-1-\alpha(t)} \tag{48}$$

which give the correct Regge behavior in view of Eq.(46), if $c(\beta) \sim \beta$. The above idea makes physical sense, perhaps better than the Veneziano model itself, because we apply it to the absorptive parti and smear the resonance peaks as in the discussion of finite energy sum rules. But the real part, when smeared, should also show a similar behavior for reasons of analyticity. Justification of the method depends on the assumption that the absorptive part grows with s ($\alpha > 0$). The formula may still be valid for $\alpha \leq 0$, but that does not follow from the above argument. Assuming the general validity of the procedure, we can also handle the high energy behavior of many particle processes. The main point is that any high energy ("hot") propagator $\sim 1/(s - H_0)$ is replaced by a Boltzmann factor $\sim \beta \exp(-\beta H_0)$

where $\beta \sim 1/s$, which would give a correct answer as far as the $s-$dependence is concerned.

At this point let us indulge in some speculations. The problems are unitarity and the Pomeron, both of which are lacking in the Veneziano model. It is generally assumed that the Pomeron (in the $t-$channel) is equivalent to non-resonant background in the $s-$channel. But the background is after all made up of hadrons, so a unitarized dual theory should naturally contain the Pomeron. Now the unitarization means taking account of dissociation and recombination of resonances among themselves. Wouldn't it be reasonable, then, to consider a grand canonical ensemble of resonances? The main problem is of course how to define the vertex operator in a scattering problem. We would like to suggest the ansatz that instead of Eq.(46), β^{-1} should be the thermodynamic temperature, or

$$\beta \approx c/\sqrt{s} \tag{49}$$

because the distribution of M^2 would be decided by the interaction among the strings in the ensemble, and not by the coupling of these states to the external channels. We would then get the result

$$A \sim s^{\alpha(t)/2} \tag{50}$$

suggesting that the Pomeron trajectory has half the universal slope of resonances. Whether this is a pole or a cut, and what the intercept is, cannot be decided in such a crude picture. It is interesting that the same behavior as Eq.(50) has been obtained by explicitly computing certain higher order diagrams of the dual model.

There is still another possibility for the Pomeron which will be discussed later.

5 Electromagnetic interactions and inelastic $e-p$ scattering

We would like to tackle the problem of inelastic $e-p$ scattering on the basis of our model. Before doing this, we have to make some remarks about the electromagnetic interactions in general. If we just have an elastic spring with a charge distribution along the string but without the quark spin, the problem of setting up the gauge principle is very simple. In the Hamiltonian (16) one makes the replacement

$$p^{\mu}(\xi) \longrightarrow p^{\mu}(\xi) - \rho(\xi)A^{\mu}(x(\dot{\xi})) \tag{51}$$

where $\rho(\xi)$ is an arbitrary distribution function. But this is not the most general form. Instead of $x(\xi)$ and $p(\xi)$, we should choose the set $\{x_n, p_n\}$, or any other orthogonal basis, and apply the gauge principle. Because $A(x)$ is nonlinear in x, we get inequivalent results.

The trouble with this method is that the form factors are all Gaussian, an undesirable characteristic of harmonic oscillators. For a pointlike charge distribution the Gaussian peak is infinitely sharp (in momentum space) since the string has an infinite zero-point length. If we renormalize away the Gaussian factor, on the other hand, what is left is in general a polynomial which is not good either.

Another popular approach to form factors is the spurion method which allows one to obtain pole dominance form factors. A difficulty here is the gauge invariance, or the conservation law, which cannot be automatically guaranteed. At any rate there is a large amount of arbitrariness in either method.

Another serious problem is dealing with external fields in a factorized dual model. It arises from the fact that one needs fifth-dimension oscillators for factorization of a general dual amplitude, but the parameters of the fifth dimension depends on the individual external masses in a non-factorizable way. Thus one cannot maintain duality and factorizability for arbitrary external fields. Full duality must then be abandoned in our model when dealing with electromagnetic interaction. For example, a virtual Compton amplitude should not be dual;

One might argue that in the pole dominance model the lines p and p', instead of the photons, may be treated as external:

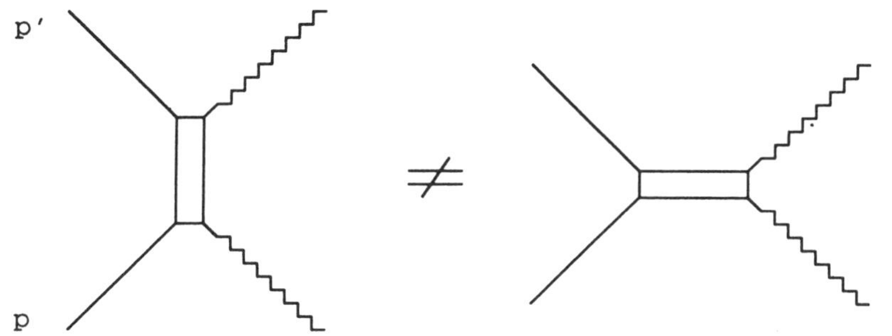

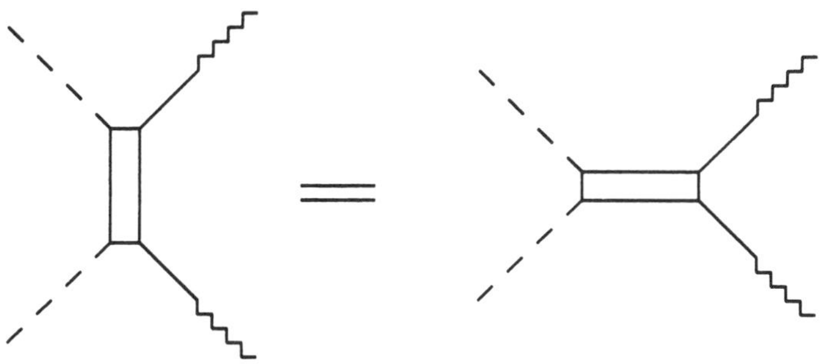

but this does not work for many photon cases.

In spite of the unpleasant Gaussian nature of form factors in the first method described above, let us see what will come out if duality is not demanded. The crossed channel singularities in the usual case arises from the singular nature of the vertex operator

$$\Gamma(q) \sim \exp[iq \cdot x(0)] = \exp\left[iq \cdot x_0 + 2iq \cdot \sum_{n=1}^{\infty} x_n\right] \tag{52}$$

In order to blunt the singularity, we assume that the charge is located at a coordinate

$$\bar{x} = \int x(\xi) f(\xi) d\xi = x_0 + 2 \sum_{n=1}^{\infty} x_n f_n \tag{53}$$

such that

$$\sum f_n^2 < \infty \tag{54}$$

According to the gauge principle, the electromagnetic interaction is obtained by the substitution $p_0^\mu \to p_0^\mu - eA^\mu(\bar{x})$, $p_n^\mu \to p_n^\mu - 2ef_n A^\mu(\bar{x})$ in the Hamiltonian. The current operator is then

$$j^\mu = e\left\{ p_0^\mu + \sum_{n=1}^{\infty} p_n^\mu f_n \ , \ e^{iq\cdot\bar{x}} \right\}_+ \tag{55}$$

For the ground state (spin zero), this gives a vertex

$$\Gamma_\mu = e(p+p')_\mu \exp\left[-q^2 \sum_{n=1}^{\infty} f_n^2/n\right] \tag{56}$$

Now let us discuss off-diagonal elements $\langle n|j_\mu|0\rangle$ corresponding to inelastic processes. We would like to compare them with the $e-p$ scattering data ignoring the effect of spin. The familiar structure functions W_1 and W_2 should be obtained from

$$\sum_{s_n=n} \langle 0|j_\nu(-q)|n\rangle\langle n|j_\mu(q)|0\rangle \tag{57}$$

where the statistical method could be applied for large s.

There still remains a problem. In the SLAC data, W_1 and W_2 seem to have an s dependence ($W_1 \sim s$, $W_2 \sim 1/s$) which is consistent with the Pomeron picture. On the other hand, Eq.(55) will not give any Regge (or power) behavior. We propose to fix this by including the fifth dimension:

$$\exp[iq\cdot\bar{x}] \longrightarrow \exp[iq\cdot\bar{x} + iq_5 x_5(0)] \tag{58}$$

where q_5 is an appropriate constant. Since $x_5(0)$ is not smeared out, it gives rise to a fixed power behavior $\sim s^\alpha$ where α is constant, corresponding to a flat trajectory. We do not know what all this means, but the Pomeron might be of this nature. At any rate we can evaluate Eq.(57) with the new ansatz, and obtain the result

$$W_1 \sim (s-m^2)^{\alpha_1} \exp\left[-2q^2 \sum_{n=1}^{\infty} \frac{1-e^{-n\beta}}{n} f_n^2\right] \tag{59}$$

and similarly for W_2. Here m^2 is the initial Hadron mass. Actually we have lost gauge invariance from a) addition of the fifth dimension and b) the statistical approximation, so we cannot get the ratio W_1/W_2 exactly. But the main feature of Eq.(59) is that for large s we have

$$\beta \sim c/(s-m^2) \tag{60}$$

so that

$$\begin{aligned}
W_i &\sim (s-m^2)^{\alpha_i}\exp[-2q^2\beta_i \textstyle\sum f_n^2] \\
&\sim (s-m^2)^{\alpha_i}\exp[-\lambda_i q^2/(s-m^2)] \\
&= (s-m^2)^{\alpha_i}\exp[-\lambda_i/(\omega-1)], \quad i=1,2 \\
\lambda_i &= 2c_i \sum_{n=1}^{\infty} f_n^2, \quad \omega = 1+(s-m^2)/q^2 \ (= 2m\nu/q^2)
\end{aligned} \tag{61}$$

Which $\alpha_1 = 1$, $\alpha_2 = -1$, we get the scaling law (for large s)

$$\begin{aligned}
(s-m^2)W_2 &\sim \exp[-\lambda_2/(\omega-1)] \\
\text{or} \qquad 2m\nu W_2 &\sim (\omega/\omega-1)\exp[-\lambda_2/(\omega-1)], \\
(s-m^2)^{-1}W_1 &\sim \exp[-\lambda_1/(\omega-1)]
\end{aligned} \tag{62}$$

The general behavior of the exponential factor in Eq.(62) is:

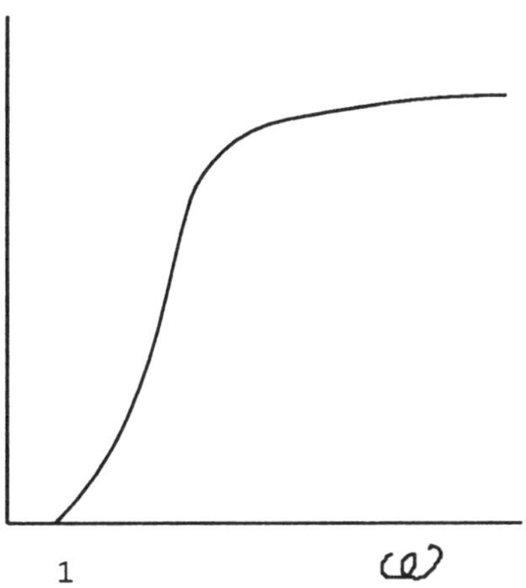

Reprinted from:

PHYSICAL REVIEW D VOLUME 7, NUMBER 8 15 APRIL 1973

Generalized Hamiltonian Dynamics*

Yoichiro Nambu

The Enrico Fermi Institute and the Department of Physics, The University of Chicago, Chicago, Illinois 60637

(Received 26 December 1972)

Taking the Liouville theorem as a guiding principle, we propose a possible generalization of classical Hamiltonian dynamics to a three-dimensional phase space. The equation of motion involves two Hamiltonians and three canonical variables. The fact that the Euler equations for a rotator can be cast into this form suggests the potential usefulness of the formalism. In this article we study its general properties and the problem of quantization.

I. INTRODUCTION

A notable feature of the Hamiltonian description of classical dynamics is the Liouville theorem, which states that the volume of phase space occupied by an ensemble of systems is conserved. The theorem plays, among other things, a fundamental role in statistical mechanics. On the other hand, Hamiltonian dynamics is not the only formalism that makes a statistical mechanics possible. Any set of equations which lead to a Liouville theorem in a suitably defined phase space will do (provided of course that ergodicity may be assumed). With this in mind, let us consider the following scheme.

Let $(x, y, z) \equiv \vec{r}$ be a triplet of dynamical variables (canonical triplet) which span a three-dimensional phase space. This is a formal generalization of the conventional phase space spanned by a canonical pair (p, q). Next introduce two functions, H and G, of (x, y, z), which serve as a pair of "Hamiltonians" to determine the motion of points in phase space. More precisely, we postulate the following "Hamilton equations":

$$\frac{dx}{dt} = \frac{\partial(H, G)}{\partial(y, z)},\quad \frac{dy}{dt} = \frac{\partial(H, G)}{\partial(z, x)},\quad \frac{dz}{dt} = \frac{\partial(H, G)}{\partial(x, y)}, \tag{1}$$

or in a vector notation

$$\frac{d\vec{r}}{dt} = \vec{\nabla}H \times \vec{\nabla}G . \tag{1'}$$

For any function $F(x, y, z)$, then, we have

$$\frac{dF}{dt} = \frac{\partial(F,H, G)}{\partial(x, y, z)} = \vec{\nabla}F \cdot (\vec{\nabla}H\times \vec{\nabla}G) . \tag{2}$$

We may call the right-hand side of (2) a generalized Poisson bracket (PB), to be denoted by $[F,H, G]$. Obviously a PB is antisymmetric under interchange of any pair of its components. As a result we have $\dot{H} = \dot{F} = 0$, i.e., both H and G are constants of motion. The orbit of a system in phase space is thus determined as the intersection of two surfaces, $H = \text{const.}$ and $G = \text{const.}$

Equation (1) or (1′) also shows that the velocity field $d\vec{r}/dt$ is divergenceless,

$$\vec{\nabla} \cdot (\vec{\nabla}H \times \vec{\nabla}G) \equiv 0 , \tag{3}$$

and this amounts to a Liouville theorem in our phase space.

The above properties immediately tempt us to construct a statistical mechanics where a canonical ensemble is characterized by a generalized Boltzmann distribution in phase space with a weight factor

$$e^{-\beta H-\gamma G} .$$

Two temperaturelike intensive parameters are thus required to specify the ensemble, much as in a grand canonical ensemble.

It is obvious that this kind of generalization can be extended to a phase space of any dimensionality, n. We would introduce an n-component vector x_i and $n-1$ Hamiltonians H_k, and postulate in lieu of Eqs. (1) and (2)

$$\frac{dx_i}{dt} = \sum_{jk\cdots l} \epsilon_{ijk\cdots l} \frac{\partial H_1}{\partial x_j} \frac{\partial H_2}{\partial x_k} \cdots \frac{\partial H_{n-1}}{\partial x_l},\quad \frac{dF}{dt} = \frac{\partial(F,H_1,H_2, \ldots ,H_{n-1})}{\partial(x_1, x_2, \ldots , x_n)}, \tag{4}$$

where $\epsilon_{ijk\cdots l}$ is the Levi-Civita tensor.

From the standpoint of physics, however, we must first examine the relevance and applicability of such generalizations. Are there real physical systems which may be described in this way? Or else can one think of these generalizations as a possible direction in which classical and quantum mechanics might develop?

In this paper we will limit ourselves to the three-dimensional case only. Then the first question can be answered affirmatively: Equation (1) is nothing but the Euler equation for a rigid rotator, if we identify $\vec{r}$ with the angular momentum $\vec{L}$ in the body-fixed frame, and G and H, respectively, with the total kinetic energy and the square of angular momentum in this frame:

$$G=\frac{1}{2}\left(\frac{L_x{}^2}{I_x}+\frac{L_y{}^2}{I_y}+\frac{L_z{}^2}{I_z}\right),\qquad H=\tfrac{1}{2}(L_x{}^2+L_y{}^2+L_z{}^2). \tag{5}$$

We believe this to be justification enough to explore further the proposed ideas at least for the three-dimensional case. Needless to say, one may in general consider a number of canonical triplets $\vec{r}_n$, $n=1,\ldots,N$ which form a $3N$-dimensional phase space, and write in place of Eq. (2)

$$\frac{dF}{dt}=\sum_n\frac{\partial(F,H,G)}{\partial(x_n,y_n,z_n)}. \tag{6}$$

For example, this will enable one to handle a model which simulates coupled spin systems.[1] For the moment, however, we are interested in the basic formalism only.

There is another direction in which Eq. (2) can be generalized. It is to assume

$$\frac{dF}{dt}=\sum_i\frac{\partial(F,H_i,G_i)}{\partial(x,y,z)}, \tag{7}$$

where (H_i, G_i) are a given set of functions. Since the Liouville theorem holds with respect to each term separately, it also holds for the sum. But the individual "Hamiltonians" H_i, G_i, are no longer constants of motion in general, though there may nevertheless be some constants of motion which are not directly related to the Hamiltonians. This may sound like an uninteresting and unnecessary digression, but we will see later that it becomes more natural when one tries quantization.

II. CANONICAL TRANSFORMATION

In this section we examine canonical transformations on the triplet (x, y, z), or more generally on the set of triplets (x_n, y_n, z_n).

We may call a mapping $(x,y,z)\to(x',y',z')$ a canonical transformation if

$$[x',y',z']=\frac{\partial(x',y',z')}{\partial(x,y,z)}=1. \tag{8}$$

Then

$$\frac{dF}{dt}=\frac{\partial(F,H,G)}{\partial(x,y,z)}=\frac{\partial(F,H,G)}{\partial(x',y',z')}, \tag{9}$$

i.e., the Hamilton equations are form-invariant if we use the new set of variables. In particular, the Hamilton equations themselves generate an infinitesimal canonical transformation in view of the Liouville theorem (3). On the other hand, not all infinitesimal canonical transformations need be generated this way. Moreover, two sets (H, G) and (H', G') generate the same transformation if there is a functional relationship $H'=h(H,G)$, $G'=g(H,G)$ such that

$$\frac{\partial(H',G')}{\partial(H,G)}=1, \tag{10}$$

because then

$$\frac{\partial(H,G)}{\partial(y,z)}=\frac{\partial(H',G')}{\partial(y,z)},\quad\text{etc.}$$

Thus the Hamiltonians are defined up to the usual type of canonical transformations (10) where (H, G) are regarded as a canonical pair of variables. In order not to confuse this with the transformations on (x, y, z) we will refer to (10) as "gauge" transformations.

Linear canonical transformations form a special class. We may use the matrix notation

$$\vec{r}'=A\vec{r}. \tag{11}$$

Equation (8) implies that the matrix A must be unimodular: $\det A=1$. In other words, linear canonical transformations form the group SL(3, R). In order to generate them we can conveniently take H and G to be, respectively, linear and quadratic forms in $\vec{r}$:

$$H=\sum_i a_i r_i,\quad G=\sum_{ij}\tfrac{1}{2}r_i B_{ij} r_j, \tag{12}$$

where B_{ij} is a symmetric 3×3 matrix. Then

$$\dot{\vec{r}}=\vec{a}\times(B\vec{r})$$

or

$$\dot{r}_i=\sum_{jkl}\epsilon_{ijk}a_j B_{kl} r_l.$$

Hence

$$A_{ik}=\sum_{jl}\epsilon_{ijl}a_j B_{lk}. \tag{13}$$

The number of parameters is 3 for H and 6 for G but there is a redundancy under the scaling $H\to\lambda H$, $G\to G/\lambda$. Therefore the degrees of freedom are 8, the correct number for the group SL(3, R).

The Euler equations for a rotator belongs to a slightly more complicated case where both H and G are quadratic forms. Let us assume in general

$$H=(\vec{r},A\vec{r}),\quad G=(\vec{r},B\vec{r}), \tag{14}$$

where A and B are symmetric matrices. It is possible, however, to bring them into a standard diagonal form

$$H=\lambda(\vec{r}\,',1\vec{r}\,'),\quad G=(\vec{r}\,',\Lambda\vec{r}\,') \tag{15}$$

by a linear canonical transformation, provided that A is positive (or negative) definite. For example, first diagonalize A by a rotation, then make it proportional to a unit matrix by a unimodular rescaling of the three coordinates. Finally, make G diagonal by a second rotation. Equation (5) for the Euler top corresponds precisely to such a form. One should note also that there is a freedom of linear gauge transformations between H and G, or equivalently between A and B:

$$\begin{aligned}&A\to\alpha A+\beta B,\\&B\to\gamma A+\delta B,\quad \alpha\delta-\beta\gamma=1\end{aligned} \tag{16}$$

which forms a group SL(2, R).

Let us now examine the case of many canonical triplets as represented by Eq. (6). Here one can define a PB to mean

$$[A,B,C]=\sum_{n=1}^{N}\frac{\partial(A,B,C)}{\partial(x_n,y_n,z_n)}. \tag{17}$$

The canonical variables then have the properties

$$\begin{aligned}&[x_l,y_m,z_n]=1\ \text{ if }\ l=m=n\\&\qquad\qquad\ \ =0\ \text{ otherwise},\\&[x_l,x_m,z_n]=[x_l,x_m,x_n]=0,\quad\text{etc.}\end{aligned} \tag{18}$$

A canonical transformation $\{r_n\}\to\{r_n'\}$ will be a transformation which leaves the PB form-invariant. In other words, Eq. (17) must hold true for $\{r_n\}$ when evaluated in terms of the new variables $\{r_n'\}$ and vice versa.

In the case of an infinitesimal canonical transformation $\vec{r}_n\to\vec{r}_n+\delta\vec{r}_n$, Eq. (18) leads to

$$\begin{aligned}\delta[x_l,y_m,z_n]&=[\delta x_l,y_m,z_n]\\&\quad+[x_l,\delta y_m,z_n]+[x_l,y_m,\delta z_n]\\&=0,\\ \delta[x_l,x_m,z_n]&=[\delta x_l,x_m,z_n]\\&\quad+[x_l,\delta x_m,z_n]+[x_l,x_m,\delta z_n]\\&=0,\\ \delta[x_l,x_m,x_n]&=[\delta x_l,x_m,x_n]\\&\quad+[x_l,\delta x_m,x_n]+[x_l,x_m,\delta x_n]\\&=0,\quad\text{etc.}\end{aligned} \tag{19}$$

Most of these conditions are satisfied trivially because the $\{r_n\}$ satisfy Eq. (18). The only nontrivial ones turn out to be

$$\delta[x_l,y_l,z_l]=\frac{\partial}{\partial x_l}\delta x_l+\frac{\partial}{\partial y_l}\delta y_l+\frac{\partial}{\partial z_l}\delta z_l=0, \tag{20a}$$

$$\delta[x_l,x_m,z_m]=-\frac{\partial}{\partial y_m}\delta x_l=0\quad(m\neq l),\quad\text{etc.} \tag{20b}$$

The condition Eq. (20b) means that $\delta\vec{r}_n$ does not depend on any r_l, $l\neq n$. If so, then the first condition (20a) is a simple statement for canonical transformations of individual triplets. We are thus led to a rather unexciting result that the only continuous canonical transformations consist of independent transformations of the individual triplets. This contrasts with the case of usual canonical doublets where a general transformation involve all the variables simultaneously.

A corollary to the above result is that a Hamilton equation like Eq. (6) cannot be regarded as generating successive canonical transformations unless H and G are simple sums

$$H=\sum_n H_n(\vec{r}_n),\quad G=\sum_n G_n(\vec{r}_n). \tag{21}$$

More general forms of H and G will satisfy Eq. (20a) but not (20b). On the other hand, Eq. (20a) alone is sufficient to guarantee that the Liouville theorem holds in the $3N$-dimensional phase space.

The problem of an infinitesimal canonical transformation can also be tackled from a different angle. The standard general solution to an equation like (20a) is

$$\delta\vec{r}=\vec{\nabla}\times\vec{A}, \tag{22}$$

where $\vec{A}$ is a vector field, determined up to the gradient of a scalar. Our Hamilton equation (2) corresponds to a special choice

$$\vec{A}=H\vec{\nabla}G\ (\text{or } -G\vec{\nabla}H). \tag{23}$$

As is well known, a transformation on (H,G) satisfying Eq. (10) has the property

$$H\delta G-H'\delta G'=\delta S \tag{24}$$

for some function $S(G,G')$. This induces a transformation on $\vec{A}$,

$$\vec{A}\to\vec{A}+\vec{\nabla}S, \tag{25}$$

which is indeed a gauge transformation in the conventional sense. On the other hand, Eq. (23) does not exhaust all possible fields $\vec{A}$, which have three independent components.

This brings us finally to the type of generalization represented by Eq. (7). Since more than two functions $\{H_n,G_n\}$ are now available, we should be able to represent an arbitrary $\vec{A}$ this way. The Hamiltonians are, however, again subject to gauge transformations $\{H_n,G_n\}\to\{H_n',G_n'\}$ such that

$$[H_i', G_j'] \equiv \sum_n \frac{\partial(H_i', G_j')}{\partial(H_n, G_n)} = \delta_{ij}\,, \qquad (26)$$

$$[H_i', H_j'] = [G_i', G_j'] = 0\,.$$

These are precisely the ordinary canonical transformations with (H_n, G_n) being regarded as canonical pairs.

III. QUANTIZATION

Can one "quantize" our system of equations just as the ordinary canonical formalism can be quantized? This is an intriguing question which we would like to investigate. The first problem is how to define a quantization. One supposes that quantization would be an algebraic mapping of the relationships which characterize the canonical formalism developed so far. As it turns out, this is not an easy task. The PB defined in (2) has two properties:

(a) Alternation law:

$$[A, B, C] = -[B, A, C] = [B, C, A] = \cdots\,.$$

In particular, $[A, A, C] = 0$, etc. (27a)

(b) Derivation law:

$$[A_1A_2, B, C] = [A_1, B, C]A_2 + A_1[A_2, B, C], \quad \text{etc.} \qquad (27b)$$

The first property guarantees that the Hamiltonians are constants of motion. The second makes the PB appropriate for a differential equation. Therefore it should be natural to characterize a PB by Eqs. (27a) and (27b). But the main problem lies in satisfying both of them simultaneously. In fact we have not been able to find a solution beside the classical one (2).

In order to gain some more insight into the situation we recall that the correct Eulerian equations for a top are obtained in quantum theory by the Heisenberg equation (with $\hbar = 1$)

$$i\dot{F} = [F, G]\,, \qquad (28)$$

where G is the kinetic energy, Eq. (7), and the angular momenta L_i satisfy the commutation relations

$$[L_i, L_j] = -i\epsilon_{ijk} L_k\,. \qquad (29)$$

The derivation law is duly satisfied by Eq. (28). [The anomalous sign on the right-hand side of Eq. (29) is well known,[2] but it is irrelevant to the discussion.] Comparing Eq. (28) with our classical form, Eqs. (2) and (5), we realize that the relations (29) were translated there as

$$[L_i, L_j] \to -i\epsilon_{ijk}\,\partial H/\partial L_k\,. \qquad (30)$$

This suggests a commutator algebra different from (29):

$$[L_i, L_j] = i\epsilon_{ijk}K_k\,, \quad [L_i, K_j] = i\delta_{ij}\,, \qquad (31)$$

and

$$\partial H/\partial L_k = i[H, K_k]\,.$$

However Eq. (31) is incompatible with the Jacobi identity

$$[[L_1, L_2], L_3] + [[L_2, L_3], L_1] + [[L_3, L_1], L_2] = 0\,.$$

These considerations seem to indicate that we had better relax the constraints on a PB in order to find other solutions. If we want to keep only the physical consequences of Eqs. (27a) and (27b) the following conditions are also admissible for a PB:

(a′) $[H, H, G] = [G, H, G] = 0\,,$ (32a)

(b′) $[A_1A_2, H, G] = A_1[A_2, H, G] + [A_1, H, G]A_2\,,$ (32b)

where H and G are fixed. Still, we have three options,

(1) $(a) + (b')$, (2) $(a') + (b)$, and (3) $(a') + (b')$.

In the following we study in some detail the case (1) because the example of Pauli spin matrices falls in this category and suggests a way to define a PB. Case (2) retains the derivation law with respect to A, B, and C. This is the most difficult condition, and probably cannot be met. Case (3) will be discussed later. Let us consider an operator algebra $\mathfrak{A}$ generated by three elements, X, Y, and Z, and define a PB by

$$[A, B, C] \equiv ABC + BCA + CAB - BAC - ACB - CBA \qquad (33a)$$

$$= A[B, C] + B[C, A] + C[A, B]$$

$$= [A, B]C + [B, C]A + [C, A]B \qquad ([A, B] \equiv AB - BA), \qquad (33b)$$

which clearly satisfies Eq. (27a). The canonical triplet (X, Y, Z) will then have the property

$$[X, Y, Z] = i\,. \qquad (34)$$

The Hamilton equation should read

$$i\frac{dF}{dt} = [F, H, G] \qquad (35)$$

for arbitrary F. According to Eq. (33), a PB of Hermitian operators is anti-Hermitian, so that Eq. (35) is consistent if all physical quantities are translated into Hermitian operators.

We will next check the derivation law (27b). First of all it requires

$$[1, H, G] = 0 \tag{36}$$

as can be seen by putting $A_2 = 1$ in Eq. (27b). But this means, from Eq. (33b),

$$[H, G] = 0 . \tag{37}$$

For general A_1 and A_2, then, Eq. (27b) can be reduced to

$$\dot{A}_1 A_2' - A_1' \dot{A}_2 = 0 ,\quad \dot{A} \equiv [A, H], \quad A' \equiv [A, G], \tag{38}$$

or

$$[A_1']^{-1}\dot{A}_1 = \dot{A}_2[A_2']^{-1} = \dot{A}_1[A_1']^{-1} = [A_2']^{-1}\dot{A}_2$$

whenever the division is possible. Thus

$$\dot{A} = \alpha A' \quad (\text{or } A' = \alpha \dot{A}) \tag{39}$$

for any A such that A'^{-1} (or $\dot{A}^{-1}$) exists, which in turn implies

$$H = \alpha G + \beta \quad (\text{or } G = \alpha H + \beta) . \tag{40}$$

Here α and β commute with all operators and therefore are c numbers. But then

$$[A, H, G] = -[A, \beta G] \quad (\text{or } [A, \beta H]) \tag{41}$$

and by redefining $\beta G \to -G$ (or $\beta H \to H$) we recover the Heisenberg equation.

We conclude thus that if Eq. (33) satisfies the conditions (27a) and (32b), one of the Hamiltonians is a c number, and Eq. (35) is equivalent to a Heisenberg equation. We may wonder at this point whether we can avoid the rather disappointing result if we go over to Eq. (7). Following the same arguments as before, we then arrive at the conditions

$$\sum_i [H_i, G_i] = 0 ,\quad \sum_i [A_1, H_i][A_2, G_i] = \sum_i [A_1, G_i][A_2, H_i] \tag{42}$$

in place of Eqs. (37) and (38). It is not clear, however, whether or not these more general relations admit nontrivial solutions.

Our next task is then to find realizations of the PB for the canonical triplet. For this purpose let us assume that the canonical variables $\vec{R} = (X, Y, Z)$ generate, under repeated commutator operations, a Lie subalgebra $\mathfrak{A}'$ of $\mathfrak{A}$ which is semisimple. Let the generators of $\mathfrak{A}'$ be $\{R_n\}$, where the first three elements coincide with (X, Y, Z). Further let

$$[R_2, R_3] \equiv i R_1' ,\quad [R_3, R_1] \equiv i R_2' ,\quad [R_1, R_2] \equiv i R_3' . \tag{43}$$

Then (R_1', R_2', R_3') belong to the vector space of $\{R_n\}$. In view of Eq. (33b) we have

$$[X, Y, Z] = i \sum_{i=1}^{3} R_i R_i' = i \sum_{i=1}^{3} R_i' R_i . \tag{44}$$

Equation (34) demands that this be a c number (in an irreducible Hilbert space generated by $\vec{R}$). Therefore it must be a Casimir operator of the Lie algebra. Then $\vec{R}$ and $\vec{R}'$ must exhaust the generators, and hence $N \leq 6$. There are only two such possibilities, $N = 6$ and $N = 3$.

(a) If $N = 6$, the algebra is either $SO(4) \approx SU(2) \times SU(2)$, $SO(3,1) \approx SL(2,c)$, or $SO(2,2)$. For the first two cases, their generators $(\vec{L}, \vec{K})$ satisfy

$$[L_1, L_2] = iL_3, \text{ etc.},\quad [L_1, K_2] = [K_1, L_2] = iK_3, \text{ etc.},\quad [K_1, K_2] = \pm iL_3, \text{ etc.}, \quad \begin{cases} SO(4) \\ SO(3,1) \end{cases} \tag{45}$$

and the Casimir operators are

$$C_1 = \vec{L}^2 \pm \vec{K}^2 ,\quad C_2 = \pm \vec{L} \cdot \vec{K} = -i(K_1[K_2, K_3] + K_2[K_3, K_1] + K_3[K_1, K_2]) . \tag{46}$$

Without loss of generality one may put $R_i = icK_i$. In this way the solution is found to be

$$\vec{R} = C_2^{-1/3}\vec{K} ,\quad \vec{R}' = \pm C_2^{-2/3}\vec{L} . \tag{47}$$

The results for the case of SO(2, 2) are obtained from that of SO(4) by a similar change of metric.

(b) If $N = 3$, the algebra is either $SO(3) \approx SU(2)$ or $SO(2,1) \approx SU(1,1)$. Their generators $\vec{L}$ satisfy

$$[L_1, L_2] = \pm iL_3 , \quad \begin{cases} SO(3) \\ SO(2,1) \end{cases} \quad [L_2, L_3] = iL_1 ,\quad [L_3, L_1] = iL_2 , \tag{48}$$

with the Casimir operator

$$C = L_1^2 + L_2^2 \pm L_3^2 = -i(L_1[L_2, L_3] + L_2[L_3, L_1] + L_3[L_1, L_2]) . \tag{49}$$

Thus the solution is

$$\vec{R} = (X, Y, Z) = C^{-1/3}(L_1, L_2, L_3) ,\quad \vec{R}' = C^{-2/3}(L_1, L_2, \pm L_3) , \quad \begin{cases} SO(3) \\ SO(2,1) . \end{cases} \tag{50}$$

We are left with the cases where $\mathfrak{G}'$ is not semisimple. We will discuss some representative cases without claiming completeness.

(c) Euclidean algebra E(3) and its noncompact version E(2, 1). These are derived from SO(4) and SO(3, 1) by group contraction. The six generators $(\vec{L}, \vec{P})$ obey the commutation relations

$$\begin{aligned} &[L_1, L_2] = \pm i L_3, \quad [L_2, L_3] = i L_1, \quad [L_3, L_1] = i L_2, \\ &[L_1, P_2] = -[L_2, P_1] = i P_3, \\ &[L_2, P_3] = \pm[L_3, P_2] = \pm i P_1, \\ &[L_3, P_1] = \pm[L_1, P_3] = i P_2, \\ &[P_i, P_j] = 0, \qquad [P_i, L_i] = 0. \end{aligned} \tag{51}$$

The Casimir operators are

$$\begin{aligned} &C_1 = P_1{}^2 + P_2{}^2 + P_3{}^2, \\ &C_2 = L_1 P_1 + L_2 P_2 + L_3 P_3. \end{aligned} \tag{52}$$

The solutions for R and R' are given by

$$\begin{aligned} &\vec{R} = (\pm C_2)^{-1/3}(L_1, L_2, P_3), \\ &\vec{R}' = \pm C_2{}^{-2/3}(P_1, P_2, L_3) \end{aligned} \tag{53}$$

or

$$\begin{aligned} &\vec{R} = C_2{}^{-1/3}(P_1, L_2, L_3), \\ &\vec{R}' = C_2{}^{-2/3}(L_1, P_2, P_3). \end{aligned}$$

[These are equivalent for E(3).]

(d) E(2) and E(1, 1) derived from SO(3) and SO(2, 1) by contraction. The generators are (L, P_1, P_2) with the commutation relations

$$[L, P_1] = i P_2, \quad [L, P_2] = \mp i P_1, \quad [P_1, P_2] = 0, \tag{54}$$

and the Casimir operator is

$$C = \pm P_1{}^2 + P_2{}^2. \tag{55}$$

The solutions are

$$\begin{aligned} &\vec{R} = C^{-1/3}(L, P_1, P_2), \\ &\vec{R}' = C^{-2/3}(0, \pm P_1, P_2). \end{aligned} \tag{56}$$

(e) This case is characterized by the commutation relations of the Galilean group in one dimension,

$$[X, Y] = i Z^{-1}, \quad [X, Z] = [Y, Z] = 0, \tag{57}$$

for which we have a representation

$$\vec{R} = \left(i \frac{d}{dq}, \frac{q}{x}, x\right), \quad \vec{R}' = \left(0, 0, \frac{1}{x}\right). \tag{58}$$

(f) By putting $x = \text{const.} = 1$ in Eq. (58) we recover the canonical formalism of quantum mechanics:

$$\vec{R} = (-p, q, 1), \quad [p, q] = -i. \tag{59}$$

IV. USE OF NONASSOCIATIVE ALGEBRAS

In this section we pose the question whether it is possible to realize the PB relations by means of nonassociative algebras.[3] As a matter of fact, we immediately see that there is some hope, because it takes three elements to characterize the nonassociative nature of an algebra, just as we have a PB involving three observables. The analog of a PB in a nonassociative algebra is the associator

$$(a, b, c) \equiv (ab)c - a(bc) \tag{60}$$

which is zero if the multiplication table is actually associative.

Let us now see whether or not one can identify the associator with the PB under some combination of conditions (27) and (32). First take the alternative (1) stated there. Equation (27a) then demands that the associator is alternative:

$$(a, b, c) = -(b, a, c) = (b, c, a) = \cdots. \tag{61}$$

That means that we are dealing with the alternative algebra of Cayley and Dickson.[3] The Cayley-Dickson algebra C is an algebra over real numbers (or any suitable field) with eight basic elements u_i, $i = 0, \ldots, 7$, where u_0 serves as the unit element; it is the only possible algebra of this kind.

Unfortunately, we run into trouble with the derivation law (32b). The associator does not in general satisfy the derivation law, and conversely any derivation algebra over the Cayley numbers is known to be generated instead by operations of the form $\sum_{a,b} D_{a,b}$ on C, where

$$\begin{aligned} D_{a,b}\, x &= a(bx) - b(ax) + (xb)a - (xa)b + (bx)a - b(xa), \\ &\equiv D(a, b; x). \end{aligned} \tag{62}$$

$D(a, b; x)$ is antisymmetric only with respect to a, b, and for associative algebras reduces to

$$D(a, b; x) = [x, [a, b]]. \tag{63}$$

Suppose we now identify the PB $[A, B, C]$ with $D(B, C; A)$, and $B = H$, $C = G$ in the Hamilton equation. We have then switched to the alternative (3), which requires Eq. (32a):

$$D(H, G; H) = D(H, G; G) = 0. \tag{64}$$

This amounts to

$$[H, (HG) - (GH)] = [G, (HG) - (GH)] = 0, \tag{65}$$

but there is no nontrivial solution to it.

On the other hand, we can give up Eq. (32a) if we are considering Eq. (9) as the classical basis. Then the form

$$\frac{dF}{dt} = \sum_i D(H_i G_i ; F) \tag{66}$$

will certainly serve as a possible Hamilton equation. Since the multiplication table is preserved under Eq. (66), it induces an automorphism on C. It is known that automorphisms on C form a Lie group of type G_2. An element of C which is left invariant under (66) will be a constant of motion.

Let us leave the Cayley algebra and next shift our attention to the Jordan algebra J, which is commutative (but not associative).[3,4] The associator has the property

$$(a,b,a)=0, \quad (a,b,a^2)=0. \tag{67}$$

(The first is an identity which follows from the definition.) A derivation operator on J is given by

$$D_{a,b}\, x = (a,b,x) - (b,a,x). \tag{68}$$

This allows one to define the PB as

$$\begin{aligned} -i[A,B,C] &= D_{B,C} A \\ &= (B,C,A) - (C,B,A) \end{aligned} \tag{69}$$

and the Hamilton equation as

$$i\frac{dF}{dt} = [F,H,G], \tag{70a}$$

or more generally

$$= \sum_i [F,H_i,G_i]. \tag{70b}$$

As before, we may add the condition (32a) in the case of Eq. (70a).

Now it is known that a Jordan algebra can be derived from an associative algebra $\mathfrak{a}$ if we define the multiplication to mean

$$axb = \tfrac{1}{2}(ab+ba), \tag{71}$$

where ab stands for a multiplication in the associative algebra $\mathfrak{a}$. Going back to this realization, we then find the PB (69) is given by

$$[A,B,C] = i[A,[B,C]]. \tag{72}$$

This is the same form as Eq. (63), and Eqs. (70a) and (70b) reduce to

$$i\frac{dF}{dt} = [F,\mathcal{H}], \quad \mathcal{H} = i[H,G] \text{ or } i\sum_i [H_i,G_i]. \tag{73}$$

In other words, it is equivalent to a Heisenberg equation with Hamiltonian $\mathcal{H}$. In addition, an example of canonical triplet $\vec{R}$ satisfying

$$-i[X,Y,Z] = [X,[Y,Z]] = 1 \tag{74}$$

can readily be found. In fact

$$\vec{R} = \left(i\frac{\partial}{\partial x}, xy, i\frac{\partial}{\partial y} \right). \tag{75}$$

Finally there exists one Jordan algebra (the exceptional Jordan algebra) which is not isomorphic to an algebra generated by Eq. (71).[3,4] It is an algebra of 3×3 matrices M with elements in a Cayley algebra. M has the typical form

$$M = \begin{pmatrix} \alpha & c & \bar{b} \\ \bar{c} & \beta & a \\ b & \bar{a} & \gamma \end{pmatrix}, \tag{76}$$

where α, β, γ are c numbers; a, b, c are Cayley numbers, where $\bar{a}$ is obtained from a by the involution $u_0 \to u_0$, $u_i \to -u_i$ $(i \neq 0)$. This should again offer another possibility of realizing the PB relations.

V. SUMMARY

Taking the Liouville theorem as a guiding principle, we have proposed a possible generalization of classical Hamiltonian dynamics to a three-dimensional phase space. The equation of motion involves two Hamiltonians and three canonical variables. A more general form may have many triplets and many Hamiltonian pairs. Such a formalism does not seem totally irrelevant to physics because the Eulerian top problem can be cast into this form, and it offers a new possibility in statistical mechanics. An attempt to find a "quantized" version of the formalism, however, has been only partially successful. In the process, the correspondence between classical and quantized versions is largely lost. One is repeatedly led to discover that the quantized version is essentially equivalent to the ordinary quantum theory. This may be an indication that quantum theory is pretty much unique, although its classical analog may not be.

On the other hand, there remains some possibility that nonassociative algebras may also be incorporated into the new formalism. Jordan was first led to what is now known as Jordan algebra in an attempt to reformulate and generalize quantum mechanics.[4] Although our starting point and motivation were different from Jordan's we have also found the potential significance of nonassociative algebras.

I would like to express my appreciation to Professor K. Husimi who kindly took an interest in my ideas when the contents of the first section were conceived more than twenty years ago. I would also like to acknowledge the inspirations I derived from recent communications with Professor F. Gürsey[5] and Dr. Pierre Ramond regarding nonassociative algebra.

*Work supported in part by the National Science Foundation (Contract No. NSF GP 32904 X) and the John Simon Guggenheim Foundation.

[1]T. Bell and Y. Nambu (unpublished report).

[2]O. Klein, Z. Physik 58, 730 (1920); J. H. Van Vleck, Rev. Mod. Phys. 23, 213 (1951); L. D. Landau and E. M. Lifshitz, *Quantum Mechanics* (Addison-Wesley, Reading, Mass. 1958), p. 373.

[3]For nonassociative algebras see, for example, R. D. Schafer, *An Introduction to Nonassociative Algebras* (Academic, New York, 1966).

[4]P. Jordan, Z. Physik 80, 285 (1933); Nachr. Ges. Wiss. Göttingen (1933), p. 209; P. Jordan, J. von Neumann, and E. Wigner, Ann. Math. 35, 29 (1934); A. Albert, Ann. Math. 35, 65 (1934).

[5]M. Günaydin and F. Gürsey, Yale University report (unpublished).

Physica 15D (1985) 147–151
North-Holland, Amsterdam

FERMION–BOSON RELATIONS IN BCS-TYPE THEORIES

Y. NAMBU
Enrico Fermi Institute and Department of Physics, University of Chicago, Chicago, IL 60637, USA

1. Introduction

Supersymmetry is a mathematical idea in search of physical relevance, but so far the search has not been successful in the area where it originated: particle physics. The present report is an outcome of my attempt to understand the physical meaning of supersymmetry and to find examples of it in more familiar low energy physics.

Iachello and collaborators [1] have observed that there is a kind of supersymmetry in nuclear physics. For certain nuclei in the mass number region of platinum, the low energy spectra of even–even nuclear species and neighboring even–odd species can be described by the same empirical formula based on group theory. I have been aware, on the other hand, that in theories of the BCS-type, there always is a simple relation between the mass (energy gap) of the basic fermion and those of the bosons (collective modes) [2]. To use the language of particle physics, the dynamically induced masses of the pion, quark and σ meson stand in the ratio 0:1:2 (subject to higher order corrections). In terms of the effective σ model (or Higgs or Ginsburg–Landau) Lagrangian, this implies that the self-coupling and the Yukawa constants are related by $\lambda = f^2$. Generic relations of this nature emerge in any BCS–Heisenberg-type four-fermion short-range interaction theory [3,11]. It is gratifying that such relations have been experimentally established in superconductors and superfluid helium 3, as I will discuss later.

My speculation, then, is that the Iachello relation in nuclear physics may also be a manifestation of the BCS mechanism which is known to account for the nuclear pairing phenomenon. An immediate question that arises is whether the BCS or Ginsburg–Landau theory has a supersymmetry of which these relations are a consequence. I do not know the answer yet. Before coming to nuclear physics, however, I will first discuss the other examples to show the origin of the relations.

2. Superconductivity

We use the two-component formalism in which the spin-up electron and spin-down hole span a τ-spin space for the quasiparticle. The free quasiparticle Hamiltonian has the form

$$H = K\tau_3 + m\tau_1, \tag{1}$$

where K is the kinetic energy measured from the Fermi surface and m is the gap parameter. (We deal only with states in the vicinity of the Fermi surface.) The charge operator is $e\tau_3$, so a τ_1 or τ_2 term will break charge conservation. The gap in the τ_1 direction will then generate a Goldstone mode in the τ_2 direction. The collective modes in the τ_1 and τ_2 directions (called amplitude and phase modes) are the analogs of σ and π mesons in particle physics.

In terms of the four-fermion interaction Lagrangian

$$G\tau_3\tau_3' \sim G\left[\left(\tau_1\tau_1' + \tau_2\tau_2' + \cdots\right)\right]$$
(under a Fierz transformation), (2)

a collective mode is determined as the pole of the

function

$$G/(1-GI)=G+G(GI)+G(GI)^2+\dots, \qquad (3)$$

where the formal expansion indicates that the function is derived by summing to infinite order chains of "bubble" perturbation theory diagrams. I is a two-point loop integral with appropriate vertex operators, τ_1 or τ_2. Its energy (E) dependence can be displayed by a dispersion integral

$$I(z)=\int \rho(x)\,\mathrm{d}x/(x-z), \quad x>4m^2,\ z=E^2, \qquad (4)$$

where the absorptive parts ρ for τ_1 and τ_2 modes satisfy the relation

$$\rho_1(x)=\rho_2(x)(1-4m^2/x). \qquad (5)$$

From this, it follows that

$$I_1(z=4m^2)=I_2(z=0). \qquad (6)$$

But $GI_2(z=0)=1$, as a result of the Ward identity corresponding to the breaking of τ_3 invariance. Thus, one sees that the τ_2 and τ_1 modes have a pole at $E=0$ and $2m$, respectively. It looks as if these states correspond to composites of two fermions with maximum and zero binding energy. I do not have a simple physical explanation for this, but technically it depends on whether the vertex operator anticommutes or commutes with the mass operator. Another way of putting it is that, in the particle theory analog, π and σ are s- and p-wave bound states of quarks, and the extra factor in the absorptive part for the latter is just the p-wave phase space factor.

The τ_1 (amplitude) mode was experimentally detected a few years ago in certain superconductors whose gap parameters are susceptible to external forces like laser beams [4]. Actually, the mode comes out as a genuine bound state somewhat below $2m$. The theoretical explanation has been worked out along the lines sketched above [5]. The massless τ_2 (phase or Goldstone) mode, on the other hand, mixes with the τ_3 (density) mode, and turns into the plasmon by the familiar Anderson–Higgs–Englert–Brout mechanism. Both τ_1 and τ_2 modes have the same canonical velocity

$$v=v(F)/\sqrt{3}, \qquad (7)$$

where $v(F)$ is the Fermi velocity. So it is a simple matter to write down an effective σ-model Lagrangian for the collective modes, which can be explicitly derived from a BCS theory. [As was mentioned in section 1, the Yukawa coupling (f) and the boson self-coupling (λ) are related as

$$\lambda=f^2=1/(\mathrm{d}I_2(z)/\mathrm{d}z)_{z=0}, \qquad (8)$$

where $\lambda/2$ is the coefficient of the quartic term in the Lagrangian.]

3. Helium 3

Superfluid helium 3, occurring at temperatures in the millikelvin range, has two phases, A and B[6]. It is believed that the Cooper pairs of two helium atoms are formed in a triplet p state in either phase. In principle, the total angular momentum j can be 0, 1, or 2, and the pairs can condense in any state in the space spanned by these 9 states; also, recall that the degeneracy of a state of angular momentum j is $2j+1$. Actually, however, the A and B phases correspond to $j=1$ and 0, respectively. The nine states may be expressed by a scalar S, vector V_i, or tensor T_{ik} order parameter transforming as

$$\tau_\alpha \sigma\cdot p, \quad \tau_\alpha(\sigma\times p)_i, \quad \text{or } \tau_\alpha\sigma_i p_k \quad (\alpha=1,2), \qquad (9)$$

respectively, where σ is the spin, and p is the relative momentum of the Cooper pair. As before, the τ_α's operate in the space of particle and hole components, so there are 18 collective modes altogether.

What symmetries are broken by the Cooper pairings? This question leads to another: what is the precise nature of the forces responsible for the pairing? The answer is not entirely clear, but it appears that they are not greatly different in different j channels, meaning that intrinsic spin-orbit interaction is small. The pairing in a particular j state will therefore break the conservation of total angular momentum j (if $\neq 0$) and spin (or equivalently, orbital) angular momentum, in addition to the usual violation of particle number. Each violation should generate a Goldstone mode (exact or approximate), whereas the remaining collective states should be massive. Again, these have a simple mass spectrum as shown in table I [7]. Note in particular that for a given j, the phase and amplitude modes satisfy a sum rule

$$m_1^2 + m_2^2 = 4m^2 \tag{10}$$

in terms of the fermion mass m. This is a consequence of a theorem which generalizes eq. (6):

$$aI_1(a) + (1-a)I_2(a) = \text{const} = I_2(0). \tag{11}$$

Vertices appearing in this problem generally give rise to the linear combination on the left-hand side, and the phase and amplitude modes are orthogonal and complementary in the sense $a \leftrightarrow 1 - a$.

The above theoretical predictions are roughly in agreement with observation. The phase and amplitude modes behave differently under particle–hole reflection relative to the Fermi surface. Since this symmetry is not exact, both modes can be excited by external density waves. Perhaps the above sum rule is to be more trusted than the individual mass values.

Table I
Collective modes in He 3, B phase. $H = K\tau_3 + m\tau_1\sigma\cdot p$

Mode	j	Mass	Loop integral	Comment
$\tau_1\sigma\cdot p$	0	$2m$	I_1	
$\tau_2\sigma\cdot p$	0	0	I_2	particle number Goldstone
$\tau_1(\sigma\times p)_i$	1	0	I_2	spin Goldstone
$\tau_2(\sigma\times p)_i$	1	$2m$	I_1	
$\tau_1\sigma_i p_k$	2	$\sqrt{2/5}(2m)$	$\frac{2}{5}I_1+\frac{3}{5}I_2$	
$\tau_2\sigma_i p_k$	2	$\sqrt{3/5}(2m)$	$\frac{3}{5}I_1+\frac{2}{5}I_2$	

4. Nuclear physics

The BCS theory has been applied with success to account for the pairing phenomenon in nuclei [8]. Typically, the ground state of a heavy even–even nucleus is filled with singlet proton pairs and neutron pairs, whereas an even–odd nucleus has one quasi-particle added or missing. The excited states of a nucleus can be due to single-particle as well as pair excitations, the latter giving rise to collective models with $j = 0, 2, \ldots$.

In the interacting boson model of Arima and Iachello [9,1], one treats the pairs as bosons, and introduces an effective Hamiltonian with quartic self-couplings, which is determined on the basis of group theory. Thus, take $j = 0$ and $j = 2$ bosons only, considered to be degenerate in the first approximation. The relevant invariance group is then $U(6) = SU(6) \times (U(1)$. Next, break it down to a chain of subgroups. There are three distinct chains:

$$\begin{aligned} &U(6) \supset U(5) \supset SO(5)\ldots, \\ &U(6) \supset SU(4) \supset SO(6) \supset SO(5)\ldots, \\ &U(6) \supset SU(3)\ldots. \end{aligned} \tag{12}$$

The question of which chain is relevant is where physics comes in, and the answer depends on the mass number region one is considering. In the platinum region, for example, it is the second chain. At any rate, one then writes down an effective Hamiltonian as a linear combination of Casimir operators corresponding to the chain of groups in terms of bose operators. This seems to work well in general. It has been pointed out [1], moreover, that in the platinum region, the even–odd nuclei have single protons in a $D_{3/2}$ shell, for which the natural invariance group is SU(4), leading to the same chain of subgroups as for the bosons. As it turns out, the same Hamiltonian can indeed describe both even–even and even–odd cases, hence suggesting a supersymmetry [1]. Real supersymmetry, however, should involve a supergroup like U(6/4) which includes fermion–boson transitions. Such a scheme still works, although not as well as before.

Now, I come back to the scenario already developed, i.e., the BCS–Ginsburg–Landau theory, as applied to the present problem. Let us say that Cooper pairs, each formed out of nucleons in the same shell, collectively condense into a ground state, which can be excited by breaking pairs up or placing them into new states. In the case of $j = 3/2$ shell, the pair can have $j = 0$ or 2.

According to the standard methods, a nucleon should be described by combining four particle and four hole (complex conjugate) states into a 8-component wave function ψ. Then one writes an analog of the Hamiltonian, eq. (1), in this space. The ψ can form a representation of SO(8) and Spin(7) (Majorana), so there are seven anticommuting Γ matrices $\Gamma_1, \Gamma_1, \ldots \Gamma_7$. One can choose them in such a way that $\Gamma_0 = +1(-1)$ for particle (hole) states, and $\Gamma_s \equiv \Gamma_1$ behaves as a scalar (or possibly a mixture of scalar and tensor). Then the Hamiltonian will take the form

$$H = K\Gamma_0 + m\Gamma_s. \tag{13}$$

A remark is in order before proceeding further. The emergence of the parameter m is a result of interactions among many nucleons in various shells, not just in the states described by the wave function under consideration. The collective modes are generated by all of them, which contribute to the loop integral I, eq. (4). For example, $j = 1/2$ shells will contribute to $j = 0$, and $j = 5/2$ states to $j = 0, 2, 4$, collective modes. The relation between m in eq. (13) and the masses of the collective modes, therefore, is no longer so clear. On the other hand, to the extent that $j = 0$ and 2 modes are approximately degenerate while others can be ignored, it would be reasonable to suppose that the symmetry group SU(4) of the fermions under consideration and the symmetry group SO(6) of the collective modes are physically identical. So the loop integrals will still maintain the same properties as before.

There should be $6 \times 2 = 12$ collective modes, which come in six pairs $\Gamma_k, \mathrm{i}\Gamma_0\Gamma_k$ $(k = 1, \ldots 6)$ of amplitude and phase modes. Thus, Γ_s is an analog of the σ with $j = 0$ at mass $2m$, whereas the zero-mass $\mathrm{i}\Gamma_0\Gamma_1$ mode is spurious because there is actually no particle number violation in a finite system. The SO(6) symmetry, if it exists in an approximate sense, is generated by $\Gamma_k\Gamma_1(K > 1)$. The Γ_s term in H breaks this down to SO(5), so there should be spurious near-zero modes $\Gamma_t (t \neq 1)$, and complementary modes $\mathrm{i}\Gamma_0\Gamma_t$ at mass $\approx 2m$, both with $j = 2$. Actually, the zero modes will in general couple to ordinary acoustic modes and get absorbed into them. These results are summarized in table II.

Table II
Collective modes in nuclei ($j = 3/2$ shell). $H = K\Gamma_0 + m\Gamma_s$

Mode	j	Mass	Baryon #(B)	Loop integral	Comment
Γ_s	0	$2m$	± 2	I_1	
$\Gamma_0\Gamma_s$	0	0	± 2	I_2	B Goldstone
Γ_t	2	0	± 2	I_2	SO(6) Goldstone
$\Gamma_0\Gamma_t$	2	$\sim 2m$	± 2	I_1	
Γ_0	0		0	$I_1 - I_2$	B generator
$\Gamma_s\Gamma_t$	2		0	$I_1 - I_2$	SO(6) generator
$\Gamma_t\Gamma_{t'}$	2		0	0	SO(6) generator
$\Gamma_0 - \Gamma_0\Gamma_s$	0		0	$(\mathrm{iE}/2\mathrm{m})\mathrm{I}_2$	off diagonal elements
$\Gamma_t - \Gamma_s\Gamma_t$	2		0	$(\mathrm{iE}/2\mathrm{m})\mathrm{I}_2$	

Γ_s: scalar, Γ_t: tensor

Finally, we will write down an effective Hamiltonian for the system. Introduce six complex boson operators $B_i(B_i^\dagger)$, transforming under the symmetry group SU(6)×U(1). It turns out that the proper Hamiltonian (ignoring the mixing with acoustic modes for the moment) is given by

$$H = \psi^\dagger K\Gamma_0\psi + f\left[\psi^\dagger(1+\Gamma_0)\Gamma_i\psi B_i + \text{h.c.}\right] + (f^2/2)\left[\left(B_i^\dagger B_i - c^2\right)^2 + |B_i^\dagger B_k - B_k^\dagger B_i|^2\right].$$

Note that the nonlinear terms for the B_i's are composed of Casimir operators of U(1) and SU(6). The terms involving the fermions, on the other hand, have the symmetry of SU(4)×U(1) ≈ SO(6) × U(1). The latter gets broken spontaneously to SO(5), giving rise to the above mentioned zero modes. It thus seems likely that an effective

Hamiltonian constructed along this line has properties similar to the one proposed by Iachello.

I have tried to uncover in this kind of system a genuine supersymmetry involving fermionic transformations, but so far I do not have a clear answer to that.

5. Particle physics

Finally, I will briefly come back to particle physics. Historically, the σ model in hadron physics is the oldest relevant example and is still of considerable interest in many respects, but our present concern is with mass relations. In accordance with our general results, it has been claimed [10] that the σ meson should have a mass roughly twice the constitutent quark mass, i.e., ≈ 700 MeV, although the actual σ resonance seems to occur at ≈ 900 MeV. Since the QCD interaction is rather different from the ones we are considering, it is not obvious that the present arguments apply here equally well.

A more interesting case should be the electroweak interaction [11]. Here the Higgs bosons in the Weinberg–Salam theory are analogs of the π and σ, but regarded as elementary objects. Their masses therefore are not predictable. But in composite models like Terazawa's [11], there emerge relations like ours between boson and constituent fermion masses. Similar situations may also exist in technicolor theories, subject to the uncertainties encountered in the hadronic case. Short of a specific model however, I will not indulge in further speculations at this moment.

Acknowledgements

I would like to thank Professors F. Iachello, A. Arima, I. Bars, A.J. Leggett, G. Baym, K. Levin, C.M. Varma, M. Scadron and A. Terazawa for enlightening conversations I have had at one time or another.

References

[1] F. Iachello, Phys. Rev. Lett. 44 (1980) 772; Physica 15D (1985) 85 (these proceedings).
A. Balantekin, I. Bars and F. Iachello, Phys. Rev. Lett. 47 (1981) 19; Nucl. Phys. A370 (1981) 284; A.B. Balantekin, I. Bars, R. Bijker and F. Iachello, Phys. Rev. C27 (1983) 1761.

[2] Y. Nambu and G. Jona-Lasinio, Phys. Rev. 122 (1961) 345.

[3] T. Eguchi and H. Sugawara, Phys. Rev. D10 (1974) 4257.

[4] R. Sooryakumar, M.V. Klein and R.R. Frindt, Phys. Rev. B23 (1980) 3213, 3222.

[5] P.B. Littlewood and C.M. Varma, Phys. Rev. B26 (1982) 4883.
C.A. Balseiro and L.M. Falicov, Phys. Rev. Lett. 45 (1980) 662.

[6] A.J. Leggett, Rev. Mod. Phys. 47 (1975) 331.
J. Wheatley, Rev. Mod. Phys. 47 (1975) 415.
D.M. Lee, Internat. J. Quant. Chem. 23 (1983) 1191.
K. Levin, Phys. Rep. 98 (1983) 1.

[7] P. Woelfle, Physica 90B (1977) 96; Prog. LTP 7A (1978) 191.

[8] A. Bohr, B.R. Mottelson and D. Pines, Phys. Rev. 110 (1958) 936.
S.T. Beliaev, Kgl. Dansk. Vid. Selsk. Mat-Fys. Medd. 31, no 11 (1959).
J.M. Eisenberg and W. Greiner, Microscopic Theory of the Nucleus, vol 3, (North-Holland, Amsterdam, 1972).

[9] A. Arima and F. Iachello, Phys. Rev. Lett. 35 (1975) 1069.

[10] R. Delbourgo and M. Scadron, Phys. Rev. Lett. 48 (1982) 379.

[11] H. Terazawa, Y. Chikashige and A. Akama, Phys. Rev. D15 (1977) 480.

Three Stages, Three Modes, and Beyond

Yoichiro Nambu
The Enrico Fermi Institute
University of Chicago

Abstract

I review the developement of particle physics from the viewpoint of theoretical methodology. Particle physics has gone through certain characteristic periods in the course of discoveries and accumulation of knowledge, both factual and theoretical. To anlyze this situation, I draw on the theses of three stages and three modes, and illustrate the points with historical examples, up to the present Standard Model and the speculative theories of unification. As an exercise in this analysis, I propose a possible regularity in the quark masses.

1 INTRODUCTION

By nature, scientists are optimists. They are forward-looking. I have been told that the main theme of the conference was not to look back, but to look forward and try to anticipate the physics of the twenty-first century. But I do not think I am up to such an ambitious task. Prediction is always difficult, especially prediction of the future, as Niels Bohr was fond of quoting. So I will first look back 50 years of particle physics. They coincide with the post-war years which also more or less covers the bulk of the most important developments of what we call now high energy or particle physics. If my tone sounds rather philosophical, it may be as much due to my age as it is due to the education I received in my early years.

Indeed the title of my talk reveals my age and my background. When I became a physics student at the University of Tokyo some 55 years ago, the Bohr atom was 28 years old, and quantum mechanics was only 15 years old. Compare this with the present day. The main achievement of particle physics is encoded in the Standard Model, which is almost 30 years old. Then the supersymmetry is 23 years old, and the superstring 15 years. Do the youngest generation of theoretical physicists look upon the Standard Model as I had looked upon the Bohr theory? Do they look upon the superstring theory as I had looked upon quantum mechanics? I do not know, but I would not be too surprised if some of them do. On the other hand, even they would admit that, unlike quantum mechanics, supersymmetry and superstring are still speculative theories. I must say that the time scale of progress has lengthened compared to those truly revolutionary days.

2 THE THREE STAGES

In my view [1], particle physics owes its origin to Lawrence and Yukawa. With his cyclotron of 1930 Lawrence supplied the basic experimental tool, and with the meson theory of 1935 Yukawa supplied one of the basic theoretical tools. They are very much alive today, and are all taken for granted. In the following I will elaborate on what I mean by theoretical tools. Immediately after the war, my generation of young theorists were strongly influenced by the thinkings of the very original Japanese school of particle physicists, which owes its origin to Y. Nishina, a theorist turned experimentalist. He laid the foundation of nuclear and cosmic ray physics at his Riken (Institute for Physical and Chemical Research) laboratories, which continues to be one of the top research establishments in Japan to this day. He nurtured theorists such as Tomonaga, Yukawa and Sakata, with whom he kept in close contact. Before the war, in 1935, Yukawa initiated the meson theory of nuclear forces. During the war years, Tomonaga, who himself was at Riken, developed the idea of renormalization. Then, with the advancement of experimental techniques, came the discovery of more and more unexpected particles and phenomena: the strange particles, a whole bunch of hadrons, resonances, and parity and CP violation.[2,3]

During these exciting times, S. Sakata and M. Taketani, collaborators of Yukawa in developing the meson theory, articulated their conscious theoretical approach and strategy to particle physics, and practiced it themselves with considerable success.[4]

The Sakata-Taketani methodology is summarized in Taketani's three stage thesis. It posits that the progress in physics, especially particle physics, goes through a cycle of three stages.

Stage 0

We start with the discovery of new phenomena which are outside of the existing physical laws.

Stage 1: Phenomenology

Our first task would be to collect data, and try to find some order or regularities in them, then arrive at some empirical laws or representations that enable us to organize them and make predictions. This is the stage of phenomenology.

Examples. The discovery of strange particles and hadron resonances fits this description. The Gell-Mann-Nakano-Nishijima law [5,6] was found by assigning isospin and strangeness to strange particles. This led to flavor SU(3) symmetry of hadrons [7,8,9,10] from which were derived various symmetry relations among masses and cross sections. The Veneziano model [11], which is an explicit formula realizing duality and straight Regge trajectories, might be called another example of this stage.

Stage 2: Model building

After the first stage of phenomenology comes the stage of model building. We want to interpret the origin of the regularities in terms of a model, in which concrete, often hypothetical, entities are introduced. Or else the model may be mathematical in nature.

The Sakata triplet (proton, neutron, and Λ) [12] and the successful quark model [13,14] are such examples of realizing the SU(3) symmetry relations in terms of material particles. For the Veneziano formula, the dual string model is its realization in terms of concrete entities called strings.

These models, however, may have large numbers of arbitrary parameters, and even so their validity is of a limited and semi-quantative nature. For different kinds of phenomena there may be different models, and they may not necessarily be mutually compatible.

Stage 3: Definitive theory

The next and the final stage is to construct or invent a theory which incorporates the various models in a precise, all-inclusive mathematical system of laws, which describes all phenomena in a quantitatively correct way, and can make precise predictions. We immediately recall the Standard Model (SM) as an example, which consitsts of quantum chromodynamics of string interactions and the Salam-Weinberg theory of electroweak interactions. In the past 15 years or so, its validity has been confirmed to an impressive degree of precision. It has all the features of a definitive theory of the three basic interactions. It is still called a model, perhaps because some aspects of it has not yet been confirmed, and perhaps also because of the large numbers of phenomenologically adjustable parameters which could be derived in a more complete theory. Nevertheless its confirmed validity is such that essentially it deserves to be called the Standard Theory.

Stage 4. Return to Stage 0

The three stages are expected to repeat themselves, when the final theory is found to break down in the face of new phenomena, which in particle physics usually happens when the available accelerator energy is raised, or the experimental precision is greatly improved. We have been so accustomed to this that we have acquired a habit always expecting it, so that as soon as a theory is established, we immediately start looking for its violations. Here is what Salam had said on this point:

"Classical physical theories are profound. Take the second law of thermodynamics, for instance: Heat cannot flow spontaneously from a colder to a hotter body. Compare this to what you have been doing. You propose some symmetry, and ten seconds later you are already trying to figure out how to break it."

—A. Salam [15]

3 THE THREE MODES

No doubt such characterization of particle physics has been most appropriate up to the 1960s. It was the period, as had been in the past, in which experiment and theory went hand in hand to explore and understand nature. In fact, the above analysis could apply equally well to the history of quantum theory. As Sakata had emphasized, the moral of it for the particle physicist is to have in one's mind at any moment a larger picture of the situation, always asking oneself in what stage one is in at the present point in time, and what is likely to follow.

But it is inevitable that even this useful general thesis about the stages of physics itself has its own limits. Indeed, starting around the seventies, there has been a sea change, so to speak, in the way particle physics has evolved. As we have successfully developed and synthesized all the theoretical ideas and experimental facts into the Standard Model, the theory further took off on its own in an exponential way, leaving experimental capabilities far behind. Here the term 'exponential' is a literal one. On the one hand, the theoretical leap is caused by the fact that physical "constants" vary with energy scales logarithmically due to renormalization, so that it becomes natural or even neccesary to envisage new phenomena to appear at exponentially high energies, whereas, on the other hand, the experimental capabilites to test such theories are limited by the worldly realities. It is true that, empirically, the accelerator energy has also grown since the 1930s exponentially with time, by a factor of 10 every 10 years, i.e., a factor of 10^6 in the past 60 years (the Livingston law [16]). But the GUT and Planck energies that theorists envisage are a factor of 10^{12} or more ahead. Without some breakthrough on the part of experimental techniques or methods, the gap looks difficult to close.

In view of this situation, some time ago I proposed [17] a different characterization of the way theories are created and developed. Again it is not a classification of the physical laws themselves but that of the theorists' modes of operation to get at them. As I see it, there are three such modes, and each physicist seems preferentially inclined to one or the other.

1. Yukawa mode

This mode can be characterized by saying that one tries to explain or understand new phenomena in terms of concrete entities, i. e., particles. One freely postulates or invents new particles if necessary. Schematically,

$$\text{new physics} \quad \rightarrow \quad \text{new particles}$$

which is to be contrasted with alternative approaches, for example,

$$\text{new physics} \quad \rightarrow \quad \text{new principles}$$

Let us illustrate the point with concrete historical examples:

Example 1. Beta decay

In order to solve the continuum spectra of beta decay, Pauli postulated the neutrino. An alternative which was considered was giving up of the energy conservation hypothesis.[18]

Example 2. Nuclear forces

In order to explain the nuclear forces, Yukawa applied the then new quantum field theory, and postulated the existence of a new particle, now called pion.[19] In order to appreciate its significance, one must recall the prevailing attitude of the leading physicists at that time. They had just succeeded in explaining atomic physics by means of a new physical principle called quantum mechanics. Nuclear phenomena involved many orders of magnitude higher energies and shorter distances. It was not unnatural for the founders of quantum mechanics to expect that to each new physics there corresponded a new mechanics or principle. On the other hand, they had taken for granted that matter consisted of the electron and the proton (perhaps also the neutron), and were reluctant to conceive of new particles just for the sake of understanding the nucleus.

6

Example 3. The cosmic ray muon puzzle

The muon discovered in the cosmic rays by Anderson and Neddermyer was at first identified erroneously but understandably with the pion which Yukawa had predicted two years before. But soon theoretical inconsistencies surfaced in the disparity between its production and interaction cross sections. The solution to this puzzle was the two-meson hypothesis[20] of Sakata who assumed that the cosmic ray 'meson' (fermion) was the daughter of the Yukawa meson (boson). On the other hand, a host of theorists tried to explain this away, unsuccessfully, in terms of the strong coupling theory of the Yukawa meson[21]. The situation is akin to example 2 above.(Incidentally, the strong coupling theory has features similar to the Skyrmion theory of the nucleon.)

Example 4. Two neutrinos and the charm quark

With the discovery of the muon, it would have seemed natural, in hindsight, to suppose the existence of its own neutrino, but in fact this was not natural or obvious in those days. When the Cabbibo theory was proposed, it would have been natural to expect a fourth quark (charm), but still this did not seem to come so naturally. I will not go into rather complicated historical details, but I would like to emphasize the fact that the Sakata school was among the earliest people to take this path, because, in my view, they did not have psychological resistance to introducing new particles[22].

Example 5. The three generations

Continuing in the same direction, it is not surprising that Kobayashi and Maskawa [23] of the Sakata school were led to anticipate the existence of three (or more) generations of fermion, in order to solve the problem of CP violation.

2. Einstein mode

In what I call the Eisntein mode, one first sets down a general physical principle to describe a physical phenom, formulate a mathematical theory that realizes it, which will then lead to predictions. Thus one may say that the Yukawa mode is inductive, whereas the Einstein mode is deductive. Of course the latter too starts from some known properties of physical phenomena, but thereafter one is guided by the logic of a theory rather than by the facts. The following examples will need no elaboration.

Example 1. Theory of gravity

The logic of Einstein's theory of gravity can be summarized by the following chart.

geometrical principle, equivalence principle ⟶ equation of gravity
⟶ predictions ⟶ (precession of the perihelion of Mercury,
the black hole, the expanding universe, etc.)

Example 2. The Gauge theory – Here the chart is

gauge principle ⟶ non-Abelian gauge (Yang-Mills) theory
⟶ QCD, the electroweak theory, grand unified theories

3. Dirac mode

The Dirac mode is more abstract and speculative than the Einstein mode. Contact with known physical phenomena becomes more indirect. The primary guiding principle appears to be mathematical beauty and elegance, based on the belief (Dirac [24]) that what is mathematically beautiful must also be physically true. Thus

mathematical possibility ⟶ physical theory ⟶ predictions

Example 1. Electric-magnetic duality

Dirac [24] invented the magnetic monopole (as well as the fiber bundle) in order to realize a symmetry between electric and magnetic forces, solely on the basis of esthetic desirabilty. Much later, the non-Abelian gauge theories were found to be natural realization of the idea in the form of topological solitons. Now this has further developed into the exciting recent developments in the new dualtity principle [25].

Example 2. Supersymmetry and superstring

Like the monopole, supersymmetry was not directly motivated by known physics, but by the conceived desirability of unifying fermions and bosons. Nevertheless, because of its mathematical elegance and richness, it has grown into a highly developed theoretical system. Superstring theory may also be placed somehow in this category. Although it is descended from the dual string model, it is a grand and speculative leap. Its concrete appeal is the possibility of unifying gravity and everything.

So far none of the above examples of the Dirac mode has been confirmed experimentally. Nevertheless they are where the center of theoretical activity lies at the moment. The reason for it is not difficult to locate: the fast mathematical developments in this mode coupled with the slow progress on the experimental side, that has been characteristic of particle physics since the 1980s.

8

To summarize the themes developed, particle physics in the past 60 years has gone through various stages, and in each stage physicists resorted to some dominant theoretical methods appropriate to that stage. We are at present at the near final point of having a complete theory of what is known up to the electroweak energy scale. We have a feeling that the Standard Model is the true theory, and it just remains to confirm it in detail. We have also theories of what to expect beyond the electroweak scale, and various pieces of indirect evidence are called upon in this respect, especially in astrophysics and cosmology, but as yet there are no clear and direct experimental signs to indicate that new physics lies ahead. In a sense, particle physics is at a self-sufficient stage now. A set of principles or paradigms may be said to have guided the physicist to achieve this success:

a. Renormalization

b. The Symmetry and the gauge principle

c. Spontaneous symmetry breaking

As for the future, we expect the following additional principles will play crucial roles, although they remain to be just theoretical expectations at the moment.

d. Supersymmetry

e. Superstring

It is also to be remarked that, theoretical speculations about new physics aside, there are some open problems within the known domain of physics that await further exploration, experimentally as well as theoretically. Outside of the Standard Model, the physics of gravity, including quantum gravity, is an outstanding one. Within the Standard Model, they concern basically the Higgs sector. Experimentally, the masses of the neutrinos are not known yet. The Higgs boson is yet to be discovered. Theoretically, they concern the origin of the masses and mass matrices of quarks and leptons, their hierarchical structure, and the nature of the Higgs field. Since my interests have been in the problem of mass, for which the symmetry breaking paradigm, c, is relevant, I will devote the remainder of the talk to discuss this problem.

4 REVIEW OF SPONTANEOUS SYMMETRY BREAKING [26]

The concept of spontaneous symmetry breaking (SSB) has been around for a long time without being recognized as such. As for the concept and use of symmetry in physics, P. Curie [27] consciously applied group theoretical considerations to crystal symmetries, and derived certain selection rules for physical phenomena. He showed that for a phenomenon to occur, for example the Wiedemann effect (magnetization caused by twisting a bar), the medium must have symmetries that are consistent with the symmetries of the effect. He did not realize, however, that the converse is not necessarily true: A symmetry can be broken spontaneously to produce an effect. P. Weiss's explanation of ferromagnetism is an example of SSB. Heisenberg gave a microscopic theory of ferromagnetism in terms of quantum mechanics. It is not surprising, then, that he later invoked an analogy to ferromagnetism in his unsuccessful attempt at a unified theory of elementary particles [28].

SSB is most relevant to the problem of fermion masses. A massless fermion has chiral symmetry, but the latter can be broken spontaneously to produce mass. In other words, mass is a dynamical quantity that is subject to theoretical explanation. The prototype of such an effect is the BCS (Bardeen-Cooper-Schrieffer) mechanism for superconductivity. Other examples are superfluidity in ^{3}He, the QCD-chiral dynamics of quark mass (the so-called constituent mass) generation, nucleon-nucleon pairing and low lying collective modes in nuclei. The electroweak symmetry breaking in the Salam-Weinberg theory may be another example if the Higgs field can be interpreted as composite in nature. [26]

The SSB in general has the following characteristics.

1. Degeneracy of the ground state of a medium.

2. Existence of the Nambu-Goldstone (NG) modes when the symmetry is continuous and the system is infinite (the thermodynamic limit)

3. Possibility of hierarchical SSB (tumbling) – This means that an SSB can trigger another in a hierarchical way. Examples of this are the chains

   a) crystal formation (continous to discrete space symmetry) $\longrightarrow$ phonon (an NG mode) $\longrightarrow$ superconductivity

b) QCD chiral SSB $\longrightarrow$ π and σ mesons (see below) as the source of nuclear forces $\longrightarrow$ formation of nuclei $\longrightarrow$ nuclear pairing and collecive modes

The BCS mechanism, in addition to the above, has the following special properties.

1. Formation of a fermion mass (energy gap)

2. Existence of the collective bosonic modes π (NG) and σ (Higgs)

3. A mass formula (approximate) for the low energy fermionic and bosonic modes

$$m_\pi : m_f : m_\sigma = 0 : 1 : 2$$

This can be shown to imply a sort of broken supersymmetry among these modes.

4. Possibility of transcribing the dynamics of these modes into an effective Ginzburg-Landau- Higgs (GLH) theory.[29] In this transcription, the Yukawa coupling f and the self-coupling λ of the effective Higgs field are given by

$$1/f^2 = C \ln(\Lambda/m_f),\ \lambda = f^2$$

where C and Λ are determined by the underlying dynamics.

5 THE STANDARD MODEL AND BEYOND: THE MASS HIERARCHIES

I come now to the main theme of the conference, i. e., to examine the present and gauge the future, as it relates here to the problem of mass. This may further be broken down to a set of three problems, although they are closely dependent on each other; the solution to one of them will influence the other.

a) Hierarchy within the SM and the details of the Cabbibo-Kobayashi-Maskawa (CKM) matrix

b) The nature of the Higgs field in the SM

c) General hierarchy problem in the GUTS and beyond

The first problem addresses what is already known experimentally: the mass spectrum of quarks and leptons. Still unknown are the neutrino masses and their mixing matrix.

In the second problem, the existence of the Higgs field has been only indirectly confirmed in the gauge sector. The Salam-Weinberg theory gives a precise account of the properties of the electroweak interaction. The existence of the Higgs boson or the content of the pure Higgs sector is an open question. Similarly, the role of the Higgs field in generating the fermion masses is theoretically expected but not experimentally confirmed.

The third problem remains a theoretical one at present. However it has arisen naturally in the process of pursuing the SM to its logical conclusion. In spite of its successes, the SM is not theortically complete. They are unsatisfactory in some respects:

1. It does not unify the strong and electroweak sectors.

2. It does not explain the fermion mass structure, including CP violation and the theta vacuum angle. The search for answers to them has led to the GUTS, but at the same time it has created a new set of problems, like proton decay, monopole, and hierarchy.

It is instructive to first look at the entire spectrum of hierarchy in the universe. In terms of length scale, it spans 60 orders of magnitude, from the size of the present universe ($10^{27} cm$), through various intermediate scales, down to the Planck length ($10^{-33} cm$). In the geometrical middle ($\sim 10^{-3}$) is the biological scale. On the small size side, the scale of the SM spans 6 orders for the charged particles, and perhaps 12 orders including the neutrinos. From the SM to the GUTS scale is again 13 orders. The nature is in fact built on layers of discrete structures and hierarchies. At each layer there is an appropriate set of physical laws which is supposed to be derived as an effective theory from the underlying sublayers by "integrating them out". In between two layers is the domain where a scaling law holds in some form or another. Although one seems to take it for granted, it is this property of natural laws that lies under the success of reductionism. Thus one can start from phenomena of a large scale and find physical laws that apply within that domain, then extend it to smaller scales as much as possible. The Hamiltonian structure of mechanics allows this: The heavenly bodies as well as the bodies on the earth obey the same laws in terms of a few degrees of freedom alone, without having to know their internal structure, and just by changing the parameters. This also holds in the quantum world since basically the Hamiltonian structure remains there. The Wilsonian scaling is another manifestation of this same general property of natural laws. Yet it is also true that such an effective theory has a finite range of validity, and will eventually cease to apply at a critical scale. The next step may be a discontinuous leap, but it still must be such that it permits the process of "integrating out", and the underlying layer decouples to a large extent when pulled back to the original layer. The challenge for particle physics then has been to probe the next layers one by one in this way.

In light of the above considerations, one may say that the three categories of problems listed above regarding the hierarchy at the current microscopic scale are all related and aims to address the same task, but differ in the emphasis or strategy that people want to follow. In concrete terms, I see three approaches to the mass problem:

1. Top down appproach

   Here one starts from a grand scheme like the GUTS at a high energy scale, which was originally motivated by an attempt for unification in the gauge sector, then comes down again to the low energy scale and try to see if it can also explain the mass problem. If successful, it will certainly reinforce our confidence in grand unification.

   Currently this seems to be the preferred strategy of theorists. Thus the various GUT theories like $SU(5), SO(10)$ and higher and richer unification groups up to the superstring level have been considered. At the moment the most promising version of gauge unification seem to be the supersymmetric one, and it is also most interesting one in that the supersymmetric partners are expected to show up in the Tev region. The nature of the Higgs fields in the SM is generally regarded as elementary, i. e., they exist already at the gauge unification scale.

2. Bottom up approach

   This may be said to be repeating anew of the process of arriving at grand unification applied to the mass problem. For this one examines the mass spectrum in the SM and tries to find some clues there without prejudice, and see how much of it can be understood within the SM, and if not, what new physics would be needed.

   The top quark condensation model[30,31,32] belongs to this category. It has achieved some success in making clear that the large top quark mass is more natural than the smallness of the other fermion masses. The Higgs boson is considered as a $t-\bar{t}$ composite here, the only input parameter is the cutoff energy, with the top and the Higgs masses satisfying the BCS relation. But it fails to explain the other lighter fermions. The real origin of the $t-\bar{t}$ binding force is left open, and to be sought eventually in a unification theory at the cuoff.

3. Intermediate approach, represented by the technicolor models[33,34]

   Here a new set of gauge fields specifically responsible for mass generation is postulated at an intermediate scale, not far above the electroweak scale, from which one comes down to account for the masses in the SM. Going up, they will be unified together with the existing forces in the SM.

It seems fair to say that so far none of these approaches has been so successful as to exclude the other possibilities. The difficulty is in giving a natural explanation for the seemingly irregular hierarchical pattern of masses from a naturally regular and simple input. One does not know yet whether the fermion masses are rooted in a relatively simple law or are a result of the whims of the renormalization group equations. In view of this situation, I will present below, as an exercise in the stage 1 of the Sakata scheme, a possible regularity in the quark masses which may have not been noticed before.

The quark masses are not directly observable quantities, but have to be inferred indirectly and are subject to theoretical uncertainties, hence also allow some degree of freedom of choice. Some of them are especially sensitive to the QCD Λ and the renormalization point referred to. With this proviso they seem to satisfy the following simple formula

$$m/m_0 = 2^n,$$

where $m_0 \approx 5$ Mev, and n runs over the following set of integers

Table 1: The quark mass formula

I		II		III
u: 0		c: 8		t: 15
	6±2		6±1	
d: 1		s: 5		b: 10

The numbers inserted between the quark entries are the differences between neighboring generations for up/down type quarks This discrete scaling law is ominously reminiscent of the Bode's law for the radii of the planetary orbits, which is not known to have a theoretical basis. This new one may not have a theoretical basis either. But it might serve as a clue in the search for a proper theoretical starting point. For example, it lends itself to the type of hierarchy theories proposed by Frogatt and Nielsen, perhaps with three different kinds of charges assigned to each quark. Or else it suggests some kind of quantization in scaling. [a]

6 CONCLUDING REMARKS

Within the last quarter century, the nature of particles physics has undergone radical changes the likes of which have not been seen before. On the theoretical side, the success of the Standard Model means that one has a working theory that enables one to describe all known phenomena in a precise and comprehensible way. But it still has some phenomenological elements; it is not a complete closed theory that is expected to hold much beyond the electroweak energy scale. Currently, the neutrino and Higgs sectors are unknown and untested. At the same time, one has also developed a class of speculative unified theories that have gone way beyond the energies conceivably accessible by traditional experimental techniques yet are able to make some statements that are subject to experimental tests. The superstring theory raises the hope of bringing gravity within the fold of quantum field theory and accomplish the ultimate unification of all known forces. On the experimental side, however, it is clear that one is slowly approaching the practical limits of the traditional accelerator-based physics. The empirical Livingston scaling law for the exponential increase of accelerator energy has served us well for the past 60 years,

[a] The leptons do not conform to such a formula as nicely as the quarks, but if one wishes, one could assign the following numbers: e: -3.5 μ: 4.5 τ: 8.5

but here again a new qualitative break has to occur if one is going to be able to continue our quest to the foreseeable future, and to be able to test the unified theories in its full content.

The first thirty years of particle physics was like the time of Renaissance. Experiment and theory went hand in hand. Since then, however, we have been getting into the baroque period, so to speak, with increasing levels of sophistication in both hardware and software, and yet with a mismatch between the two. This is the way the things are, whether one likes it or not. According to theory, particle physics has a bright future. Supersymmetry may be around the corner. Superstring may be able to make a connection with reality. Still, as a person from the older era, I feel an urge to close this talk with a couple of quotes

"As a mathematical discipline travels far from its empirical source, or still more, if it is second and third generation only indirectly inspired by ideas coming from 'reality', it is beset with very grave dangers. It becomes more and more purely aesthetizing, more and more purely *l'art pour l'art.* This need not be bad, if the field is surrounded by correlated subjects, which still have closer empirical connections, or if the subject is under the influence of men with an exceptionally well-developed taste. But there is a grave danger that the subject will develop along the line of least resistance, that the stream, so far from its source, will separate into a multitude of insignificant baranches, and that the discipline will become a disorganized mass of details and complexities. In other words, at a great distance from its empirical source, or after much' 'abstract' inbreeding, a mathematical subject is in danger of degeneration. At the inception the style is usually classical; when it shows signs of becoming baroque, then the danger signal is up. — In any event, whenever this stage is reached, the only remedy seems to be the rejuvenating return to the source; the reinjection of more or less directly empirical idea. I am convinced that this was a necessary condition to conserve the freshness and the vitality of the subject and that this will remain equally true in the future."

—— John von Neumann [35]

" Eppure si muove" — Galileo

ACKNOWLEDGEMENTS

This work was supported in part by the National Science Foundation, Grant #PHY 91-23780, and the University of Chicago. I also thank Prof. K. Kikkawa of Osaka University for his hospitality.

16

1. Y. Nambu, Prog. Theor. Phys. Suppl. 85 (1985) 104.
2. Articles in Prog. Theor. Phys. Suppl. 105 (1991) 105.
3. L. M. Brown, Historia Scientiarum #36 (1989) p.1;
4. See the articles by Sakata and Taketani in Prog. Theor. Phys. Suppl. 50 (1971).
5. M. Gell-Mann, Phys. Rev. 92 (1953) 833.
6. T. Nakano and K. Nishijima, Prog.Theor. Phys. 10 (1953) 581.
7. M. Gell-Mann, CalTech Report CTSL/20 (1964).
8. Y. Ne'eman, Nucl. Phys. 26 (1961) 222.
9. M. Ikeda, S. Ogawa, and Y. Ohnuki, Prog. Theor. Phys. 22 (1959) 715.
10. Y. Yamaguchi, Prog. Theor. Phys. Suppl. N0.11 (1959) 1, 37.
11. G. Veneziano, Nuovo Cimento 57A (1968) 190.
12. S. Sakata, Prog. Theor. Phys. 16 (1956) 686.
13. M. Gell-Mann, Phys. Lett. 8 (1964) 214.
14. G. Zweig, CERN reports 8182/TH 401.
15. Quoted in J. J. Sakurai, Ann. Phys., 11 (1960), 1.
16. M. S. Livingston, High Energy Accelerators (Interscience Publishers, 1958), p. 149.
17. Y. Nambu, Prog. Theor. Phys. Suppl. 85 (1985) 104.
18. See A. Pais, in *Twentieth Century Physics*, eds. L. Brown, A. Pais, and B. Pippard, IOP Publishing Ltd, and AIP Press, Inc., 1995), p. 43
19. H. Yukawa, Proc. Phys.-Math. Soc. Japan 20 (1937) 712.
20. S. Sakata and T. Inoue, Prog. Theor. Phys. 1 (1946), 189.
21. See L. M. Brown, in *Twentieth Century Physics*, eds. L. Brown, A. Pais, and B. Pippard, IOP Publishing Ltd, and AIP Press, Inc., 1995), p. 357.
22. See V. Fitch and J. Rosner, in *Twentieth Century Physics*, eds. L. Brown, A. Pais, and B. Pippard, IOP Publishing Ltd, and AIP Press, Inc., 1995), 635; ?? Z, Maki and T. Nakagawa, Prog. Theor. Phys. 31 (1964) 115; Z. Maki, Prog. Theor. Phys. 31 (1964) 331, 333.
23. M. Kobayashi and T. Maskawa, Prog. Theor. Phys. 49 (1973) 652.
24. P. A. M. Dirac, Proc. Roy. Soc. A133 (1931) 60.
25. N. Seiberg and E. Witten, Nucl. Phys. B426 (1994) 19; B431 (1994) 484.
26. See Y. Nambu, in *Eolutionary Trends in Physical Sciences* (Springer Proceedings in Physics, Vol. 57), eds. M. Suzuki and R. Kubo (Springer-Verlag Berlin Heideberg 1991), p. 51; Proc. ICTP Workshop on Quarks and Leptons, 1996, to be published, (EFI preprint 96-39).
27. P. Curie, J. Phys. 3 (1907) 393.
28. H. P. Duerr,W. Heisenberg, H. Mitter, S. Schlieder, and K. Yamazaki, Z. f. Naturforschung 14A (1959) 441.

29. V. L. Ginzburg and L.D. Landau, Zh. Exp. Teor. Fiz. 20 (1940) 1064. P. W. Higgs, Phys. Lett. 13 (1964) 132; Phys. Rev. Lett. 13 (1964) 508.
30. Y. Nambu, in *New Theories in Physics, Proc. XI Int. Symposium on Elementary Particle Physics*, Eds.: A. Ajduk et al., (World Scientific, Singapore, 1989)
31. A. Miransky, M. Tanabashi, and K. Yamawaki, Mod. Phys. Lett. A4 (1989) 1043; Phys. Lett. B221 (1989) 177
32. W. A. Bardeen, C. T. Hill and M. Lindner, Phys. Rev. D41 (1990) 1647.
33. S. Weinberg, Phys. Rev. D13 (1976) 974, D19 (1979) 1277.
34. D. Susskind, Phys. Rev. D20 (1979) 2613.
35. John von Neumann, in *The Works of the Mind*, (University of Chicago Press, 1947), p. 196

SOME ANOMALIES RELATED TO SPONTANEOUS SYMMETRY BREAKING

YOICHIRO NAMBU
University of Chicago, Illinois, USA
E-mail: nambu@theory.uchicago.edu

I discuss what I call Nielsen, Lorentz, and Newton anomalies accompanying some cases of spontaneous symmtry breaking (SSB). The first refers to an anomaly in the number and the dispersion law of the Nambu-Goldstone modes in the SSB of Lie algebras; the second to those in the case of Lorentz symmetry breaking, in a medium or in vacuo; and the third to the peculiar behavior of quasiparticle and other familiar excitations in a medium when they are regarded as classical particles. Speculations are made in regard to an SSB of Lorentz symmetry *in vacuo*.

1. Nielsen anomaly

In general the number of the Nambu-Goldstone (NG) bosons associated with a spontaneous symmetry breaking (SSB) $G \to H$ is equal to the number of symmetry generators Q_i in the coset G/H. In the absence of a gauge field, their energy Ω goes as a power k^γ of wave number. In a relativistic theory, $\gamma = 1$ necessarily unless Lorentz invariance is broken. There are, however, exceptions to the above 'theorem.'[1] Recently a few new examples have been found to occur in connection with color superconductivity in high density quark matter. I would like to give a general explanation to the phenomenon.[2]

Suppose a set of local operators $O = \{O_i\}$ forms a representation of a symmetry group G, under which the O_i transform into each other. The generators $\{Q_i\}$ of G themselves are an example. If one member of the elements, say O_0, develops a vacuum expectation value $< O_0 >= C$, then the other elements O_k will generate the zero modes in their low energy part. If two of them are such that their commutator closes on O_0: $[O_j, O_k] \sim iO_0$, or $< [O_j, O_k] >\sim iC$. Then their low-energy components behave like canonical conjugates of each other, therefore they belong to the same dynamical degree of freedom, and the number of NG bosons is thereby reduced to one per each such pair. The set of generators $\{Q_i\}$ are precisely of this nature. The dispersion law $\gamma = 2$ is obtained by a more detailed analysis. Basically one constructs a low-energy effective quadratic Hamiltonian in terms of the deviations $\{\widetilde{O}_i\}$ and their canonical conjugate $\{\widetilde{\Pi}_i\}$ by expanding around the expectation value $< O_0 >= C$:

$$H_{eff} = a\Sigma\widetilde{\Pi}_i^2 + b\widetilde{O}_0^2 + ck^2\Sigma\widetilde{O}_i^2 \tag{1}$$

The zero modes are $\{O_j, j \neq 0\}$ and have no b terms since they are in the flat directions from $< O_0 >$. So from the a and c terms one gets a phonon-like spectrum with $\gamma = 1$. This is the normal case. However, if O_j and O_k are canonical conjugates, then $\Pi_k \sim O_j$, $\Pi_j \sim O_k$, hence the a terms for them should also be missing, and their c terms directly lead to a dispersion law $\gamma = 2$. (The Higgs mode is $\widetilde{O}_0$, with $b_0 \neq 0$, and this gives a finite gap with k^2 corrections.)

Ferromagnetism and antiferromagnetism present the simplest examples to illustrate the normal and anomalous cases. Their relevant operators are the total spin (density) and the graded (staggered) sum

$$O_i = \frac{1}{N}\Big(\sum_{\text{even}} \sigma_i/2 - \sum_{\text{odd}} \sigma_i/2\Big), \tag{2}$$
$$\langle O_z \rangle = C, \qquad \langle S_z \rangle = 0.$$

S and O together form an SO(4) algebra. The zero modes are expected to be S_x, S_y and O_x, O_y respectively. But since $< [S_x, S_y] >=< [O_x, O_y] >= i < S_z >$, the ferromagnetic case has only one zero mode, whereas the antiferromagnetic case has two. In the former, the modes $S_x \pm iS_y \sim S_x \pm i\Pi_x$ serve respectively as the annihilation and creation operators of a single chiral mode behaving as $\exp(\mp i\omega t)$, but in the latter $O_x \pm i\Pi_x$ and $O_y \pm i\Pi_y$ generate two nonchiral modes.

I will mention two other examples of the anomalous case. They have been found in connection with color superconductivity. In one of them,[4] the SSB of a complex scalar multiplet (Higgs field) ϕ_i is induced by a chemical potential instead of a negative mass term: its potential term is of the form

$$V = \frac{\lambda}{4}(\phi^\dagger \cdot \phi)^2 - i\mu(\pi \cdot \phi - \phi^\dagger \pi^\dagger) \tag{3}$$

The chemical potential destabilizes one of the ϕ_i away from zero. Since $\dot{\phi} \neq \pi$, both ϕ and π acquire nonzero expectation values, and this leads to noncommutativity among different zero modes. In still another example,[6] one deals with a set of multicomponent Cooper pairs $\{\psi_i\psi_j\}$

and $\{\psi_j^\dagger \psi_i^\dagger\}$, some combination of which will condense. Since the commutators $[\psi_i \psi_j, \psi_k^\dagger \psi_l^\dagger]$ of the expected zero modes yield the bilinears $\{\psi \psi'^\dagger\}$, they may not be independent if one of the latter quantities has a nonzero expectation value because of a chemical potential.

2. Breaking of Lorentz Symmetries

The above considerations about the anomalous zero modes had to do with internal symmetries, except for the case of ferromagnetism since the spin is part of the Lorentz group. So the next natural problem would be to examine the spacetime symmetries. Can spacetime symmetries be spontaneously broken *in vacuo*? What would be the properties of the ensuing zero modes? The first question has been taken up by many people in the past, so I will concentrate here on the second one, in particular, the case of rotations and boosts. Here, one immediately realizes that the generators of the Lorentz group, namely $L_{\mu\nu} = x_\mu \partial_\nu - x_\nu \partial_\mu + \Sigma_{\mu\nu}$, are not homogeneous, i.e., their orbital part depends on the coordinates explicitly. Hence if a set of field operators transforming under $L_{\mu\nu}$ develops a vacuum expectation value, one cannot necessarily apply the usual argument and claim that the rest of the members would represent the zero modes. The spin part $\Sigma_{\mu\nu}$ of $L_{\mu\nu}$ is free of the problem, but it alone does not generate Lorentz invariance.

To illustrate the point, let us consider a relativistic analogue of the ferromagnet in the vacuum. The spin density of a fermion field is the spatial part of the chiral current $j_{5\mu} \equiv \bar{\psi}\gamma_5\gamma_\mu\psi = \psi^\dagger(\rho_1, \sigma_i)\psi$, and it is also a part of the angular momentum density M_{0ik} satisfying the $su(2)$ (current) algebra relations.[a] Consider as a model the following nonlinear interaction[2,3]

$$L_{\text{int}} = -(g/2) j_{5\mu} j_5^\mu,$$
$$(\gamma \cdot p - m + g V_\mu \gamma_5 \gamma^\mu)\psi = 0, \tag{4}$$

leading to an SSB by an axial vector constant $V_\mu = < j_{5\mu} >$. Here a Dirac mass is assumed but the result is similar for a Majorana mass. In general

[a]It is interesting to note that the relations among the spatial components of axial (σ_i) and vector ($\rho_1\sigma_i$) currents (the latter corresponding to imaginary Lorentz boosts), stand in direct correspondence to those for ferro- and antiferromagnetism.

one may consider three possible cases.

I. $g > 0$, spacelike: $V = (0, 0, 0, v)$, $\quad Z = (Z_t Z_x, Z_y, -)$,
vacuum has net spin in the z direction, $T^- C^+ P^+$,
$\omega^2 = p_x^2 + p_y^2 + (\sqrt{p_z^2 + m^2} \pm gv)^2$.
II. $g < 0$, timelike: $V = (v, 0, 0, 0)$, $\quad Z = (-, Z_x, Z_y, Z_z)$,
vacuum has net chiral charge, $T^+ C^+ P^-$,
$\omega^2 = m^2 + (|p| \pm gv)^2$.
III. lightlike: $V = (\pm' v, 0, 0, v)$, $\quad Z = (\pm' Z_0, Z_x, Z_y, Z_0)$,

$$(\omega \mp gv)^2 = p_x^2 + p_y^2 + (p_z \pm \pm' gv)^2 + m^2. \tag{5}$$

Case I seems the more natural one since it corresponds to an exchange of

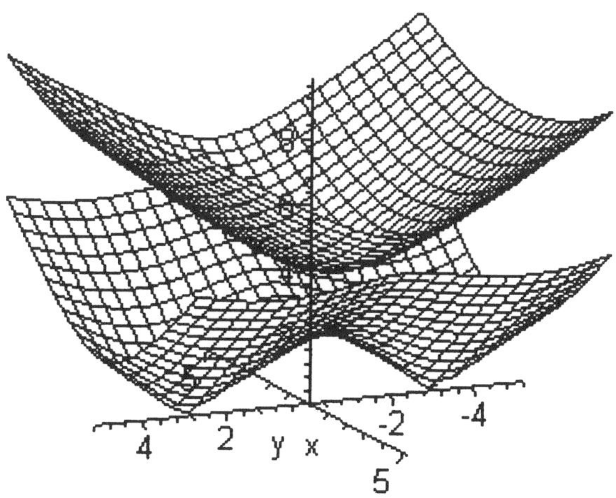

Figure 1.

The group velocity does not exceed 1, but the branch with the negative sign has various peculiarities, such as spacelike states, $\omega < k$, and negative group velocity relative to the momentum. I will come to them later, but our concern now is the properties of the associated zero modes Z, which follow from the loop integral $< j_{5\mu}, j_{5\nu} >$. For example, the effective Lagrangian

[b]In the present dynamical model, a stable solution exists if $gv > m$ for Case I.

$\mathcal{L}$ and the dispersion law for Case I are found to be of the form

$$\begin{aligned}\mathcal{L} &= L_{\mu\nu}Z^{\mu}Z^{\nu},\\ L_{\mu\nu} &= Ag_{\mu\nu}p^2 + Bp_{\mu}p_{\nu} + Cg_{\mu\nu}p_z^2 + Dg_{z\mu}g_{z\nu}p^2 + Eg_{z\mu}g_{z\nu}p_z^2\\ &\quad +F(g_{z\mu}p_zp_{\nu} + g_{z\nu}p_zp_{\mu}) + iG\varepsilon_{z\mu\nu\lambda}p^{\lambda}\\ &\quad -M^2g_{\mu\nu} - \Delta M^2 g_{z\mu}g_{z\nu},\end{aligned} \tag{6}$$

$$(M^2 + \Delta M^2 - p^2)^2 - g^2p^2 = 0, \qquad (M^2 - p^2)^2 - g^2p^2 = 0. \tag{7}$$

The points to be noted are: (1) as was anticipated, the would-be zero modes are actually massive, and (2) there is a Chern-Simons term in the subspace (t, x, y).[5c]

3. Quasiparticles as classical objects[d]

As we have seen above, the dispersion law for the fermion after an SSB can have a shape which leads to a group velocity opposite to the momentum. But this is not a new phenomenon. A similar behavior can be found in some familiar examples. One of them is the Laudau roton-phonon in superfluid ^{4}He. Another is the Bogoliubov-Valatin (BV) quasiparticle in superconductors, which is a superposition of an electron and a hole:

$$\Psi(p) = \alpha\psi_{\uparrow}(p) + \beta\psi_{\downarrow}^{\dagger}(-p) \tag{8}$$

having the dispersion law

$$\omega^2 = (p - p_F^2)^2/4m^{*2} + \Delta^2, \tag{9}$$

where p_F is the Fermi momentum. My goal is now to study the implications of these peculiar dispersion laws.

The relation between a dispersion law $\omega(k)$ and the group velocity $v = d\omega/dp$ is equivalent to a Hamiltonian equation of motion: $\omega \sim H$, $k \sim p$, $dx/dt = v = \partial H/\partial p$. So let us regard a quasiparticle as a classical point particle and study its motion in an external field, assuming that the medium is of sufficiently large extent and the acceleration is sufficiently small to

[c]The Chern-Simons term has been discussed in connection with Lorentz violation in electromagnetism. [7,3]

[d]First reported at a symposium in honor of P. Ramond, University of Florida, Jan. 2003, and at another in honor of O.W. Greenberg, University of Maryland, May 2003.

avoid the constraints of the uncertainty principle. Consider for example the following system:

$$\begin{aligned} H &= H_0(p) + V(x), \quad H_0 = ap^4 + bp^2 + c, \\ dx/dt &= 3ap^3 + 2bp, \\ dp/dt &= -\partial V/\partial x. \end{aligned} \tag{10}$$

Qualitatively H_0 is similar to Eq. (9) if $a > 0$, $b < 0$. To have a clear space-time picture, one would like to go back to the corresponding Lagrangian $L = p\dot{x} - H$ by solving p in terms of $\dot{x}$. But $\dot{x}$ is not a single-valued function of p, and it has in general three real solutions. In other words, to a Hamiltonian given in the coordinates and the momenta does not correspond a unique Lagrangian in the coordinates and the velocities. This is at odds with the usual Newtonian causality, according to which the position and the velocity of a particle at a given time uniquely determines its motion at all times. I will call this a Newton anomalv.

$$\omega = ((p^2 - p_F{}^2)^2/4m^{*2} + \Delta^2)^{1/2}$$

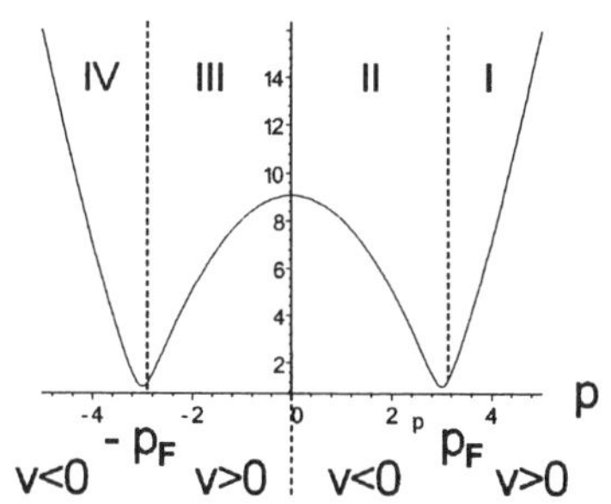

Figure 2.

The behavior (9) of the BV quasiparticle spectrum ω is shown in Fig. 2. There are four regions, I through IV, depending on the signs of the pair (p, v). Regions I and IV correspond to the outside of the Fermi surface where the quasiparticle is mostly in the electron state $\psi(p)$ ($|\alpha| > |\beta|$) with positive kinetic mass, whereas regions II and III correspond to the inside where it is mostly in the hole state $\psi(-p)^\dagger$ ($|\alpha| < |\beta|$) with negative kinetic mass. There is a degeneracy of energy in the range between Δ and $\sqrt{\Delta^2 + p_F^2/4m^{*2}}$.

7

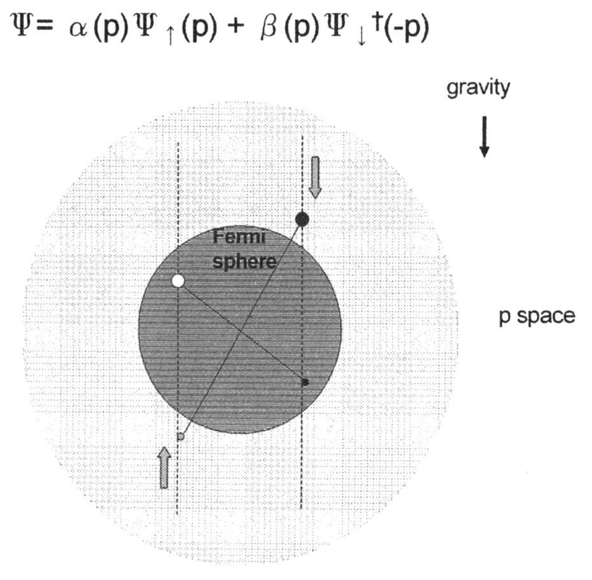

Figure 3.

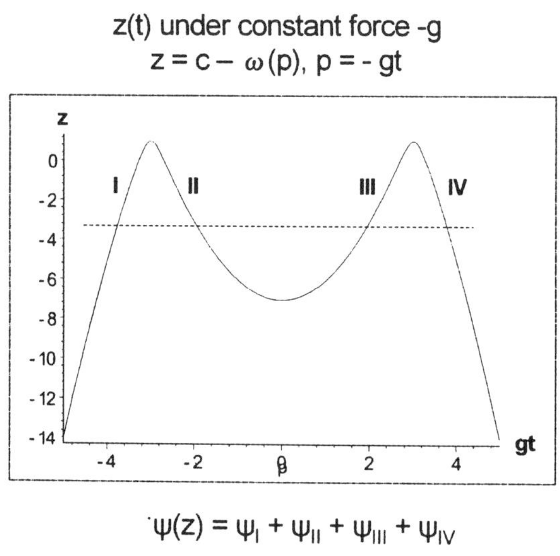

Figure 4.

It is interesting to follow the motion of a BV particle as a classical partice in the field of weak uniform gravity.[e] Given the kinetic part of the Hamiltonian H_0, the equivalence principle tells us that $V = H_0 gz/c^2$ for nonrelativistic systems. The qualitative behavior, however, may be gleaned by taking V to be of the simple form $V = \lambda z$, with some constant λ. The momentum p_z keeps decreasing with time. Starting from a sufficiently large positive value outside of the Fermi surface, the particle component $\psi(p)$ will go through the Fermi sphere smoothly downwards, whereas the hole component $\psi(-p)^\dagger$ will move upwards (Fig. 3).

In real space the motion is more interesting (Fig. 4). If one tosses the BV particle upwards with sufficiently large velocity, it rises up to a maximum height $z_{\text{max}} = z$, where $p = p_F$ but $v = 0$, then starts falling down, stops once at z_{min} when $p = 0$ and $v = 0$, rises up again to z_{max}, where $p = -p_F$, $v = 0$, and thereafter falls steadily. The violation of Newtonian causality is seen by noting that if the initial upward velocity is within the values between z_{min} and z_{max}, one cannot be sure if it is on the first upward swing or the second one.[f] In reality, the gravity is too weak to be able to exhibit the whole trajectory for a BV quasiparticle. But going back to quantum mechanics, the wave spatial function $\psi(z)$ between z_{max} and z_{min} will consist of components from four regions, two corresponding to states below the Fermi surface, and two from above.

4. Speculative Remarks

In Section 3 a model of Lorentz symmetry violation in the vacuum by an axial vector was considered in analogy with ferromagnetism. In the real world I speculate that this could be a possibility, for example, for the neutrinos since they are the least known sector of the Standard Model. The violation occurs in the space component (in a certain Lorentz frame), and the vacuum has a spontaneous 'magnetization' of non-electromagnetic nature (C-even), but it could be also in the time component, in which case the vacuum has a net (non-spontaneous) chirality, perhaps due to the presence of a chemical potential.

[e]Gravity is considered because of its universal nature. One has to assume that the medium is held against gravity, but the quasiparticles carry extra energy and should be able to respond to gravity freely. In real systems there will be complications such as the change of the density of the medium with the height. These effects are ignored.

[f]One could say that in the case of the BV paticle the nonuniqueness is due to the neglect of the spin degree of freedom in the classical description, but it is difficult to apply a similar argument in the case of the Landau roton.

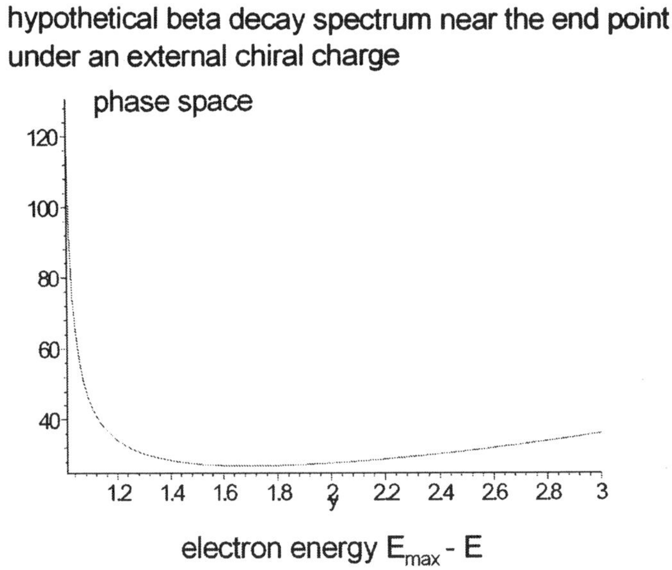

Figure 5.

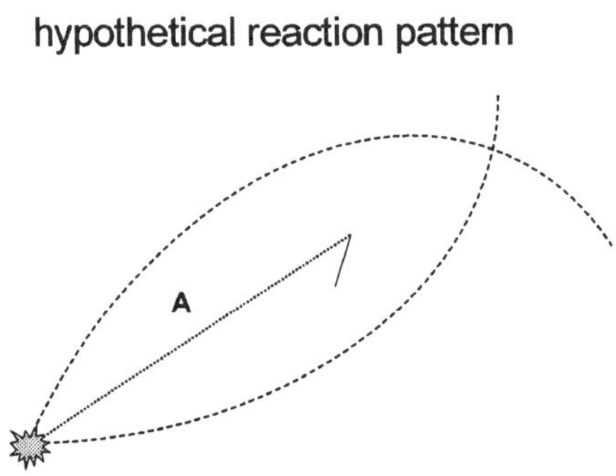

Figure 6.

One of the obvious manifestations of the effects would be in the beta decay spectrum. The presence of the peculiar dispersion law for a neutrino would be reflected in the electron energy spectrum near the end point. As is illustrated in Fig. 5, the degeneracy of the spectrum is responsible for an enchancement towards the end. The fact that the momentum and velocity can be in opposite directions causes some paradoxical phenomena if the particle in question is involved in scattering or decay (Fig. 6). There is an apparent lack of momentum conservation when an anomalous particle (A) is produced, or absorbed.

Acknowledgment. This work was supported by the University of Chicago.

References

1. H.B. Nielsen and S. Chadha, Nucl. Phys. **B105**, 445 (1976).
2. Y. Nambu, J. Statist. Phys. **115**, 7 (2004), and references therein.
3. D. Colladay and V.A. Kostelecký, Phys. Rev. D **55**, 6760 (1997); Phys. Rev. D **58**, 116002 (1998); V.A. Kostelecký and R. Lehnert, Phys. Rev. D **63**, 065008 (2001).
4. V.A. Miransky and I.A. Shovkovy, hep-ph/0108178; Phys. Rev. Lett. **88**, 11601 (2002).
5. D. Ebert, V.Ch. Zhukovsky, and A.S. Razmovsky, hep-ph/0401241 v1.
6. D. Blaschke, D. Ebert, K.G. Klimenko, M.K. Volkov, and V.L. Yudichev, hep-ph/0403151.
7. S.M. Carroll, G.B. Field, and R. Jackiw, Phys. Rev. D **41**, 1231 (1990).

International Journal of Modern Physics A
Vol. 24, No. 13 (2009) 2371–2377

World Scientific
www.worldscientific.com

SPONTANEOUS SYMMETRY BREAKING IN PARTICLE PHYSICS: A CASE OF CROSS FERTILIZATION*
Nobel Lecture, December 8, 2008

YOICHIRO NAMBU
University of Chicago, USA

I will begin by a short story about my background. I studied physics at the University of Tokyo. I was attracted to particle physics because of the three famous names, Nishina, Tomonaga and Yukawa, who were the founders of particle physics in Japan. But these people were at different institutions than mine. On the other hand, condensed matter physics was pretty good at Tokyo. I got into particle physics only when I came back to Tokyo after the war. In hindsight, though, I must say that my early exposure to condensed matter physics has been quite beneficial to me.

Particle physics is an outgrowth of nuclear physics which began in the early 1930's with the discovery of the neutron by Chadwick, the invention of the cyclotron by Lawrence, and the "invention" of meson theory by Yukawa.[a] The appearance of an ever increasing array of new particles in the subsequent decades, and the advances in quantum field theory gradually led to our understanding of the basic laws of nature, culminating in the present standard model.

When we faced those new particles, our first attempts were to make sense out of them by finding some regularities in their properties. They invoked the symmetry principle to classify them. A symmetry in physics leads to a conservation law. Some conservation laws are exact, like energy and electric charge, but these attempts were based on approximate similarities of masses and interactions.

Nevertheless, seeing similarities is a natural and very useful trait of the human mind. The near equality of proton and neutron masses and their interactions led to the concept of isospin SU(2) symmetry.[2] On the other hand, one could also go in the opposite direction, and elevate a symmetry to a more elaborate gauged symmetry. Then symmetry will determine the dynamics as well, a most attractive possibility. Thus the beautiful properties of electromagnetism was extended to the

*Published with permission from the Nobel Foundation.

[a]See, for example, Ref. 1.

SU(2) non-Abelian gauge field.[3] But strong interactions are short range. Giving a mass to a gauge field destroys gauge invariance.

Spontaneous symmetry breaking (SSB), which is the main subject of my talk, is a phenomenon where a symmetry in the basic laws of physics appears to be broken. In fact, it is a very familiar one in our daily life, although the name SSB is not.[b] For example, consider a elastic straight rod standing vertically. It has a rotational symmetry; it looks the same from any horizontal direction. But if one applies increasing pressure to squeeze it, it will bend in some direction, and the symmetry is lost. The bending can occur in principle in any direction since all directions are equivalent. But you do not see it unless you repeat the experiment many times. This is SSB.

The SSB in quantum mechanics occurs typically in a uniform medium consisting of a large number of elements. It is a dynamical effect. A symmetry allows some freedom of action to each of them but the interaction among them forces them, figuratively speaking, to line up like a crowd of people looking into the same direction. Then it is not easy to change the direction wholesale even if it is allowed by the symmetry hence does not take energy, because the action is not local operator. So the symmetry *appears* to be lost. It is still possible to recover the lost symmetry by a global operation, but it would amount to a kind of phase transition. Some of the examples are:

Physical system	Broken symmetry
Ferromagnets	Rotational invariance (with respect to spin)
Crystals	Translational and rotational invariance (modulo discrete values)
Superconductors	Local gauge invariance (particle number)
Superfluid ^{4}He	Global gauge invariance (particle number)

SSB in a medium then has the following characteristic properties:

(i) The ground state has huge degeneracy. A symmetry operation takes one ground state to another.
(ii) Only one of the ground state and a whole spectrum of excited states built on it are realized in a given situation.
(iii) SSB is in general lost at sufficiently high temperatures.

In relativistic quantum field theory, this phenomenon becomes also possible for the entire space–time, for the "vacuum" is not void, but has many intrinsic degrees of freedom. In this context, it may play an important role in cosmology. As the universe expands and cools down, it may undergo one or more SSB phase transitions

[b]The name is due to S. Glashow.

from states of higher symmetries to lower ones, which change the governing laws of physics.

I will now recall the chain of events which led me to the idea of SSB and its application to particle physics. One day in 1956 R. Schrieffer gave us a seminar on what would come to be called the BCS theory[4] of superconductivity. I was impressed by the boldness of their ansatz for the state vector, but at the same time I became worried about the fact that it did not appear to respect gauge invariance. Soon thereafter Bogoliubov[c] and Valatin[7] independently introduced the concept of quasiparticles as fermionic excitations in the BCS medium. The quasiparticles did not carry a definite charge as they were a superposition of electron and hole, with their proportion depending on the momentum. How can one then trust the BCS theory for discussing the electromagnetic properties like the Meissner effect? It actually took two years for me to resolve the problem to my satisfaction. There were a number of people who also addressed the same problem, but I wanted to understand it in my own way. Essentially it is the presence of a massless collective mode, now known by the generic name of Nambu–Goldstone (NG) boson, that saves charge conservation or gauge invariance.

The Bogoliubov–Valatin (BV) quasiparticles are described by the equations[8]

$$\begin{aligned} E\psi_{\mathbf{p},+} &= \varepsilon_{\mathbf{p}}\psi_{\mathbf{p},+} + \Delta\psi^{\dagger}_{-\mathbf{p},-}\,, \\ E\psi^{\dagger}_{-\mathbf{p},-} &= -\varepsilon_{\mathbf{p}}\psi^{\dagger}_{-\mathbf{p},-} + \Delta\psi_{\mathbf{p},+}\,, \\ E &= \sqrt{\varepsilon_{\mathbf{p}}^2 + \Delta^2}\,. \end{aligned} \tag{1}$$

Here, $\psi_{\mathbf{p},+}$ and $\psi^{\dagger}_{-\mathbf{p},-}$ are the wave functions for an electron and a hole of momentum p and spin $+$ or $-$, ε_p is the kinetic energy relative to the Fermi energy, and 2Δ is the energy gap. In terms of spinlike matrices τ_i, the corresponding Hamiltonian and the charge-current are

$$H_0 = \varepsilon_{\mathbf{p}}\Psi^{\dagger}\tau_3\Psi + \Delta\Psi^{\dagger}\tau_1\Psi\,, \quad \rho_0 = \Psi^{\dagger}\tau_3\Psi\,, \quad \mathbf{j}_0 = \Psi^{\dagger}(\mathbf{p}/m)\Psi\,. \tag{2}$$

The BV ground state is $\Psi^{\dagger}_{\mathbf{p}}|0\rangle = 0$ for all $\mathbf{p}$. The charge does not commute with H_0, and the continuity equation does not hold, which is the problem. But it has turned out that the same interaction that led to the BCS-BV ground state also leads to collective excitations f, which contributes to the charge-current and restores the continuity equation. The correct expression is

$$\rho \simeq \rho_0 + \frac{1}{\alpha^2}\partial_t f\,, \quad \mathbf{j} \simeq \mathbf{j}_0 - \boldsymbol{\nabla} f\,, \quad \left(\boldsymbol{\nabla}^2 - \frac{1}{\alpha^2}\partial_t^2\right) f \simeq -2\Delta\Psi^{\dagger}\tau_2\Psi\,. \tag{3}$$

f represents the NG mode. Physically it corresponds to excitations that tend to restore the lost symmetry. Its energy goes to zero in the long wavelength limit as it corresponds to the global symmetry operation. It also happens that the above

[c]Reference 5. This is a fermionic version of transfomation which he first introduced in a description of superfluidity, Ref. 6.

NG mode also mixes with the Coulomb interaction among the electrons because of their common long-range nature, and turns into the well-known plasmons with

$$\omega^2 = k^2 + e^2 n\,, \tag{4}$$

where n is the electron density.

The formal similarity of the BV equation to the Dirac equation naturally led me to transport the BCS theory to particle physics.[9] The gap Δ goes over to the mass M, which breaks chirality $\sim \gamma_5$ rather than the ordinary charge ~ 1. The axial current is the analog of the electromagnetic vector current in the BCS theory. If chirality is a broken symmetry, the matrix elements of the axial current between nucleon states of four-momentum p and p' should have the form

$$\Gamma_{\mu 5}(p', p) = (\gamma_\mu \gamma_5 - 2M\gamma_5 q_\mu / q^2) F(q^2)\,, \quad q_\mu = p'_\mu - p_\mu\,. \tag{5}$$

So chiral symmetry is compatible with a finite nucleon mass M provided that there exists a massless pseudoscalar NG boson. In reality, there is a pseudosclar pion, and the vector and axial vector interactions that appear in weak decays of the nucleons and the pion had the properties

$$g_V \simeq g_A\,, \qquad g_\pi \simeq \sqrt{2} M g_A / G\,, \tag{6}$$

where g_V and g_A are vector and axial vector couplings of the nucleon, g_π is axial coupling for the pion, G is the pion–nucleon coupling, and M is the nucleon mass. The second relation was known as the Goldberger–Treiman relation,[10] and it implies that the matrix element of the axial vector part of nucleon decay is

$$\Gamma_{\mu A} \simeq (\gamma_\mu \gamma_5 - 2M\gamma_5 q_\mu)/(q^2 - m_\pi^2) \tag{7}$$

which differs from Eq. (5) by the presence of pion mass. In view of the smallness of m_π compared to M, I made the hypothesis that the axial current is an approximately conserved quantity, the nucleon mass is generated by an SSB of chirality, and the pion is the corresponding NG boson which should become massless in the limit of exact conservation. (Proton and neutron masses should also become equal.)

The model system[d] I worked out subsequently with Jona-Lasinio is a concrete realization of the proposed SSB. It has the form similar to the BCS model

$$L = -\bar{\psi}\gamma^\mu \partial_\mu \psi + g[(\bar{\psi}\psi)^2 - (\bar{\psi}\gamma_5\psi)^2] \tag{8}$$

which is invariant against the particle number and chiral transformations

$$\begin{aligned} \psi &\to \exp(i\alpha)\psi\,, & \bar{\psi} &\to \bar{\psi}\exp(-i\alpha)\,, \\ \psi &\to \exp(i\gamma_5\alpha)\psi\,, & \bar{\psi} &\to \bar{\psi}\exp(i\gamma_5\alpha)\,. \end{aligned} \tag{9}$$

[d] References 11 and 12. The first presentation of the model was made at Midwest Conference on Theoretical Physics, entitled "A superconductor model of elementary particles and its consequences" (Ref. 13).

After SSB, the "nucleon" acquires a mass $M \sim 2g\langle\bar{\psi}\psi\rangle$. Although the model is nonrenormalizable, it is easy to demonstrate the SSB mechanism. The generated mass M is determined by the "gap equation"

$$\frac{2\pi^2}{g\Lambda^2} = 1 - \frac{M^2}{\Lambda^2}\ln\left(1 + \frac{\Lambda^2}{M^2}\right), \tag{10}$$

where Λ is a cutoff. Bound states of nucleon–antinucleon (meson) and nucleon–nucleon (dibaryon) pairs of spin 0 and 1 were also found. In particular, the masses of 0^- ($\sim \bar{\psi}\gamma_5\psi$) and 0^+ ($\sim \bar{\psi}\psi$) mesons are found to be 0 and $2M$ respectively. A more realistic two-flavor model was also considered by generalizing Eq. (8) to

$$L = -\bar{\psi}\gamma^\mu\partial_\mu\psi + g[(\bar{\psi}\psi)^2 - \Sigma_i(\bar{\psi}\gamma_5\tau_i\psi)(\bar{\psi}\gamma_5\tau_i\psi)] \tag{11}$$

with a similar gap equation. We get an isovector 0^- pion and isoscalar 0^+. The actual pion mass was generated by a small explicit bare mass in the Lagrangian of the order of 5 MeV. This also induced a change of axial coupling constant g_A in the right direction.

Other examples of the BCS-type SSB are ^{3}He superfluidity and nucleon pairing in nuclei.[14,15] In general there exist simple mass relations among the fermion and the bosons in BCS-type theories (e.g. Refs. 16 and 17).

The BCS theory also accounts for the generation of the London mass for the electromagnetic field. This problem is made simple in terms of the Higgs scalar field.[18–20] The relativistic analog of the London relation is, in momentum space,

$$\begin{aligned} j_\mu(q) &= K_{\mu\nu}(q)\Lambda^k\,, \\ K_{\mu\nu} &= (\delta_{\mu\nu} - q_\mu q_\nu/q^2)K(q^2)\,, \\ K(q^2) &\simeq q^2/(q^2 - m^2)\,. \end{aligned} \tag{12}$$

The third relation shows the massless NG boson turning into a massive "plasmon," a process corresponding to Eq. (4). This was successfully applied to weak gauge field in the Weinberg–Salam (WS) theory[21,22] of electroweak unification. The fermion masses are also generated and break chiral invariance. These so-called current masses for the up and down quarks play the role of the bare mass in the NJL model.

In the current standard model of particle physics, the NJL model may be regarded as an effective theory for the QCD with respect to generation of the so-called constituent masses. One is interested in the low energy degrees of freedom on a scale smaller than some cutoff $\Lambda \sim 1$ GeV. The short distance dynamics above Λ as well as the confinement may be treated as a perturbation. The problem has been extensively studied by many people. The Lagrangian adopted by Hatsuda and Kunihiro[23] is

$$L = L_{\mathrm{QCD}} + L_{\mathrm{NJL}} + L_{\mathrm{KMT}} + \delta L\,. \tag{13}$$

$L_{\rm NJL}$ is for the quarks, and contains "current mass" terms. $L_{\rm KMT}$ refers to the Kabayashi–Maskawa–'t Hooft chiral anomaly

$$L_{\rm KMT} = g_D(\det[\bar{q}_i(1-\gamma_5)q_j] + \text{h.c.})\,. \tag{14}$$

Both of them contributes to the explicit breaking of chiral invariance. (δL contains the effects of confinement and one gluon exchange.) The WS theory resembles the Ginzburg–Landau[24] description of superconductivity which was shown to follow from the BCS theory by Gor'kov.[25] In the same way the NJL model goes over to the model of Gell-Mann and Lévy.[26] If this analogy turns out real, the Higgs field might be an effective description of underlying dynamics.

Finally I will end this lecture with a comment on the mass hierarchy problem. Hierarchical structure is an outstanding feature of the universe. The masses of known fundamental fermions also make a hierarchy stretching 11 orders of magnitude. Mass is not quantized in a simple regular manner like charge and spin. Mass is a dynamical quantity since it receives contributions from interactions. But we do not see yet a pattern like those in the hydrogen atom which led to quantum mechanics, or the Regge trajectories which led to the dual string picture.

The BCS mechanism seems relevant to the problem, as was remarked earlier. It generates a mass gap for fermions, plus the Goldstone and Higgs modes as low-lying bosons. The bosons may act in turn as an agent for further SSB, leading to the possibility of hierarchical SSB or "tumbling."[27] In fact, we already have examples of it:[28]

(i) The chain atoms–crystal–phonon–superconductivity. The NG mode for crystal formation is the phonon which induces the Cooper pairing of electrons to cause superconductivity.

(ii) The chain QCD–chiral SSB of quarks and baryons–π, σ and other mesons–nuclei formation and nucleon pairing–nuclear collective modes. No further elaboration would be required.

I am greatly thankful to G. Jona-Lasinio for his help in the planning of the lecture.

References

1. Y. Nambu, *J. Phys. Soc. Japan* **76**, 111002 (2007).
2. N. Kemmer, *Nature* **140**, 192 (1937).
3. C. N. Yang and R. L. Mills, *Phys. Rev.* **96**, 191 (1954).
4. J. Bardeen, L. N. Cooper and J. R. Schrieffer, *Phys. Rev.* **108**, 1175 (1957).
5. N. N. Bogoliubov, *J. Exp. Theor. Phys.* (*U.S.S.R.*) **34**, 33 (1958).
6. N. N. Bogoliubov, *J. Phys.* (*U.S.S.R.*) **11**, 23 (1947).
7. G. Valatin, *Nuovo Cimento* **7**, 843 (1958).
8. Y. Nambu, *Phys. Rev.* **117**, 648 (1960).
9. Y. Nambu, *Phys. Rev. Lett.* **4**, 380 (1960).
10. M. L. Goldberger and S. Treiman, *Phys. Rev.* **110**, 117 (1958).
11. Y. Nambu and G. Jona-Lasinio, *Phys. Rev.* **122**, 345 (1961).

12. Y. Nambu and G. Jona-Lasinio, *Phys. Rev.* **124**, 246 (1961).
13. F. J. Biedenhahn *et al.* (eds.), *Proc. of the Midwest Conference on Theoretical Physics*, Purdue University, Lafayette, Indiana, April 1–2 1960, p. 1.
14. A. Arima and F. Iachello, *Phys. Rev. Lett.* **35**, 1069 (1975).
15. A. Arima and F. Iachello, *Ann. Phys.* **99**, 253 (1976).
16. Y. Nambu, *Physica D* **15**, 147 (1985).
17. Y. Nambu, *Phys. Lett.* **201**, 1 (1988).
18. P. W. Anderson, *Phys. Rev.* **130**, 439 (1963).
19. F. Englert and R. Brout, *Phys. Rev. Lett.* **13**, 321 (1964).
20. P. W. Higgs, *Phys. Rev. Lett.* **13**, 508 (1964).
21. S. Weinberg, *Phys. Rev. Lett.* **19**, 1264 (1967).
22. A. Salam, in *Elementary Particles* (Almquist, Forlag AB, 1968), p. 167.
23. T. Hatsuda and T. Kunihiro, *Phys. Rep.* **247**, 221 (1994).
24. V. I. Ginzburg and L. D. Landau, *J. Exp. Theor. Phys.* (*U.S.S.R.*) **20**, 1064 (1950).
25. L. P. Gor'kov, *J. Exp. Theor. Phys.* (*U.S.S.R.*) **36**, 1918 (1959).
26. M. Gell-Mann and M. Lévy, *Nuovo Cimento* **16**, 705 (1960).
27. S. Dimopoulos, S. Raby and L. Susskind, *Nucl. Phys. B* **169**, 373 (1980).
28. Y. Nambu, Masses as a problem and as a clue, presented at the *International Centre for Theoretical Physics*, Trieste, 2004.

List of Publications

1948

1. On the Relativistic Formulation of the Perturbation Theory, *Prog. Theor. Phys.* **3**, 444.

1949

2. The Level Shift and the Anomalous Magnetic Moment of the Electron, *Prog. Theor. Phys.* **4**, 82.

3. Second Configuration Space and Third Quantization, *Prog. Theor. Phys.* **4**, 96.

4. Effect of the C-Meson Field on the Anomalous Magnetic Moment of the Electron (with *Z. Koba and T. Tati*), *Prog. Theor. Phys.* **4**, 99.

5. On the Method of Third Quantization, I and II, *Prog. Theor. Phys.* **4**, 331; 339.

1950

6. A Note on the Eigenvalue Problem in Crystal Statistics, *Prog. Theor. Phys.* **5**, 1.

7. The Use of the Proper Time in Quantum Electrodynamics I, *Prog. Theor. Phys.* **5**, 82.

8. On the Electromagnetic Properties of Mesons (with *T. Kinoshita*), *Prog. Theor. Phys.* **5**, 307.

9. Derivation of the Interaction Potential from Field Theory, *Prog. Theor. Phys.* **5**, 321.

10. Force Potentials in Quantum Field Theory, *Prog. Theor. Phys.* **5**, 614.

11. On the Interaction of Mesons with the Electromagnetic Field II (with *T. Kinoshita*), *Prog. Theor. Phys.* **5**, 749.

1951

12. On the Nature of V-Particles I and II (with *K. Nishijima and Y. Yamaguchi*), *Prog. Theor. Phys.* **6**, 615; 619.

13. Meson–Nucleon Scattering (with *Y. Yamaguchi*), *Prog. Theor. Phys.* **6**, 1000.

1952

14. On Lagrangian and Hamiltonian Formalism, *Prog. Theor. Phys.* **7**, 131.

15. An Empirical Mass Spectrum of Elementary Particles, *Prog. Theor. Phys.* **7**, 595.

1954

16. The Collective Description of Many-Particle System (A Generalized Theory of Hartree Fields) (with *T. Kinoshita*), *Phys. Rev.* **94**, 598.

1955

17. Structure of the Scattering Matrix, *Phys. Rev.* **98**, 803.

18. Structure of Green's Function in Quantum Field Theory I, *Phys. Rev.* **100**, 394.

1956

19. Structure of Green's Function in Quantum Field Theory II, *Phys. Rev.* **101**, 459.

20. Renormalization Constants, *Phys. Rev.* **101**, 1183.

21. Application of Dispersion Relations to Low Energy Meson–Nucleon Scattering (with *G. F. Chew, M. L. Goldberger and F. E. Low*), *Phys. Rev.* **106**, 1337.

22. Relativistic Dispersion Relation Approach to Photomeson Production (with *G. F. Chew, M. L. Goldberger and F. E. Low*), *Phys. Rev.* **106**, 1345.

1957

23. Possible Existence of a Heavy Neutral Meson, *Phys. Rev.* **106**, 1366.

24. Dispersion Relations for Nucleon–Nucleon Scattering (with *M. L. Goldberger and R. Oehme*), *Ann. Phys.* **2**, 226.

25. Parametric Representations of General Green's Functions, *IL Nuovo Cimento X* **6**, 1064.

1958

26. Dispersion Relations for Form Factors, *IL Nuovo Cimento X* **9**, 610.

27. Dispersion Theory Treatment of Pion Production in Electron–Nucleon Collisions (with *S. Fubini and V. Wataghin*), *Phys. Rev.* **111**, 329.

1960

28. Quasi-Particles and Gauge Invariance in the Theory of Superconductivity, *Phys. Rev.* **117**, 648.

29. Axial Vector Current Conservation in Weak Interactions, *Phys. Rev. Lett.* **4**, 380.

30. Electro-production of *pi*-Mesons (with *R. Blankenbecler, S. Gartenhous and R. Huff*), *IL Nuovo Cimento X* **17**, 775.

31. Anomalous Thresholds in Dispersion Theory I (with *R. Blankenbecler*), *IL Nuovo Cimento X* **18**, 595.

1961

32. Odd $\Lambda\Sigma$ Parity and the Nature of the $\pi\Lambda\Sigma$ Coupling (with *J. J. Sakurai*), *Phys. Rev. Lett.* **6**, 377.

33. A 'Superconductor' Model of Elementary Particles and Its Consequences, in the *Proceedings of the Midwest Conference on Theoretical Physics*, eds. F. J. Belin-

fante, S. G. Gartenhaus and R. W. King, Purdue University, Lafayette, Indiana, April 1–2, 1960, p. 1.

34. Dynamical Model of Elementary Particles based on an Analogy with Superconductivity I (with *G. Jona-Lasinio*), *Phys. Rev.* **122**, 345.

35. Dynamical Model of Elementary Particles based on an Analogy with Superconductivity II (with *G. Jona-Lasinio*), *Phys. Rev.* **124**, 246.

36. Possible Bound $\Sigma - \Lambda$ System (with *E. Shrauner*), *IL Nuovo Cimento X* **21**, 864.

1962

37. Chirality Conservation with Soft Pion Production (with *D. Lurié*), *Phys. Rev.* **125**, 1429.

38. Soft Pion Emission Induced by Electromagnetic and Weak Interactions (with *E. Shrauner*), *Phys. Rev.* **128**, 862.

39. Magnetic Field Dependence of the Energy Gap in Superconductors (with *San Fu Tuan*), *Phys. Rev.* **128**, 2622.

40. Rare Decay Modes of the $\omega(\eta)$ Meson (with *J. J. Sakurai*), *Phys. Rev. Lett.* **8**, 79.

1963

41. Double Phase Representation of Analytic Functions (with *M. Sugawara*), *Phys. Rev.* **131**, 2335.

42. High-Energy Behavior of Total Cross-Sections (with *M. Sugawara*), *Phys. Rev.* **132**, 2724.

43. Magnetic Field and Phase Transition in Thin Film Superconductors (with *San Fu Tuan*), *Phys. Rev. Lett.* **11**, 119.

44. Non-Shrinking Diffraction Scattering (with *M. Sugawara*), *Phys. Rev. Lett.* **10**, 304.

45. $K^*(725)$ and the Strangeness-changing Currents of Unitary Symmetry (with *J. J. Sakurai*), *Phys. Rev. Lett.* **11**, 42.

46. Self-Consistent Models of Strong Interactions with Chiral Symmetry (with *P. Pascual*), *IL Nuovo Cimento* **30**, 354.

1964

47. Considerations on the Magnetic Field Problem in Superconducting Thin Films (with *San Fu Tuan*), *Phys. Rev.* **133**, A1.

48. Axial Vector Mesons (with *P. Freund*), *Phys. Lett.* **12**, 248.

49. Broken SU(x) × SU(3) × SU(3) × SU(3) Symmetry of Strong Interactions (with *P. Freund*), *Phys. Rev. Lett.* **12**, 714.

50. Mass and Coupling Constant Formulas in Broken Symmetry Schemes (with *P. Freund*), *Phys. Rev. Lett.* **13**, 221.

51. Magnetic Field and Phase Transitions in Superconducting Thin Films (with *San Fu Tuan*), *Rev. Mod. Phys.* **36**, 288.

52. Quasi-elementary Massless Bosons Associated with the Quantum Electrodynamics of Johnson, Baker and Willy, *Phys. Lett.* **9**, 214.

53. Outlook of Elementary Particle Physics, in *Nature of Matter,* ed. Luke C. L. Huan, BNL F88 (T-360), p. 55.

1965

54. Three-Triplet Model with Double SU(3) Symmetry (with *M.-Y. Han*), *Phys. Rev.* **139**, B1006.

55. Dynamical Symmetries and Fundamental Fields, in the *Proceedings of Coral Gables Conference 1965*, eds. B. Kursunoglu, A. Perlmutter and I. Sakmar (W. H. Freeman and Co.), p. 274.

56. Broken SU(3) × SU(3) × SU(3) × SU(3) Symmetry (with *P. Freund*), *Ann. Phys.* **32**, 201.

57. Triplets, Static SU(6), and Spontaneously Broken Chiral SU(3) Symmetry, in the *Proceedings of the International Conference on Elementary Particles, in Commemoration of the Thirtieth Anniversary of Meson Theory*, Kyoto, 1965, ed. Y. Tanikawa,

p. 131; General Discussions, p. 327.

58. Axial Vector Meson and the Hyperfine Structure of Hydrogen (with *S. Fenster and R. Köberle*), *Phys. Lett.* **19**, 513.

59. Baryon Structure and Electromagnetic Properties (with *S. Fenster*), *Prog. Theor. Phys. Suppl.* Extra No. 250.

60. Electromagnetic Properties of the Baryon (Hyperfine Structure of Hydrogen), in *1965 Tokyo Summer Lecture, Part II*, ed. G. Takeda (Shokabō Publishing), p. 87.

1966

61. A Systematics of Hadrons in Subnuclear Physics, in *Preludes in Theoretical Physics,* eds. A. De-Shalit, H. Feshbach and L. van Hove (North-Holland), p. 133.

62. Coupling Constant Relations for $1^{\pm}$ and Induced $0^{\pm}$ Mesons (with *J. Cronin*), *IL Nuovo Cimento A* **41**, 380.

63. Nonleptonic Decays of K-Mesons (with *Y. Hara*), *Phys. Rev. Lett.* **16**, 875.

64. Nonleptonic Decays of Hyperons (with *Y. Hara and J. Schechter*), *Phys. Rev. Lett.* **16**, 380.

65. Relativistic Wave Equations for Particles with Internal Structure and Mass Spectrum, *Prog. Theor. Phys. Suppl.* **37** and **38**, 368.

1967

66. Infinite Multiplets, in the *Proceedings of 1967 International Conference on Particles and Fields*, eds. C. R. Hagen *et al.* (Interscience), p. 347.

67. Infinite Component Wave Equations with Hydrogenlike Mass Spectra, *Phys. Rev.* **160**, 1177.

1968

68. Magnetic Moments and Charge Radii for States Described by an Infinite Component Wave Equation (with *S. P. Rosen*), *Prog. Theor. Phys.* **40**, 5.

69. Quantum Electrodynamics in Nonlinear Gauge, *Prog. Theor. Phys. Suppl.* Extra No. 190.

70. Scalar Fields Coupled to the Trace of the Energy Momentum Tensor (with *P. Freund*), *Phys. Rev.* **174**, 1741.

71. *S*-Matrix in Semiclassical Approximation, *Phys. Lett. B* **26**, 626.

72. Relativistic Groups and Infinite Component Fields, talk at *Nobel Symposium on Elementary Particle Theory,* Goteberg, ed. N. Svaartholm (Almquist and Wiskell), p. 105.

1970

73. Asymptotic Behavior of Particle Widths in the Veneziano Model of Scattering Amplitudes (with *P. Frampton*), in *QUANTA, A Collection of Scientific Essays Dedicated to Gregor Wenzel,* eds. P. G. O. Freund, C. Goebel and Y. Nambu (University of Chicago Press), p. 33.

74. Quark Model and the Factorization of the Veneziano Amplitude, in the *Proceedings of the International Conference on Symmetries and Quark Models,* Wayne University, ed. R. Chand (Gordon and Breach), p. 269.

75. Statistical Approach to the Veneziano Model (with *L.-N. Chang and P. Freund*), *Phys. Rev. Lett.* **24**, 628.

76. Axial-Vector Form Factor of Nucleon Determined from Threshold Electropion Production (with *M. Yoshimura*), *Phys. Rev. Lett.* **24**, 25.

77. Symmetry Breakdown and Small Mass Bosons, in *Fields and Quanta* **1**, 33, also published in *The Past Decade in Particle Theory,* eds. E. C. G. Sudarshan and Y. Ne'eman (Gordon and Breach, 1973), p. 33.

78. Duality and Hadrodynamics, note prepared for Copenhagen High Energy Symposium, August 1970 (unpublished).

1971

79. Electromagnetic Currents in Dual Hadrodynamics, *Phys. Rev. D* **4**, 1195.

1972

80. Gauge Conditions in Dual Resonance Models (with *F. Mansouri*), *Phys. Lett. B* **39**, 375.

81. Use of Regulator Fields in Dual Resonance Models (with *J. Willemsen*), submitted to *XVI International Conference on High Energy Physics, 1972*, eds. J. D. Jackson and A. Roberts, National Accelerator Laboratory, 1972, No. 110, unpublished.

82. Chiral Symmetries and Current Algebras, *Lecture at International Summer School,* Erice, Italy, 1972 (unpublished).

1973

83. Generalized Hamiltonian Dynamics, *Phys. Rev. D* **7**, 2405.

84. Models Concerning the Chemical Structure of Hadrons, in the *Proceedings of Tokyo Symposium on High Energy Physics*, July.

85. Elementary Particle Physics in Perspective, *Butsuri* **28**, 452 (in Japanese).

1974

86. Three Triplets, Paraquarks and "Colored" Quarks (with *M.-Y. Han*), *Phys. Rev. D* **10**, 675.

87. Quarks, Strings and Gauge Fields, in the *Proceedings of Johns Hopkins Workshop on Current Problems in High Energy Particle Theory,* 1973.

88. Strings, Monopoles and Gauge Fields, *Phys. Rev. D* **10**, 4262.

1975

89. Dynamics of the Zweig–Iizuka Rule and a New Vector Meson below 2 GeV/c^2 (with *P. Freund*), *Phys. Rev. Lett.* **34**, 1645.

1976

90. Magnetic and Electric Confinement of Quarks, talk given at *Topical Conference on*

Extended Systems, Ecole Normale Superieure, June 1975, *Phys. Rep. C* **23**, 250.

91. Diquark Color Excitation and the Narrow Resonance (with *M.-Y. Han*), *Phys. Rev. D* **14**, 1459.

1977

92. Strings, Vortices and Gauge Fields, in *Quark Confinement and Field Theory*, eds. D. R. Stump and D. H. Weingarten (John Wiley and Sons), p. 1.

93. Description of Hadronic Structure, in the *Proceedings of the International School on Electro- and Photonuclear Reactions, 1976*, eds. S. Costa and C. Shaerf; also published in *Quark Models and Hadronic Structure,* Physics Workshop Series, 1, ed. G. Morpurgo (Plenum Press), p. 281.

94. Monopoles, Strings and Instantons, in *Five Decades of Weak Interactions, Proceedings of CCNY Conference, 1977,* ed. N. P. Chang, *Ann. of the N.Y. Acad. of Sci.* **74**, 294.

95. Elementary Particles, *Butsuri* **32**, 11 (in Japanese).

96. String-Like Configurations in the Weinberg–Salam Theory, *Nucl. Phys. B* **130**, 505.

1978

97. Some Topological Configurations in Gauge Theories, *Int. J. Theor. Phys.* **17**, 287.

98. Remarks on the Topology of Gauge Fields, in *New Frontiers in High Energy Physics, Proceedings of the Orbis Scientiae,* University of Miami, January 1978, eds. Kursunoglu and Perlmutter (Plenum Press), p. 632.

1979

99. Topological Problems in Gauge Theories, talk at *UCLA Symposium in Honor of Julian Schwinger on the Occasion of His 60th Birthday, Physica A* **96**, 89.

100. QCD and the String Model, *Phys. Lett. B* **80**, 372.

101. Concluding Talk at the *XIX Conference on High Energy Physics, Tokyo, 1978;* in the *Proceedings of the Conference*, eds. S. Homma *et al.* (Phys. Soc. Japan), p. 971.

1980

102. Quark Confinement: The Case For and Against, in *To Fulfill the Vision: Jerusalem Einstein Centennial Symposium on Gauge Theories and Unification of Physical Forces,* ed. Y. Ne'eman (Addison and Wesley), p. 118.

103. Hamilton–Jacobi Formalism for Strings, *Phys. Lett. B* **92**, 327.

1981

104. Effective Action Gauge Fields, *Phys. Lett. B* **102**, 149.

105. Strings and Vortices, in *VPI Conference on Weak Interactions as Probes of Unification,* AIP Conference Proceedings, Particles and Fields Subseries No. 23, p. 633.

106. *Quarks* (Kodansha Publishers), in Japanese.

107. Theory of Strings, in the *Proceedings of the 1981 INS Symposium on Quark and Lepton Physics,* Tokyo, Japan, eds. K. Fujikawa *et al.* (Institute for Nuclear Study, University of Tokyo), p. 347.

1982

108. Many-Quark Problem in Two-Dimensional Gauge Theory (with *B. Bambah*), *Phys. Rev. D* **26**, 2871.

109. One-Dimensional Quark Gas (with *B. Bambah and M. Gross*), *Phys. Rev. D* **26**, 2875.

1983

110. Concluding Remarks, in *Lecture Notes in Physics,* Vol. 176; *Proceedings of the International Symposium on Gauge Theory and Gravitation,* Nara, Japan, 1982, eds. K. Kikkawa *et al.* (Springer), p. 277.

111. Magnetic Monopoles and Related Topics, in the *Proceedings of the Topical Conference,* June, 1982, eds. T. Eguchi and Y. Yamaguchi (World Scientific), p. 3.

112. The Activity of the Tomonaga Group up to the Time of the 1947 Shelter Island Conference (with *K. Nishijima*), in *Shelter Island II, Proceedings of the 1983 Shelter Island Conference on Quantum Field Theory and the Fundamental Problems of Physics,* eds. R. Jackiw *et al.* (MIT Press), p. 367.

1984

113. Concluding Remarks, in the *Proceedings of the Solvay Conference on High Energy Physics,* Austin, Texas, November 1982, *Phys. Rep.* **104**, 237.

114. Quantum Mechanics; Prospects and Problems, in the *Proceedings of the Symposium on Foundations of Quantum Mechanics in Light of New Technology,* Tokyo, August 1983, eds. Kamefuchi *et al.* (Physical Society of Japan), p. 363.

115. Superconductivity and Particle Physics, talk at *International Conference on Low Temperature Physics*, Karlsruhe, Germany, August 1984; *Physica B* **126**, 328.

116. The Turbulent Ether, in *Symmetries in Particle Physics, Symposium in Honor of F. Gürsey's 60th Birthday,* eds. I. Bars, A. Chodos and C.-H. Tze (Plenum Press), p. 9.

117. Fermions Living in a Space of Lie Groups, in *From $SU(3)$ to Gravity, Festschrift in Honor of Y. Ne'eman*, eds. E. Gotzman and G. Tauber (Cambridge Univ. Press), p. 45.

1985

118. Topological Excitations in Physics, Lectures in Applied Mathematics Vol. 21, 3, *Proceedings of 1982 Mathematical Seminar* (Amer. Math. Soc.).

119. Fermion–Boson Relations in BCS-Type Theories, in *Supersymmetry in Physics, Proceedings of the Los Alamos Workshop on Supersymmetry in Physics,* December 1983, eds. A. Kosteleck'y and D. K. Campbell (North-Holland); also published as *Physica, D* **15**, 147.

120. Some Theoretical Problems in Particle Physics, *Phys. Blätter* **41**, 173.

121. Directions of Particle Physics, in the *Proceedings of the Kyoto International Symposium, MESON50,* 1985, eds. M. Bando, R. Kawabe and N. Nakanishi, *Prog. Theor. Phys. Suppl.* **85**, 104.

122. Gauge Principles, Vector Meson Dominance and Spontaneous Symmetry Breaking, in *International Symposium on Particle Physics in the 1950s,* Fermilab, May.

1986

123. Supersymmetry and Superconductivity, in *Rationale of Beings, Festschrift in Honor of Gyo Takeda,* eds. K. Ishikawa *et al.* (World Scientific), p. 3.

1987

124. Field Theory of Galois' *Fields, in Quantum Field Theory and Quantum Statistics, Essays in Honor of the Sixtieth Birthday of E. S. Fradkin,* eds. I. A. Batalin *et al.* (Adam Hilger), p. 625.

125. Thermodynamic Analogy in Quantum Field Theory, in *Wanderings in Physics, Festschrift in Honor of K. Nishijima,* eds. K. Kawarabayashi and A. Ukawa (World Scientific), p. 167.

1988

126. BCS Mechanism and the Interacting Boson Model (with *M. Mukherjee*), *Phys. Lett. B* **209**, 1.

127. Dynamical Symmetry Breaking in Nuclei, in *Fest-Val, Festschrift in Honor of Valentine Telegdi's 60th Birthday,* ed. K. Winer (North-Holland), p. 181.

128. BCS Theory, Quasi-Supersymmetry and Nuclear Physics (Encuentro Latino-Americo de Fisica de Alta Energia, Universidad Federal Santa Maria, Valparaiso, Chile, December 1987), eds. N. Bralic *et al., Sciencia* **165-166**, 141.

1989

129. The BCS Theory and the Sigma Model Revisited, in *Themes in Particle Physics, Dedicated to 70th Birthday of Julian Schwinger,* eds. S. Deser and R. J. Finkelstein (World Scientific).

130. BCS Mechanism, Quasi-Supersymmetry and Fermion Masses, in the *Proceedings of XI Warsaw Symposium on Elementary Particle Physics, New Theories in Physics,* Kazimierz, Poland, May 1988, eds. Z. Ajduk *et al.* (World Scientific), p. 1.

131. Quasi-Supersymmetry, Bootstrap Symmetry Breaking and Fermion Masses, in the *Proceedings of the 1988 International Workshop on New Trends in Strong Coupling Gauge Theories,* Nagoya, August 1988 (World Scientific), p. 3.

132. BCS and IBM (with *M. Mukherjee*), *Ann. Phys.* **191**, 143.

1990

133. Particle Physics since Lawrence and Yukawa, in *Particle Accelerators, Proceedings of the 14th International Conference on High Energy Accelerators, Part I,* eds. Y. Kimura *et al.* (Gordon and Breach), Vol. 26, p. 1.

134. Model Building Based on Bootstrap Symmetry Breaking, in the *Proceedings of the 1989 Workshop on Dynamical Symmetry Breaking,* Nagoya, December 1989, eds. T. Muta and K. Yamawaki (Nagoya University), p. 1.

135. Mass Formulas and Dynamical Symmetry Breaking, in *From Symmetries to Strings: Forty Years of Rochester Conferences, A Symposium in Honor of Susumu Okubo on His 60th Year,* May 1990, ed. A. Das (World Scientific), p. 1.

136. New Bootstrap and the Standard Model, in *Plots, Quarks and Strange Particles, Proceedings of the Dalitz Conference*, Oxford, July 1990, eds. I. J. R. Aitchison, C. H. Llewellyn Smith and J. E. Paton (World Scientific), p. 56.

137. More on Bootstrap Symmetry Breaking, in the *Proceedings of the 1990 International Workshop on Strong Coupling Gauge Theories and Beyond,* Nagoya University, Japan, July 1990, eds. T. Muta and K. Yamawaki (World Scientific), p. 3.

138. Summary of Personal Recollections of Tokyo Group, in *Elementary Particle Theory in Japan, 1935–1960: Japan–US Collaboration on the History of Particle Physics*, Lake Yamanaka, 1985, eds. L. Brown *et al., Prog. Theor. Phys. Suppl.* **105**, 111.

1991

139. Supersymmetry and Quasi-supersymmetry, in *Elementary Particles and the Universe, Essays in Honor of Murray Gell-Mann,* ed. J. H. Schwarz (Cambridge Univ. Press), p. 89.

140. Theoretical Perspective in High Energy Physics, in the *Proceedings of the 25th International Conference on High Energy Physics, Singapore,* eds. K. K. Phua and Y. Yamaguchi (South East Theoretical Association), p. 51.

141. Dynamical Symmetry Breaking, in *Evolutionary Trends in the Physical Sciences, Proceedings of the Yoshio Nishina Centennial Symposium,* Tokyo, Japan, December 1990, eds. M. Suzuki and R. Kubo (Springer-Verlag), p. 51.

142. Majorana's Infinite Component Wave Equation, A Commentary to be published in the Collected Papers of E. Majorana, ed. V. Telegdi.

143. Old Wine in New Bottles: Mass Spectra, Bootstrap and Hierarchy, talk given at *E. C. G. Sudarshan's Contributions to Theoretical Physics,* ed. A. Gleeson.

1992

144. Dynamical Symmetry Breaking: An Overview, in the *Proceedings of the International Workshop on Electroweak Symmetry Breaking,* Hiroshima, Japan, November 1991, eds. W. A. Bardeen *et al.* (World Scientific).

145. Cosmic Rays and Particle Physics, talk at the *International Workshop on Super-high Energy Hadronic Interactions at Waseda University,* October 1991.

1993

146. Spontaneous Symmetry Breaking and the Origin of Mass, in *Some New Trends on Fluid Mechanics and Theoretical Physics, in Honor of Professor Pei-Yuan Chou's 90th Anniversary,* 1992, eds. C. C. Lin and Ning Hu (Peking University Press), p. 97.

147. Nonlocal Separable Solutions of the Inverse Scattering Problem (with *T. Gherghetta*), *Int. J. Mod. Phys. A* **8**, 3163.

1994

148. Feedback Effects in Superconductors (with *T. Gherghetta*), *Phys. Rev. B* **49**, 49.

1997

149. Three Stages, Three Modes, and Beyond, in the *Proceedings of XX Nishinomiya–*

Yukawa Symposium, 1996 (World Scientific), p. 1.

1998

150. Symmetry Breaking, Chiral Dynamics, and Fermion Masses, talk given at *Quark Matter 97,* Osaka University, *Nucl. Phys. A* **629**, 3c.

151. Particle Physics in Perspective, talk given at *Qulen 97,* Osaka University, *Nucl. Phys. A* **638**, 35c.

2000

152. Aharonov-Bohm Problem Revisited, *Nucl. Phys. B* **579**, 590.

2002

153. CPT, SSB, Ether and All That, in the *Proceedings of CPT01,* Indiana University, (World Scientific), p. 1.

2004

154. Spontaneous Breaking of Lie and Current Algebras, *J. Stat. Phys.* **115**, 7.

2005

155. Some Anomalies Related to Spontaneous Symmetry Breaking, in the *Proceedings of CPT04,* Indiana University (World Scientific), p. 1.

2007

156. From Yukawa's Pion to Spontaneous Symmetry Breaking, *J. Phys. Soc. Jpn.* **76**, 111002.

157. My Memories of Sakata-sensei, *Prog. Theor. Phys. Suppl.* **167**, 175.

158. Legacies of Yukawa and his Disciples, in the *Proceedings of INPC2007* (International Nuclear Physics Conference).

159. From S-Matrix to the String Theory, Conference on *"The Birth of String Theory"*, Florence, May 2007 (Cambridge Univ. Press).

160. Nobel Lecture, *Rev. Mod. Phys.* **81**, 1019, also in *Chem. Phys.* **10**, 1718, and *Int. J. Mod. Phys. A* **24**, 2371.

2010

161. Energy Gap, Mass Gap, and Spontaneous Symmetry Breaking, *Int. J. Mod. Phys. A* **25**, 4141.

Biographical Data

Yoichiro Nambu

Date of birth	:	18 January 1921
Place of birth	:	Tokyo, Japan
U.S. Citizenship	:	1970

Education

B.S. University of Tokyo 1942

Ph.D. University of Tokyo 1952

Positions

1950–1956	Osaka City University,	Professor
1952–1954	Institute for Advanced Study,	Member
1954–1956	University of Chicago,	Research Associate
1956–1958	University of Chicago,	Associate Professor
1958–1990	University of Chicago,	Professor
1971–1976	University of Chicago,	Distinguished Service Professor
1976–1991	University of Chicago,	Harry Pratt Judson Distinguished Service Professor
1991-present	University of Chicago,	Emeritus Professor

Honors

Dannie Heineman Przie for Mathematical Physics,	1970
Member, National Academy of Sciences,	1971
Member, American Academy of Atrs and Sciences,	1971
J. Robert Oppenheimer Prize, University of Miami,	1976
Order of Culture, Government of Japan,	1978
Honorary Degree, Osaka City University,	1980
United States National Medal of Science,	1982
Honorary Member, Japan Academy,	1984
Max Planck Medal, German Physical Society,	1985
Honorary Degree, Northwestern University,	1985
P.A.M. Dirac Medal, International Center for Theoretical Physics, Trieste,	1986
Sakurai Prize, American Physical Society,	1994
Wolf Prize, Government of Israel,	1995
Gian Carlo Wick Medal, World Federation of Scientists Lausanne,	1996
Honorary Degree, Osaka University, Japan,	1997
Fukui Prefectural Award, Japan,	2003
N. N. Bogoliubov Prize, Joint Institute for Nuclear Research, Russia,	2003
Benjamin Franklin Medal,	2005
Pomeranchuk Prize, Institute of Theoretical and Experimental Physics, Russia,	2007
Nobel Prize in Physics,	2008
Honorary Member, Physical Society of Japan,	2009